Satiation: From Gut to Brain

Satiation:
From Gut to Brain

EDITED BY

Gerard P. Smith

Bourne Behavioral Research Laboratory
New York Hospital-Cornell Medical Center
Cornell University Medical College

New York Oxford
Oxford University Press
1998

Oxford University Press

Oxford New York
Athens Auckland Bangkok Bogota Bombay
Buenos Aires Calcutta Cape Town Dar es Salaam
Delhi Florence Hong Kong Istanbul Karachi
Kuala Lumpur Madras Madrid Melbourne
Mexico City Nairobi Paris Singapore
Taipei Tokyo Toronto Warsaw

and associated companies in
Berlin Ibadan

Published by Oxford University Press, Inc.
198 Madison Avenue, New York, New York 10016

Oxford is a registered trademark of Oxford University Press

Library of Congress Cataloging-in-Publication Data
Satiation: from gut to brain
edited by Gerard P. Smith.
p. cm. Includes bibliographical references and index.
ISBN 0-19-510515-X
1. Hunger. 2. Stimulus satiation.
I. Smith, Gerard P.
[DNLM: 1. Satiation—physiology.
2. Endocrine Glands—physiology.
3. Neurosecretory Systems—physiology.
4. Feeding Behavior—physiology. WL 103 S926 1998]
QP138.S77 1998
612.3′91—dc21
DNLM/DLC
for Library of Congress 97-14182

9 8 7 6 5 4 3 2 1

Printed in the United States of America
on acid-free paper

For Barbara

PREFACE

This book is concerned with the fundamental biological problem of how ingested food stops one eating. That process is called satiation. The ten chapters of the book provide critical reviews of the recent progress and current problems in the scientific analysis of satiation, and they demonstrate the effective deployment of ideas and techniques from molecular, cellular, endocrine, neural, and behavioral biology. The inclusion of a chapter on the satiation of alcohol is novel in a book on food intake, and this signals the convergence of the investigations of satiation in these two areas.

This book is unusual in that all of its contributors except one have done experiments on satiation in the Edward W. Bourne Behavioral Research Laboratory. This has shaped the book and accounts for the experimental and interpretive themes that ripple through the individual chapters.

This background of shared experience may account for why the authors have been unusually good-humored and responsive to my editorial prods. Whatever the reason, I am grateful to all of them.

This book, like the other work at the Bourne Laboratory, would not have been possible without the intellectual and financial support of William T. Lhamon, Robert Michels, and Jack Barchas, the three Chairmen of the Department of Psychiatry here at Cornell since the Bourne Laboratory opened in 1968. I hope they see this book as a good return on their investment.

Finally, I thank James Gibbs, my constant colleague, for discussions about what this book should be and for his help in editing some of the chapters; Berkeley Cooley, my administrative assistant, for careful word-processing of apparently endless revisions; and Jeffrey House, my editor, for helping to make this book better.

White Plains, N.Y. G. P. S.
May, 1997

CONTENTS

CONTRIBUTORS

John D. Davis
Professor of Psychology in Psychiatry
Cornell University Medical College

Nori Geary
Associate Professor of Psychology in Psychiatry
Cornell University Medical College

James Gibbs
Professor of Psychiatry
Cornell University Medical College

Danielle Greenberg
Associate Professor of Psychology in Psychiatry
Cornell University Medical College

Paul J. Kulkosky
Professor of Psychology
University of Southern Colorado

Ellen E. Ladenheim
Assistant Professor of Psychiatry
Johns Hopkins University School of Medicine

Timothy H. Moran
Professor of Psychiatry
Johns Hopkins University School of Medicine

Kenny J. Simansky
Professor of Pharmacology
Medical College of Pennsylvania and
Hahnemann University

Gerard P. Smith
Professor of Psychiatry
Cornell University Medical College

Dennis A. VanderWeele
Professor of Psychology
Occidental College

Satiation: From Gut to Brain

1

Introduction

GERARD P. SMITH

"What stops eating?" The scientific answers to that question are surprising, encouraging, and incomplete: surprising in their complexity, encouraging in how much has been learned in the past two decades, and incomplete at the level of mechanism.

Unless we make a conscious choice to halt food intake, the physiological process that stops eating is unconscious. This process is called *satiation* (Blundell, 1979). It begins early in a meal and terminates when eating ends. As a process that occurs during a meal, satiation is different from postprandial satiety, a state of noneating that begins at the end of one meal and lasts until the beginning of the next meal.

Satiation is a Negative-Feedback Process

Satiation is an example of the kind of active inhibitory process that Sherrington (1947) emphasized as necessary for central neural integration. Because satiation is stimulated by ingested food and inhibits eating, satiation is a negative-feedback process. Identifying satiation as a negative-feedback process is helpful because whenever negative-feedback control has been identified in modern biology, analysis of it has led to insights into the mechanisms of control. This has been true of control at all levels of biological organization, e.g., gene transcription, enzyme activity, neuroendocrine secretion, and somatic movement.

Preabsorptive Sites of Action

As the following chapters show, numerous techniques have been used to analyze the negative feedback produced by ingested food. They include surgery of the brain and visceral nerves, vascular system, and gastrointestinal tract; peptide and monoamine physiology and pharmacology; receptor measurement and function; and various behavioral measurements, such as the microstructure of licking, the expression of associative mechanisms based on prior experience, behavioral sequences, choice, and taste reactivity. What have these techniques revealed about the process of satiation that occurs during a meal?

The first thing is that satiation is elicited primarily, perhaps exclusively, by mechanical and chemical food stimuli acting on specific receptors in the mucosa lining the mouth, esophagus, stomach, and upper small intestine. All of these sites of action are *preabsorptive* in the sense that the nutrient stimuli act prior to absorption into the blood. This includes any effects produced by nutrients during transport through the small intestinal mucosa. That satiation is activated by preabsorptive stimuli is a fundamental change from the traditional perspective that satiation depends primarily on the postabsorptive repletion of metabolic fuels.

A Problem of Sensory Neurophysiology

Recognizing that preabsorptive stimuli stop eating means that understanding the negative-feedback control of the size of meals is primarily a neurological problem rather than a nutritional or metabolic one. Furthermore, it is a neurological problem of the sensory type, demanding answers concerning the site of adequate stimuli, the types of receptors and transduction mechanisms, the specific nerve fibers that carry the peripheral information to the brain, and the components of the central networks that process and integrate this information into inhibition of the motor output for eating (Smith, 1997).

The analysis of satiation appears to be a typical problem in behavioral neuroscience because it is concerned with understanding the mechanisms of the sensory control of behavior. But the analysis of satiation differs from many other problems in behavioral neuroscience in two ways. First, all of the adequate, unconditioned, chemical, and mechanical stimuli for the satiation of eating are derived from the ingestion and digestion of nutrients. Second, the central processing and integration of the preabsorptive chemical and mechanical stimuli are affected by the neural and endocrine effects of a variety of biological and behavioral processes and phenomena. These include metabolic state, growth, pregnancy, lactation, diurnal and ovarian rhythms, temperature, ecological constraints of access to food, prior experience with foods, and, in the human, cultural preferences and aversions.

Direct and Indirect Controls

To emphasize the distinction between the control of satiation by food stimuli acting directly on peripheral preabsorptive receptors to stimulate visceral afferent nerves and other factors acting centrally to modulate the processing and integration of the peripheral information, I recently proposed that the controls of meal size be categorized as *direct* and *indirect.* This highlights the fact that all indirect controls of satiation are expressed by changing the potency of the direct controls of satiation (see Smith, 1996, for details).

Synergistic Interactions

The second revelation from recent work is that satiation involves two types of synergistic interaction — spatial and mechanistic. Spatial synergistic interactions arise from the normal sequential penetration of food from the mouth to the esophagus, stomach, and small intestine. As food stimuli pass through these organs, the mechanical and chemical stimuli of the ingested and digested food play on the mucosal receptors. Because the initial phase of gastric emptying occurs rapidly, simultaneous activation of pregastric, gastric, and small intestinal receptors occurs during a meal. There is now experimental evidence for synergistic satiating interactions between food stimuli acting at pregastric and gastric receptors (Chapter 2), pregastric and small intestinal receptors (Chapter 3), and gastric and small intestinal receptors (Chapter 2).

Satiation also involves mechanistic synergisms for the peripheral transduction of the mechanical and chemical stimuli and for transmission of this information to the brain. These mechanistic synergisms include synergisms between visceral afferent nerves and gut peptides (Chapters 3, 5, and 6), gut peptides with other gut peptides (Chapter 7), and the gut peptides and monoamines (Chapter 5).

Theme and Organization of the Book

The theme of this book is that satiation during a meal is a process of negative-feedback control and depends on peripheral and central mechanisms for the detection, transmission, processing, and integration of the chemical and mechanical stimuli of the ingested food. This theme is expressed in different ways by the authors of the following chapters.

Chapters 2 and 3 by Smith and Greenberg, respectively, review the evidence for preabsorptive sites of satiating stimuli. They emphasize the importance of experimental control of chemical and mechanical stimuli in the mouth, stomach, and small intestine. These chapters contain two major surprises. The first is how little we know about the satiating effect of food in the stomach, and the second is how much we have learned about the satiating effect of food in the small intestine.

In Chapter 4, Davis shows how measuring the rate and pattern of licking within a meal of liquid food, rather than measuring only the total volume ingested, has enriched the investigation of satiation. His microstructural analysis has revealed the time during a meal that conditioned and unconditioned satiating mechanisms act, and he has measured the potency of these mechanisms.

Chapter 5 moves beyond stimulus control and microstructure of the licking response to the peripheral peptides as mechanisms of satiation. Smith and Gibbs review cholecystokinin and the bombesin-like peptides gastrin-releasing peptide and neuromedin B. The work with cholecystokinin that led to its identification as the first molecule to have a satiating effect under physiological conditions is discussed in detail because it provided the experimental framework in which other peptides have been investigated and the criteria by which their physiological function has been judged. It is also a case study of how a satiating effect of an endogenous peptide can be distinguished from nonspecific or toxic effects.

Chapter 6 by Moran and Ladenheim complements Chapter 5, providing a critical discussion of techniques for measuring peptide receptor function and demonstrating the successful use of these techniques to identify and investigate receptors for cholecystokinin and bombesin-like peptides. It shows how the satiating effect of cholecystokinin and bombesin-like peptides has been successfully pursued to the level of cellular function.

The satiating effect of pancreatic glucagon is evaluated by Geary in Chapter 7 and of pancreatic insulin by VanderWeele in Chapter 8. The conclusion is that the prandial secretion of both these peptides has physiological satiating effects under certain conditions.

Chapter 9 takes up the roles of central and peripheral serotonin in satiation. Simansky shows that current pharmacological techniques are powerful tools for analyzing physiological function.

The final chapter by Kulkosky reviews recent work that attempts to apply the analysis of satiation for food to the satiation of alcohol intake. The application has been successful so far, revealing apparently common peptide mechanisms.

The book can be read in two ways. Readers with little background in this subject matter should read the chapters sequentially because the information in the first three chapters concerned with behavioral analysis and identification of peripheral sites and stimuli is essential for understanding the detailed mechanistic information presented in the subsequent chapters.

The other way to read the book is to start with a chapter of interest. This is feasible because each chapter is a review of its topic and contains cross-references to relevant information in other chapters.

This book is concerned with the current knowledge of satiation; it is not a complete account of the control of eating. This would require discussion of the many factors controlling the initiation of eating (Weingarten, 1983; Campfield et al., 1995) and the positive-feedback processes stimulated in the mouth that oppose satiation and act to maintain eating once it has begun (Kirkham and Cooper, 1988; Smith, 1995). It would also need to review the mechanisms of conditioned preferences and aversions (Braveman and Bronstein, 1985; Bolles, 1991) and how and where the brain compares the potency of positive and negative feedbacks.

The recent discoveries of the molecular basis of six forms of genetic obesity in rodents (Leibel et al., 1996) are also not discussed. Given the attention this important work has received, this requires comment. Increased meal size is a phenotype of the forms of genetic obesity that are caused by a defective gene product in adipose tissue in ob/ob mice (Zhang et al., 1994) or by a defect in the receptor for the normal protein from adipose tissue, called leptin, in db/db mice and fa/fa rats (Chua et al., 1996; Tartaglia et al., 1995). Because the lack of normal action of leptin results in abnormally large meals and the administration of leptin to normal mice decreases food intake, leptin has been considered a negative-feedback signal with a satiating effect (Campfield et al., 1995; Zhang et al., 1994). But a decrease in meal size is the beginning, not the end, of the experimental analysis. This follows from the fact that a decrease in meal size is not only produced by an increase in negative feedback, but also by a decrease in positive feedback, or by a combination of these effects (Smith, 1996).

If leptin is shown to have a satiating effect, it would act by modulating the central processing of one or more of the satiating mechanisms discussed in this book so that the potency of those mechanism(s) is enhanced. Thus, the chapters in this book contain essential information for filling in the mechanistic gap between these genetic mutations that result in the lack of normal leptin activity and the abnormally large meal size that is one of the phenotypic expressions of these mutations.

The information contained in these chapters is derived almost entirely from experiments in rodents. We would have liked to discuss the mechanisms of satiation in humans, but almost nothing is known (Blundell et al., 1996).

Although the book does not provide a comprehensive account of the control of food intake, it is the first to deal with the fundamental topic of satiation in a systematic and critical way. It shows how a "top-down" strategy that proceeds from behavioral analysis through systems physiology to cellular and molecular mechanisms has led to increasingly detailed information about the normal structure and mechanisms of satiation in rodents. Because much of this information can be searched for in humans, it will facilitate better understanding and treatment of the hyperphagia that occurs in obesity and other eating disorders.

REFERENCES

Blundell, J. *Hunger, appetite and satiety—Constructs in search of identities.* In Turner, M. (ed.): *Proceedings of the British Nutrition Foundation,* First Annual Conference. London: Applied Science Pub. Ltd., 1979, pp. 21–42.

Blundell, J. E., C. L. Lawton, J. R. Cotton, and J. I. Macdiarmid. Control of human appetite: Implications for the intake of dietary fat. *Annu. Rev. Nutr.* 16:285–319, 1996.

Bolles, R. C. *The Hedonics of Taste.* Hillsdale, NJ: Lawrence Erlbaum Associates, 1991.

Braveman, N. S., and P. Bronstein (eds.): *Experimental Assessments and Clinical Applications of Conditioned Food Aversions. Vol. 443.* New York: Annals of New York Academy of Science, 1985.

Campfield, L. A., F. J. Smith, Y. Guisez, R. Devos, and P. Burn. Recombinant mouse OB protein: Evidence for a peripheral signal linking adiposity and central neural networks. *Science* 269:546–549, 1995.

Chua, S. C., W. K. Chung, X. S. Wu-Peng et al. Phenotypes of mouse *diabetes* and rat *fatty* due to mutations in the OB (leptin) receptor. *Science* 271:994–996, 1996.

Kirkham, T. C. and S. J. Cooper. Attenuation of sham feeding by naloxone is stereospecific: Evidence for opioid mediation of orosensory reward. *Physiol. Behav.* 43:845–848, 1988.

Leibel, R. L., S. C. Chua, Jr., and W. K. Chung. Animal models of genetic obesity. In Angel, A., H. Anderson, C. Bouchard, D. Lau, L. Leiter, and R. Mendelson (eds.): *Progress in Obesity Research: 7.* London: John Libbey & Company, 1996, pp. 263–271.

Sherrington, C. *The Integrative Action of the Nervous System. 2nd Ed.* New Haven: Yale University Press, pp. 329–330, 1947.

Smith, G. P. *Dopamine and food reward.* Prog. Psychobiol. Physiol. Psychol., 16:83–144, 1995.

Smith, G. P. The direct and indirect controls of meal size. *Neurosci. Biobehav. Rev.* 20:41–46, 1996.

Smith, G. P. *Feeding: control of eating.* In Adelman, G., and B. Smith (eds.):

Encyclopedia of Neuroscience CD-ROM. 2nd Ed. New York: Elsevier Science Inc., 1997, (in press).

Tartaglia, L. A., M. Dembski, and X. Weng et al. Identification and expression cloning of a leptin receptor, OB-R. *Cell* 83:1263–1271, 1995.

Weingarten, H. P. Conditioned cues elicit feeding in sated rats: A role for learning in meal initiation. *Science* 220:431–433, 1983.

Zhang, Y., R. Porcina, M. Maffei, M. Barone, L. Leopold, and J. M. Friedman. Positional cloning of the mouse obese gene and its human homologue. *Nature* 372:425–432, 1994.

2

Pregastric and Gastric Satiety

GERARD P. SMITH

In 1885, Pavlov devised a new operative technique to study the control of gastric secretion in dogs. The technique consisted of making a longitudinal incision in the anterior wall of the esophagus and then suturing the edges of the esophagus to the edges of the overlying incision. This produced a chronic esophageal fistula. The purpose of the esophageal fistula was to allow ingested food to drain out it so that gastric secretion during eating could be collected from another, chronic gastric, fistula without being mixed with food particles. The dogs were maintained in good nutrition by introducing food through the chronic gastric fistula.

In a classic series of experiments, Shumova-Simonovskia and Pavlov used this technique to demonstrate that food in the mouth was a much more potent stimulus for gastric acid secretion than food in the stomach and that this effect of food in the mouth was a cephalic reflex mediated by the efferent fibers of the vagal nerves (Pavlov, 1910).

In the course of these experiments, Pavlov made a fundamental discovery about satiation. When ingested food drained out of the esophageal fistula, dogs ate for hours. Satiation apparently never occurred! The amount of food eaten when it drained out of the esophageal fistula depended on the type of food eaten and on the duration of food deprivation. Pavlov called this *fictional feeding,* but in the English literature it has always been referred to as *sham feeding* (Smith, 1995b).

Sham feeding is central to the modern analysis of satiation and satiety. This analysis has been carried out mostly in the rat prepared with a chronic gastric fistula (Young et al., 1974). Less frequently, it has been carried out in the rat with a chronic esophageal fistula (Mook, 1963) or in the rat with a chronic gastric catheter that permits continuous withdrawal of gastric contents during eating (Davis and Campbell, 1973).

This analysis has revealed unsuspected satiation effects during sham feeding. Because ingested food contacts receptors in the mouth and the esophagus during sham feeding,* the satiating effect of sham-fed food is referred to as *pregastric satiety* to distinguish it from gastric and intestinal satiety (Smith and Gibbs, 1979).

Pregastric Satiation

To reveal pregastric satiating effects of food that are part of the normal control of food intake, satiety should be measured during the first time rats sham feed the maintenance diet after an interval of food deprivation that is within the range of the normal intermeal interval. Kraly et al. (1978) satisfied these requirements by testing the intake of a maintenance diet by gastric fistula rats after 3 hours of food deprivation in the light phase. Time-sampled observations of behavior were used to generate three measures of satiety: (1) meal size, (2) latency to rest, and (3) duration of the intermeal interval. These measures were based on the identification of the behavioral sequence of satiety. This sequence consists of the termination of eating, followed by exploratory behaviors and grooming, and culminates in resting behavior or apparent sleep (Antin et al., 1975; Bolles, 1960; Richter, 1922). Although its specificity is controversial (Deutsch, 1990), this sequence is a reliable marker of postprandial satiety when rats and rhesus monkeys (Gibbs and Falasco, 1978) are housed singly with only food and water available, i.e., the usual test conditions for investigating the control of food intake.

Kraly and co-workers used the onset of resting behavior to denote postprandial satiety. All food eaten before the onset of resting constituted the size of the test meal. The time from the beginning of the test meal to the onset of resting was the latency to rest, and the time from the onset of resting to the beginning of the next meal was the intermeal interval.

The behavioral sequence of satiety occurred after the first sham-fed meal, but meal size doubled and the intermeal interval was about 50% shorter than was observed in the preceding real feeding test (Table 2.1). Because the direction of these changes would increase food intake, Kraly et

* Although ingested food also always contacts gastric receptors during sham feeding in the gastric fistula rat, and a small amount may, in some experiments, contact intestinal or postabsorptive receptors (Nissenbaum and Sclafani, 1987; Sclafani and Nissenbaum, 1985; Grill et al., 1984), such contact is not sufficient to affect the rate or duration of sham feeding (Nissenbaum and Sclafani, 1987) or the hormonal (Berthoud and Jeanrenaud, 1982) or metabolic (Weingarten and Powley, 1980) responses elicited by it.

TABLE 2.1.
Measures of Satiety During Real and Sham Feeding[a]

	Real	Sham 1	Sham 2
Meal size (ml)	4.9 ± 0.5	9.4 ± 1.8*	10.1 ± 1.5 **
Latency to rest (min)	12.1 ± 0.9	12.6 ± 1.8	15.5 ± 2.0
Intermeal interval (min)	56.5 (22.0–190.5)	25.0[†] (2.5–129.0)	23.5[†] (5.0–48.0)

[a]Data are mean ± SE for meal size and latency to rest and are median (range) for intermeal interval. Meal size was significantly larger during the first (Sham 1) and second (Sham 2) sham-feeding tests than during the last of at least five consecutive real-feeding tests (Real) $*p < 0.05$, $**p < 0.01$, matched-pairs t-test. Intermeal interval was significantly smaller during both sham-feeding tests than during the real-feeding test, $^{\dagger}p < 0.01$, Wilcoxon signed-ranks test. From Kraly et al. (1978), with permission of the publisher.

al. concluded that the liquid food was efficacious for satiety during sham feeding, but its potency was reduced. When a second sham-feeding test was performed, the results were the same (Table 2.1).

Although the satiating potency of the test diet was less during sham feeding than real feeding, these experiments demonstrated that, when pregastric satiating mechanisms were activated in the absence of gastric or small intestinal satiating mechanisms, pregastric mechanisms stopped eating, elicited the behavioral sequence of satiety, and produced a substantial intermeal interval. What modulates the potency of the pregastric satiating effects of food stimuli?

Modulation of Satiating Potency of Pregastric Stimuli

Four factors are known to modulate the satiating potency of pregastric stimuli: (1) diurnal rhythm, (2) duration of food deprivation, (3) kind and concentration of liquid food stimuli, and (4) ingestive experience with food stimuli.

Diurnal Rhythm

When Kraly et al. (1980) used the same liquid food and 3 hours of food deprivation to study sham feeding in the dark phase rather than in the light phase, they obtained different results. Sham-fed meal size was larger, the latency to rest was longer in the dark phase than in the light phase, but the intermeal interval was not different (Fig. 2.1). The larger meal followed by the appearance of the normal behavioral sequence of satiety is evidence that the satiating potency of the high-carbohydrate test diet decreased

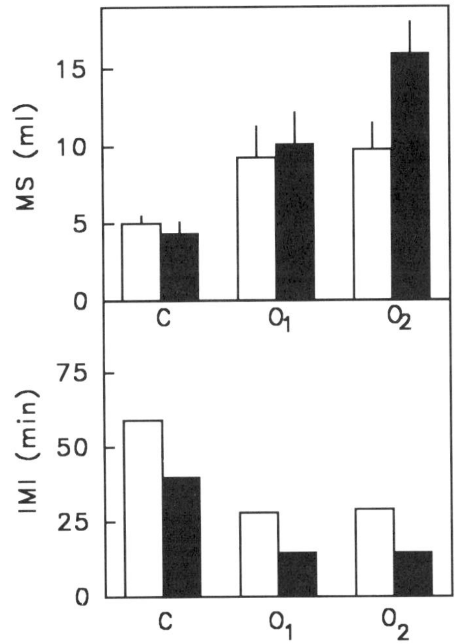

FIGURE 2.1. Mean meal size (MS) and postprandial median intermeal interval (IMI) for rats eating after 3-hour food deprivation at the midpoint of the day (open bars) or night (closed bars) phase of the diurnal cycle. Rats ate with the gastric cannula closed (normal feeding) on test C and with the cannula open (sham feeding) on tests 0_1 and 0_2. Vertical lines are SE. From Kraly et al. (1980), with permission of the publisher.

significantly under these conditions. Furthermore, the failure of diurnal rhythm to change the intermeal interval suggests that meal size is more sensitive to changes in the satiating potency of pregastric stimuli than the intermeal interval is. These results and the other evidence contained in their paper suggested that decreased satiation is the critical change in the controls of meal size that accounts for larger meals observed in the dark phase than in the light phase during spontaneous feeding. It is remarkable that no further experiments concerned with this issue have appeared. This is a fundamental problem wide open to investigation.

Duration of Deprivation

Increasing the duration of food deprivation from the 3–4 hours to 17 or more hours markedly decreases the satiating potency of pregastric stimuli. For example, after 17 hours of food deprivation, sham feeding continued for 7.5 hours in rats without the occurrence of the behavioral sequence of satiety (Young et al., 1974). Continuous sham feeding has also been observed in rhesus monkeys after overnight food deprivation (Gibbs et al., 1981; Gibbs and Falasco, 1978; Falasco et al., 1979).

Although the satiating potency of pregastric stimuli was reduced after 17 hours of food deprivation, subsequent experiments demonstrated that it was not abolished. To search for subthreshold satiating effects, Antin et al. (1977) combined duodenal infusions of the liquid diet being sham fed with sham feeding. The duodenal infusion had been shown to stop sham feeding and elicit the behavioral sequence of satiety after 17 hours of food deprivation (Liebling et al., 1975) (see Chapter 3). Antin et al. (1977) fixed the duration of the duodenal infusion at 6 minutes and then varied the time of the infusion in relationship to the onset of sham feeding (Table 2.2). When the 6-minute infusion began 12 minutes before sham feeding, it had *no* effect. When it began 6 minutes before, at the same time, or 6 minutes after sham feeding began, the infusion produced a significant decrease in intake. When the infusion began 12 minutes after the onset of sham feeding, the percentage of inhibition of intake was significantly larger. Because the content and duration of the duodenal infusion were constant, the change in its satiating potency must have been due to temporal summation

TABLE 2.2.
Sham Feeding by five Rats after Duodenal Infusions of Liquid Food or Saline that Preceded, Coincided with, or Followed Presentation of Liquid Food at Time 0

Time Infusion Began (min)	Intake After 0.15 M NaCl Infusion	Intake After Liquid Food Infusion	Percent Inhibition of Intake for Group	No. of Rats Inhibited	Duration of Inhibition
−12	63 ± 4	64 ± 7	−3	1	7 ± 8
−6	68 ± 7	41 ± 3**	39	5	23 ± 3
0	61 ± 6	43 ± 9*	30	5	23 ± 4
+6	63 ± 7	34 ± 9**	45	5	30 ± 5*
+12	63 ± 5	15 ± 11**	76	5	38 ± 3**

Measurement of sham intakes during a 30-minute interval began after a 12-minute delay following the beginning of each duodenal infusion. This delay permitted comparison of the same interval relative to the beginning of each infusion. It was the minimum delay allowed by the temporal constraint of including the −12-minute infusion.

Inhibition is defined as an intake during any one 6-minute interval below the range of intakes for the same rat on an adjacent saline-infusion day.

Intake after liquid infusion was significantly smaller than intake after saline infusion, $^{*}p < 0.05$, $^{**}p < 0.01$, matched-pairs t-test, one-tailed.

Duration of inhibition after infusion at +6 minutes was significantly longer than duration after infusions at −6 and −12 minutes, $^{*}p < 0.05$; duration of inhibition after infusion at +12 was significantly longer than all other durations, $^{**}p < 0.025$, matched-pairs t-test, one-tailed.

From Antin et al. (1977), with permission of the publisher.

of the satiating effects of intestinal and pregastric stimuli. Furthermore, the synergism was not identical at all times. The satiating potency of pregastric stimuli was significantly larger after 12 minutes of sham feeding than after 6 minutes of sham feeding. Very similar results occurred when cholecystokinin (CCK), a mechanism of intestinal satiety (see Chapter 3), was injected intraperitoneally at the same times in relation to the onset of sham feeding (Forsyth et al., 1985; Antin et al., 1978).

Thus, the temporal summation between duration of sham feeding and intestinal infusion or CCK demonstrates that overnight deprivation of food reduces the satiating potency of pregastric stimuli below the threshold for eliciting satiety, but it does not abolish it. Much remains to be learned about the quantitative aspects of the relation(s) between duration of food deprivation and the satiating potency of pregastric stimuli. For example, deprivation affects the sham-fed intake of oils much more than the intake of carbohydrates (Joyner et al., 1985; Mindell et al., 1990). Furthermore, nothing is known about the neural site and mechanisms that mediate these relation(s).

Kind and Concentration of Stimulus

If phase of the diurnal rhythm and duration of food deprivation are fixed, the kind and concentration of the orosensory stimulus control the rate of sham feeding and, thus, the amount ingested during a test. For both carbohydrate solutions and oil emulsions, the rate of sham feeding increases linearly as concentration increases until the rate plateaus or decreases at high concentrations (Grill and Kaplan, 1992; Mindell et al., 1990; Joyner et al., 1985; Nissenbaum and Sclafani, 1987; Weingarten and Watson, 1982; Mook, 1963). In the linear, ascending part of the concentration-ingestion function, it is clear that increasing the concentration of the orosensory stimulus produces a larger increase in the positive-feedback, afferent stimuli that maintain sham feeding than in the negative-feedback, satiating stimuli that slow and terminate sham feeding (Smith, 1996). Because rats ingest more of most of these concentrations during sham feeding than real feeding, the larger intake during sham feeding results from the lack of the usual negative-feedback effects.

Intake is not always larger during sham feeding than real feeding. Rats ingest equivalent volumes of dilute carbohydrate solutions (0.1 M) during sham and real feeding (Grill and Kaplan, 1992; Nissenbaum and Sclafani, 1987; Mook and Wagner, 1988). This is the only situation that has been identified in which the satiating effect of pregastric stimuli alone is sufficient to control meal size in the normal range.

Because a significant volume of dilute carbohydrate is ingested during sham feeding, the initial difference between the positive-feedback effect and the negative-feedback effect of pregastric stimuli must decrease during the meal until a central comparator mechanism judges the magnitude of the feedbacks to be equal and it commands eating to stop (Smith, 1996). Two time-dependent processes, metering and habituation, have been proposed to account for the increase in the relative satiating potency of pregastric negative-feedback stimuli under these conditions.

Metering has been proposed by Mook and his colleagues (see Mook, 1990). They found that the number of licks elicited by an ingested liquid, rather than its chemosensory properties, correlated with the satiating potency of the liquid. Thus, they suggested that sensory negative feedback associated with oral or esophageal movements is the neural substrate for the pregastric satiating effect under these conditions. Mook believes that the complexity of this metering requires central mechanisms to determine a target number of licks, to "count" the licks, to compare the number of licks elicited with the target number, and to stop eating when the target number is achieved, but no relevant experiments have been done.

Habituation provides a very different explanation because it focuses on a loss of potency of orosensory stimuli to maintain eating as a result of repetitive chemosensory stimulation rather than an increase in negative feedback. Swithers-Mulvey and Hall (1992) and Swithers and Hall (1994) demonstrated habituation by observing the diminishing response to *intermittent* delivery of carbohydrates or milk through chronically implanted oral cannulas that could not be accounted for by orosensory fatigue or adaptation because disinhibition occurred when a different flavored liquid was infused. This observation was first made in preweanling pups and then in adult rats. Habituation of the orosensory stimuli that maintain eating could contribute to the termination of eating because the decreasing rate of eating that culminates in the termination of eating is a dynamic interaction between positive- and negative-feedback effects of ingested food (Smith, 1996). Thus, habituation could contribute to the *relative* increase in satiating potency of pregastric stimuli that occurs during ingestion of a dilute carbohydrate solution.

Sensory-specific satiety, i.e., the decreased intake in animals or ratings of pleasantness in humans that occurs after prior ingestion of a specific food, involves pregastric stimuli (Rolls, 1986; Hetherington et al., 1989). The relationship of habituation of pregastric stimuli to sensory-specific satiety is not clear.

The relative importance of metering, sensory-specific satiety, and habituation to the pregastric control of meal size remains to be determined. They are not mutually exclusive and could operate simultaneously. In fact,

Swithers-Mulvey and Hall (1992) suggest that metering and sensory-specific satiety are instances of oral habituation because oral habituation is the basic mechanism by which ongoing oral experience can control ingestive behavior. That suggestion deserves further testing.

Experience

During the first sham-feeding test, rats eat more than during real feeding (except in the case of dilute carbohydrate solutions; see above). The increased intake, however, is not maximal. Intake increases on subsequent sham-feeding tests until it plateaus after three tests when carbohydrates or milk are the test diets (Mook et al., 1983; Young et al., 1974; Davis and Campbell, 1973) and after about five tests when oils are sham fed (Mindell et al., 1990). Similar results have been observed in rhesus monkeys (Gibbs and Falasco, 1978). This is not simply a novelty effect of sham feeding because it occurs when a new liquid is sham fed after a rat has had numerous sham-feeding tests with a different liquid (Mook et al., 1983; Davis and Campbell, 1973). Instead, the progressive increase in sham feeding during the first three or more tests is the result of extinguishing a conditioned inhibition that is due to the association of taste and postingestive negative feedback formed during prior real-feeding tests Weingarten and Kulikovsky (1989). Two real-feeding tests interposed between sham-feeding tests are sufficient to maintain this conditioned inhibition (Davis and Smith, 1990). Microstructural analysis revealed that this conditioned inhibition occurs during the first 5 minutes of a meal (see Chapter 4 for details). Gastric and postpyloric stimuli function as unconditioned stimuli for this association between taste and postingestive stimuli (Davis et al., 1993; Davis and Smith, 1990), but increases in plasma glucose do not (Gowans and Weingarten, 1991).

If extinction of conditioned inhibition accounts for the gradual rise to a maximum of intake during sham feeding, what accounts for the increase of intake on the first sham-feeding test? The answer is the removal of negative-feedback, satiating stimuli acting in the stomach and beyond the pylorus, probably in the small intestine (Davis et al., 1993; Davis and Smith, 1990). Microstructural analysis revealed that removal of these negative-feedback stimuli produced an increase in intake after about 4–6 minutes during a meal of 0.8 M sucrose, i.e., after the time when conditioned inhibition operates, and that the effect is maximal on the first test (see Chapter 4 for details). Because repeated experience did not change the effect of the removal of these negative-feedback stimuli on intake, Davis et al. concluded that these stimuli elicited unconditioned, negative-feedback controls of intake.

Cranial nerves 2, 5, 7, 9, and 10 are candidate pathways for carrying the effects of pregastric stimuli to the brain for processing as negative-feedback controls of eating, but the effect of deafferentation on sham feeding has not been investigated. In contrast, there are four studies of the effects of central lesions. Bilateral lesions of the ventromedial hypothalamic nucleus (VMH) increased sham-fed intake of sucrose (Weingarten, 1982; Black and Weingarten, 1988) and of sweet milk (Cox and Smith, 1986). Bilateral knife cuts of the stria terminalis (ST) also increased sham feeding of sucrose (Black and Weingarten, 1988), but large bilateral ibotenic acid lesions of the basolateral amygdala decreased sham-fed intake of sucrose (Siegel et al., 1988). There was an interaction between the effect of the lesions and the concentration of sucrose sham fed. The VMH and ST lesions increased intake of concentrated solutions more than dilute solutions, while the amygdaloid lesions reduced the intake of 2.5% and 5% sucrose, but not of 10% or 40% sucrose.

Because pregastric food stimuli stimulate positive-feedback mechanisms that maintain eating as well as negative-feedback, satiating mechanisms, the changes in sham-fed intake produced by these neural lesions could result from an increase in positive feedback, a decrease in negative feedback, or both (Smith, 1996). Only further analysis can determine if any of these changes can be attributed to a change in the potency of pregastric satiating mechanisms.

It is likely that the caudal brainstem has sufficient complexity to process pregastric stimuli into satiating effects of terminating meal size and producing an intermeal interval, because chronic decerebrate rats show apparently normal satiation and postprandial satiety (Grill, 1986). Furthermore, Swithers-Mulvey and Hall (1992) recently reported normal oral habituation after acute decerebration in preweanling rat pups.

Recent work has implicated central and peripheral serotonergic mechanisms in pregastric satiety (see Chapter 9). Although dopamine and opioid antagonists decrease sham feeding, current evidence suggests that this represents attenuation of the potency of orosensory positive feedback rather than an enhancement of pregastric negative feedback (Smith, 1995a; Kirkham and Cooper, 1988).

Gastric Satiation

Although the stomach has figured prominently in ideas about the control of eating for a long time, modern analysis of its contribution to the satiating effect of food began when Hall (1973) developed a technique for acute and reversible blockade of gastric emptying so that gastric stimuli could be investigated in the absence of small intestinal stimuli. Until that time, all liquid preloads of the stomach changed stimuli in the stomach and small intestine because stomach contents begin emptying quickly after a gastric preload. Removal of stomach contents also changed stimuli in the small intestine because it interrupted or reduced gastric emptying. The only stimulus confined to the stomach was inflation of a balloon, but the relevance of these results to the normal controls of eating was not clear (for details of this problem, see Smith and Gibbs, 1979).

Hall's technique (1973) was a noose that was chronically implanted around the pyloric sphincter. The ends of the noose were exteriorized at the rear of the neck. This permitted the noose to be tightened remotely to close the pylorus before a feeding test. When the animal ate, all of the ingested food remained in the stomach. At the end of the experiment, the noose was loosened. This permitted gastric emptying of ingested food to occur outside the test situation so that animals maintained their normal nutritional state. This technique was modified by Young and Deutsch (1981). They implanted a small, inflatable cuff around the pylorus. When the cuff was inflated, the pylorus was closed. Because the cuff can be used for a longer series of experiments than the noose, it has become the standard technique for closing the pylorus acutely.

Acute closure of the pylorus permitted the satiating effect of food in the stomach to be measured in the absence of postpyloric stimuli. By comparing intake when ingested food accumulated in the stomach with the pylorus closed to intake during sham feeding when ingested food drained out of the stomach, the satiating effect of food in the stomach could be measured.

In 1978, the first reports appeared using this technique to study the control of meal size. Kraly and Smith (1978) used the same conditions (3 hours of food deprivation in the light phase) as Kraly et al. (1978) used to study sham feeding. Recall that meal size doubled when it was sham fed under these conditions in the earlier study (Kraly et al., 1978). But when the pylorus was closed, meal size was not different from meal size when rats ate normally with the pylorus open. Furthermore, the latency to resting behavior in the behavioral sequence of satiety was also normal.

The results of this experiment suggested that gastric stimuli made a significant contribution to the satiating effect of ingested food, because, when pregastric and gastric stimuli occurred simultaneously in the absence of postpyloric stimuli, meal size was normal. Intragastric volume at the end of the meal in cuff-closed rats, however, was almost 50% larger than the intragastric volume in cuff-open rats. This suggested that gastric stimuli were abnormally large in cuff-closed rats.

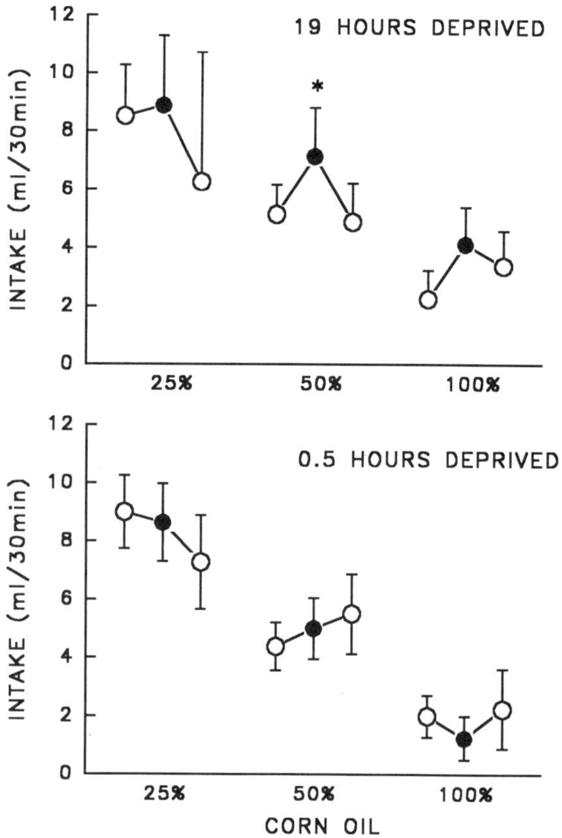

FIGURE 2.2. Mean ± SE ml/30 minutes of 25%, 50%, or 100% corn oil ingested with a pyloric cuff open (open circles) and a pyloric cuff closed (closed circles) on sequential test days after 19 or 0.5 hours of food deprivation. *Intake with a pyloric cuff closed was significantly larger than intake with a pyloric cuff open, $p < 0.05$. From Rauhofer et al. (1993), with permission of the publisher.

Deutsch et al. (1978) obtained the same result when rats ingested milk. Furthermore, they showed that excessive intragastric pressure due to the accumulation of ingested milk in the cuff-closed stomach was not necessary for the control of meal size by equipping rats with an overflow valve that maintained intragastric pressure equal to that observed at the termination of a meal when the cuff was open. This blunts the criticism that excess intragastric volume (Seeley et al., 1995; Kraly and Smith, 1978) is necessary for the normal meal size observed in cuff-closed rats. Deutsch et al. (1978) also provided data that appeared to eliminate the possibility that the inflated cuff produced aversive stimuli.

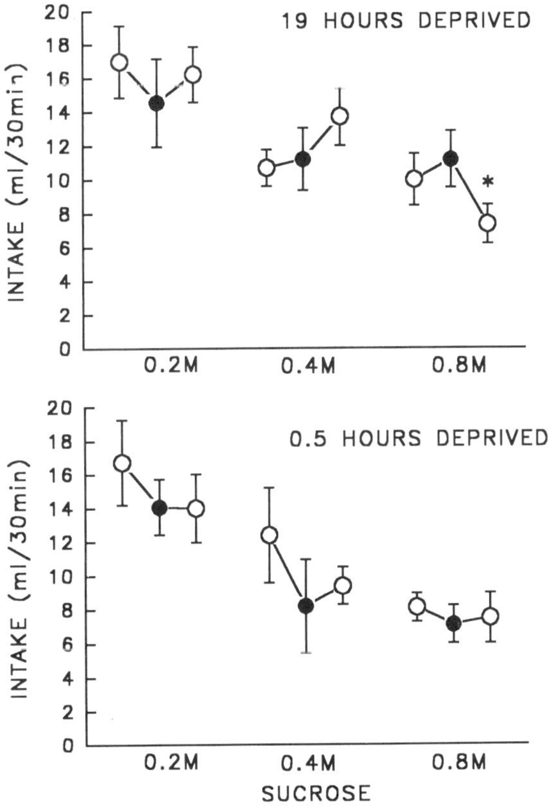

FIGURE 2.3. Mean ± SE ml/30 min of 0.2, 0.4, or 0.8 M sucrose ingested with a pyloric cuff open (open circles) and with a pyloric cuff closed (closed circles) on sequential test days after 19 or 0.5 hours of food deprivation. *Intake with a cuff open was significantly smaller than intake with a cuff closed, $p < 0.05$. From Rauhofer et al. (1993), with permission of the publisher.

The ability of the combination of pregastric and gastric stimuli that occur during eating when the pylorus is closed to control meal size equivalent to meal size when the pylorus is opened has now been demonstrated with milk (Gonzalez and Deutsch, 1981; Deutsch et al., 1978), three nutritionally complete liquid diets (Phillips and Powley, 1996; Deutsch et al., 1980; Kraly and Smith, 1978), glucose (Seeley et al., 1995), and three concentrations of sucrose and of corn oil after short or long durations of food deprivations (Rauhofer et al., 1993) (Figs. 2.2, 2.3).

The only exception occurred in the experiments of Deutsch et al. (1980). All 16 rats ingested less milk when the cuffs were closed than when they were open (see control intakes in their Table 1). Note that the difference between intake under the two conditions was larger when intake with cuffs open was larger than 20 ml.

Although these results demonstrate that the combination of pregastric and gastric stimuli is sufficient to control meal size, they do not reveal the relative contribution of pregastric and gastric satiating mechanisms. To attempt to gain more evidence about the potency of gastric satiating mechanisms, investigators measured the effect on intake of withdrawing gastric contents after a meal or of infusing saline or nutrients during the meal.

Gastric Withdrawal Experiments

Do gastric stimuli merely provide a sensory cue to the central satiety system that is necessary for the efficacy of pregastric satiating stimuli, or do gastric stimuli provide a separate, quantitative, negative feedback related to the mechanical or chemical aspects of gastric contents? One way to answer this question is to remove gastric contents during or immediately after a meal. If the rat eats an additional quantity of food that is correlated with the amount withdrawn from the stomach, this would be consistent with a quantitative gastric signal.

Davis and Campbell (1973) withdrew 3.5–11.0 ml of gastric contents in 0.5-ml increments 20 minutes after eating began on different test days. They found that milk intake over the next 30 minutes was quite close to the volume of gastric contents removed. The withdrawal of gastric contents decreased gastric volume directly, but it also had indirect effects on small intestinal nutrient stimulation because the decreased gastric volume probably decreased gastric emptying.

Deutsch et al. (1978) attempted to gain more direct evidence for the role of the stomach in the mediation of intake after gastric withdrawal by withdrawing gastric contents from the stomach isolated from the small intestine by an inflated pyloric cuff. Four rats were allowed to drink milk for 20 minutes with cuffs open (intakes during this interval were not presented). Then the milk was removed for 5 minutes. During that 5 minutes, the pyloric cuffs were closed for two tests or left open for two tests. In one of the two tests with each cuff condition, 10 ml of gastric content was withdrawn. In the other tests, there was no gastric withdrawal. At the end of the 5 minutes, rats were given access to milk for 25 minutes and intakes were measured. The results of the four different tests are presented in Table 2.3. When cuffs were closed, rats ingested less milk than when cuffs were open. This occurred when gastric contents were withdrawn and when gastric contents were left in place. This is quite different from the equal intakes when the cuff is closed or open during a single test (see above) and

TABLE 2.3.
Compensatory Intake after Gastric Withdrawal with Pyloric Cuffs Open or Closed

Pyloric Cuff	Volume Withdrawn	Intake	Net Compensatory Intake
Closed	10	8.9	7.0
Closed	0	1.9	
Open	10	15.1	7.4
Open	0	7.7	

Four rats were allowed to drink milk for 20 minutes; the pyloric cuffs were open. Then milk was removed, the cuffs were closed or left open, and 10 ml of gastric content was withdrawn in two tests or not withdrawn in two tests. At the end of 5 minutes, rats had access to milk for 25 minutes and the intakes (ml) were measured. Net compensatory intake was calculated by subtracting the intake after no gastric withdrawal from the intake after 10 ml was withdrawn from the stomach. Although intakes when the cuffs were closed were smaller than intakes when the cuffs were open, the calculated net compensatory intakes were similar. Data are from Table 1 in Deutsch et al. (1978).

suggests that the postpyloric stimulation provided by the gastric contents that emptied into the small intestine during the 20 minutes of access to milk prior to cuff inflation increased the satiating potency of pregastric or gastric stimuli during the second intake interval.

This interesting and potentially important result was interpreted differently by the authors. They suggested that, when the cuff is open, the gastric contents emptied into the duodenum are functionally equivalent to the withdrawal of gastric contents from the stomach so that the rats drink more to compensate. They support this interpretation by calculating the net compensatory intake by subtracting the mean intake after no withdrawal from the mean intake after withdrawal of 10 ml in both cuff conditions (no estimate of variability was provided). The results were very similar (Table 2.3).

They tested their interpretation in another experiment using the same four rats. After ingesting milk for 20 minutes with cuffs closed, the cuffs were opened for 30 minutes and one of two conditions was imposed. In the first condition, rats had access to milk during the last 20 minutes of the 30-minute period. The mean intake was 12.8 ml. In the other condition, rats had no access to milk during the 30 minutes. Then the cuffs were closed and rats had access to milk for 20 minutes. The mean intake (12.7 ml) was not significantly different from the intake during 20 minutes of ingestion when the cuff was open. This result was considered evidence for the interpretation that gastric emptying was functionally equivalent to gastric withdrawal.

Note that Deutsch and his colleagues never measured gastric emptying or gastric volume in any of these experiments. Both of these measures are relevant to their interpretation of the results. Despite this important omission, the authors made a strong inference from these experiments and concluded that "these results demonstrate that signals arising from the stomach itself (as distinct from the duodenum and jejunum) regulate meal size and produce satiety" (Deutsch et al., 1978).

Deutsch and Gonzalez pursued the problem in two further experiments. In the first experiment (Deutsch and Gonzalez, 1980), rats ingested a mixed diet (Nutrico) for 10 minutes with cuffs open. Then cuffs were inflated. On separate days, gastric contents were withdrawn or left in place, and the rats were given access to Nutrico for 30 minutes. Rats showed compensatory intake after the withdrawal of gastric contents. This appears to provide strong evidence for the hypothesis, because withdrawal and compensatory intake occurred with the cuff closed so that small intestinal stimuli were not changed by withdrawal and could not be affected by the ingested diet during compensatory intake. The possibility that postpyloric nutrient stimuli produced during the 10 minutes of ingestion with cuffs open could affect the gastric signals produced by ingestion when the cuffs were closed was not considered. As we shall see, this was a mistake.

The second experiment (Gonzalez and Deutsch, 1981) used three cuff-closed rats eating milk after overnight food deprivation. When the first 5 ml of ingested milk was withdrawn, rats showed compensatory intake. This experiment excluded all postpyloric stimuli. Although these results are constrained by the small number of rats, they are the best evidence available for the hypothesis that compensatory intake in response to withdrawal that only affects gastric stimuli can be controlled by gastric and pregastric stimuli without any simultaneous postpyloric stimuli. This shows that gastric and pregastric stimuli are sufficient for compensatory intake after withdrawal, but it does not exclude an effect of postpyloric stimuli when they occur during compensatory intake with cuffs open.

These results provoked other experiments in monkeys and rats. Gibbs et al. (1981) demonstrated that compensatory intake occurred after withdrawal from the stomach or duodenum of rhesus monkeys. The compensatory intake after duodenal withdrawal was apparently not consistent with the hypothesis that the withdrawal of only gastric stimuli is necessary for compensatory intake. It was possible to save the hypothesis, however, by saying that duodenal withdrawal increased gastric emptying; thus, its effect was due to withdrawing gastric contents, although not as completely as gastric withdrawal did. The observation that intake after duodenal withdrawal was less than after gastric withdrawal in these experiments is consistent with that interpretation.

Other experiments came from McHugh's laboratory. The first experiment used a combination of gastric preload, withdrawal, and infusion in monkeys eating chow (McHugh, 1979). Twenty minutes before the feeding period, 150 ml of glucose (0.25 g/ml) was infused into the stomach through a chronic gastric catheter. Just before the feeding period, three gastric treatments were performed on separate days: (1) gastric contents were withdrawn; (2) gastric contents were withdrawn and 150 ml of isotonic saline was infused into the stomach; and (3) gastric contents were left in place. Monkeys ate about twice as much in the first 30 minutes after withdrawal of gastric contents than after withdrawal plus saline infusion or after no withdrawal. The equivalent effect on intake of leaving glucose in the stomach or of replacing it by an equal volume of saline suggested that volume was the sufficient gastric stimulus and that glucose in the gastric contents was not necessary to prevent this intake after withdrawal of gastric contents.

It is important to note that the inhibitory effect on intake of 150 ml of saline in the stomach required some glucose to have emptied from the stomach into the small intestine. When the gastric infusion of saline was given without a prior glucose preload, the saline infusion had no effect on intake (McHugh and Moran, 1985).

The authors also pointed out that the glucose emptied from the stomach could not prevent intake in response to gastric withdrawal when there was no gastric stimulus. This is the situation when gastric contents were withdrawn and not replaced with saline. Thus, gastric volumetric stimuli required postpyloric glucose to be efficacious. The requirement for postpyloric glucose was attributed to its slowing of gastric emptying, which maintained the gastric volume of the saline load longer than when no glucose was in the small intestine.

Wirth and McHugh (1983) refined this approach to the problem. Monkeys ingested glucose for 15 minutes. The glucose contained phenol red, a marker for the measurement of gastric emptying. On different test days, intake of glucose was measured in a sequence of 15-minute tests (two to nine tests). Each test was preceded by one of three gastric manipulations: (1) withdrawal of gastric contents (glucose and secretions); (2) withdrawal of gastric contents and immediate replacement with an equal volume of isotonic saline; and (3) no withdrawal, leaving the gastric contents including the ingested glucose undisturbed.

By measuring the gastric emptying of glucose, Wirth and McHugh (1983) showed that intake varied according to the state of the stomach when equivalent amounts of glucose had emptied from the stomach. When the stomach was empty, six of seven monkeys ate significantly more. This response to gastric withdrawal could be abolished if the stomach contained ingested glucose or 150 ml of saline. The equivalent effect on intake of leaving glucose in the stomach or of replacing it by an equal volume of saline demonstrated again that volume was the sufficient gastric stimulus for controlling the magnitude of the intake after gastric withdrawal.

But, as in the preceding experiment, gastric volume did not act alone in these experiments, because glucose was present in the small intestine in all of the experiments. Furthermore, as the authors noted, the satiating potency of the postpyloric glucose alone or interacting synergistically with pregastric stimuli increased during the sequence of four or nine bouts of ingestion because the mean intake of glucose in each bout gradually declined during successive bouts even when gastric contents were removed and not replaced before each bout. This is evidence for the action of nongastric satiating mechanisms.

Thus, gastric volumetric stimuli have a satiating effect under these conditions, but the relative contributions of gastric, pregastric, and postpyloric stimuli to the control of intake were not determined.

Kaplan et al. (1994) recently carried out a series of experiments in the rat similar in design to those of Wirth and McHugh (1983). In the first experiment, total gastric volume was removed or left in place after an intake test in which glucose (12.5%) was infused through an intraoral catheter. A rat terminated intake by allowing the infused fluid to drip out of its mouth or by active rejection that consisted of head shakes or forelimb flails. After 10 minutes, rats were given a second intraoral intake test. In the second experiment, rats had about 50% of the gastric volume removed after the first three of four sequential intake tests. Both experiments confirmed previous results—intake was equivalent to the volume of gastric contents withdrawn.

These experiments, however, added an important dimension to the withdrawal studies by measuring the amount of glucose in the volume withdrawn and using this to compute the amount of glucose retained in the stomach and the amount emptied from the stomach during the initial and compensatory intake tests. These measurements indicated that the gastric volume and amount of glucose in the stomach after the compensatory intake test(s) were significantly smaller than the gastric volume and amount of glucose withdrawn after the initial intake test. Thus, when the pylorus is not occluded, as in these experiments and in those of Wirth and McHugh (1983), some postpyloric stimuli contribute to the control of compensatory intake after gastric withdrawal.

These results demonstrating a synergistic interaction between postpyloric and gastric stimuli in the control of compensatory intake when the pylorus was open do not support the hypothesis (Deutsch et al., 1978) that only gastric stimuli control the intake response to the withdrawal of gastric contents. When the cuff was closed, however, the results from one experiment in three rats was consistent with the hypothesis (Gonzalez and Deutsch, 1981).

Note that all of the experiments are consistent with some role for gastric stimuli in the control of intake after gastric withdrawal. In two experiments with monkeys (Wirth and McHugh, 1983; McHugh, 1979), gastric volume rather than nutritional content appeared to be the stimulus for gastric satiating effects. Further evidence for this came from experiments in which gastric infusions were administered during a meal in cuff-closed rats.

Gastric Infusion Experiments

Saline Infusions

Deutsch et al. (1980) were the first to do such an experiment. They infused saline into the stomach with the cuff closed during the ingestion of milk. The infusion was yoked to ingestion rate so that 1 ml of saline was infused for each 1 ml of milk ingested. The control intake was the volume of milk ingested in the 30-minute test when no gastric infusion was given. Gastric infusion decreased intake in 15 of 16 rats (rat 16 was the exception; Table 2.4). The magnitude of the decrease varied from 13.5 ml (rat 2) to 2.5 ml (rat 9). Most rats compensated for the volume infused because the mean (± SE) calculated difference (1.9 ± 1.3 ml) was not significantly different from zero. Three rats (9, 12, and 16) made only a small or no compensatory response.

These results support the hypothesis that gastric volume in the absence of postpyloric stimuli can decrease intake. Three rats that showed the least inhibitory effect of infusions had control intakes below the median intake of this group of rats. This led Deutsch et al. (1980: 55) to suggest that "rats that are lighter, spontaneous drinkers regulate their intake by using some other index exclusively or in addition." This other index was postulated to be the amount of nutrient in the stomach, and a series of experiments was performed to test this idea (see below).

Gonzalez and Deutsch (1981) extended the evidence for the volume of gastric infusion decreasing intake in cuff-closed rats by using the same yoked-infusion design except that 0.5 ml of 0.9% saline was infused for each milliliter of milk ingested. Thus, this experiment investigated the inhibition of intake when the volume entering the stomach was only 50%

TABLE 2.4.
Effect of Gastric Infusion During Ingestion on Intake in Rats with Pyloric Cuffs Closed

Rat	Intake: Control	Intake: During Infusion	Intake: Difference in Intake	Saline Infused	Difference in Gastric Volume
1	24.5	11.5	13.0	11.5	−1.5
2	23.0	9.5	13.5	9.5	−4.0
3	22.0	9.0	13.0	9.0	−4.0
4	20.5	10.5	10.0	10.5	+0.5
5	19.5	10.5	9.0	10.5	+1.5
6	19.5	11.0	8.5	11.0	+2.5
7	17.5	10.0	7.5	10.0	+2.5
8	17.5	10.0	7.5	10.0	+2.5
9	17.5	15.0	2.5	15.0	+12.5
10	17.5	10.5	7.0	10.5	+3.5
11	17.5	9.5	8.0	9.5	+1.5
12	17.0	13.5	3.5	13.5	+10.0
13	16.5	7.5	9.0	7.5	−1.5
14	14.5	8.0	6.5	8.0	+1.5
15	12.5	2.5	10.0	2.5	−7.5
16	9.5	10.0	−0.5	10.0	+10.5

Intakes (ml/30 min) during control (no gastric infusion) and during gastric infusion of 0.9% saline yoked to the rate of ingestion (1 ml saline infused for 1 ml of milk ingested). Difference in intake is the remainder when intake during gastric infusion was subtracted from intake during control. Difference in gastric volume is the calculated difference between intake during control and the intake during gastric infusion plus an equal volume of saline infused. If rats adjusted intake to match the volume infused, the calculated difference would be zero. The actual gastric volumes were not measured in either condition and were probably larger due to salivary and gastric secretions. Recalculated from data in Table 1 of Deutsch et al. (1980).

more than the volume being ingested; in the previous experiment, the volume entering the stomach was 100% more than the volume ingested. All three rats decreased intake significantly, and the volume decreased was similar to the volume infused (see right panel in Fig. 1 of Gonzalez and Deutsch, 1981).

Recent experiments by Phillips and Powley (1996) provide impressive support for the inhibition of intake by gastric volume in cuff-closed rats. They demonstrated that gastric infusions of isotonic saline that were equivalent to 50%, 100%, 150%, and 200% of control intake and administered at the beginning of 30 minutes of access to liquid food decreased intake significantly in cuff-closed rats. The inhibition of intake was not linearly related to volume. Only the largest volume (200%) produced a significantly larger inhibition than the equivalent inhibitions produced by the three smaller volumes.

These three experiments provide compelling evidence that increasing gastric volume by infusing isotonic saline during ingestion decreases intake in cuff-closed rats. Because this effect was obtained in cuff-closed rats, the result is strong evidence that infusing saline produced gastric stimuli without changing postpyloric stimuli.

Although these results are clear evidence for a gastric inhibitory control of intake sensitive to intragastric volume, several aspects of this phenomenon need to be clarified before the physiological significance of this negative-feedback mechanism can be ascertained.

The first issue is sensitivity. Thus far, the threshold volume that decreases intake significantly is 50% of the control (noninfused) intake. Furthermore, in the only study that varied volume of infusion, the inhibition of intake was equivalent for infusions that were 50%, 100%, or 150% of control intake. Thus, there is much less accurate inhibition of intake for increased gastric volume during the meal than for the increase of intake that compensates for the withdrawal of gastric contents at the end of the meal. This asymmetry in the effects on intake suggests that gastric controls of intake revealed by infusion and withdrawal experiments are not identical. This conclusion is reinforced by the differential results of vagotomy in such experiments (see Mechanisms of Gastric Satiation, below).

It is relevant to the issue of sensitivity that the gastric volume in infusion experiments is probably underestimated, because the volume of salivary and gastric secretions has not been measured and there is no reason to assume that they remain constant across the control and infusion conditions.

The second issue is whether the inhibition of intake has the characteristics of normal satiation and lacks an aversive component. Although it seems clear that aversion is not involved when rats eat a meal with cuffs closed, this has not been evaluated when gastric infusions are made during ingestion with the cuffs closed. The analysis of the microstructure of licking, measures of conditioned aversion, and appearance of the behavioral sequence of satiety should be used to settle this issue.

Thus, the physiological significance of this volumetric mechanism is uncertain. Its apparent insensitivity when volume infused is varied argues against it. Even more worrisome is the observation by Phillips and Powley (1996) that when a volume of isotonic saline equal to 100% of control intake was infused, no inhibition of intake occurred when the pyloric cuffs were open.

Nutrient Infusions

Recall that Deutsch et al. (1980) used the failure of 3 of 16 rats to show little or no decrease of intake when saline was infused into the stomach to postulate that gastric nutrient stimuli had satiating effects.

Deutsch and his colleagues tested this idea in a series of experiments that demonstrated an inhibitory effect on intake by intragastric infusions of partially digested corn oil (Deutsch and Gonzalez, 1980, 1981) and oleic acid (Deutsch and Tabuena, 1986), but not by infusions of corn oil, lysine, tryptophan, or a mixture of lysine and threonine (Deutsch and Tabuena, 1986). Unfortunately, the positive results were obtained in rats with the pyloric cuffs open. This permitted emptying partially digested corn oil and oleic acid from the stomach into the duodenum so that the site of their inhibitory action on intake was not clear. Certainly, in the case of oleic acid, this is a significant problem of interpretation because duodenal infusions of oleic acid produce a potent inhibition of intake during sham feeding (see Chapter 3).

Recent experiments by Phillips and Powley (1996) confirm this concern. They showed that intragastric infusions of sucrose (40%), fructose (40%), or glucose (10%, 20%, and 40%) inhibited intake significantly more than the volume infused when pyloric cuffs were open, but not when pyloric cuffs were closed. These results provide no evidence for the hypothesized efficacy of gastric nutrient stimuli for inhibiting intake, and they shift the burden of proof for a satiating effect of gastric nutrient stimuli to those who hold that hypothesis.

Oral and Gastric Interactions

In two reviews, Deutsch (1983, 1990) cited unpublished data from his laboratory demonstrating that the introduction of a novel flavor abolished the compensatory intake after gastric withdrawal and the inhibition of intake by gastric infusion in rats with pyloric cuffs open. From these observations, Deutsch suggested that gastric signals control intake only when rats eat familiar food.

This hypothesis predicts that if a diet is familiar to a rat, gastric satiating mechanisms are the dominant, negative-feedback control of meal size. The experiment of Davis et al. (1993) is a clear test of this prediction. They adapted rats to 30-minute intake tests of 0.8 M sucrose after 4 hours of food deprivation. The pylorus was closed by an inflated cuff in all of the tests. Thus, only pregastric and gastric satiating stimuli were produced during ingestion. When intakes were stable, a series of intake tests were done in which ingested sucrose drained out of a gastric cannula during every third test. Thus, a sham-feeding test occurred after two real-feeding tests. The hypothesis proposed by Deutsch predicts that drainage of gastric contents removes most, if not all, of the negative feedback occurring during the meal, so meal size should become much larger than normal. The results demonstrated that meal size doubled the first time the gastric cannula was open; this is consistent with the hypothesis, although not as large an effect as predicted. But the hypothesis did not predict the significant satiating effect of the pregastric stimuli of the familiar diet. There was a rapid decline in the rate of licking in the first 4–6 minutes of the sham-fed meal (see Chapter by 4 for details). As discussed above (see Experience above), this pregastric, negative-feedback effect was learned presumably by associating orosensory stimuli with the postingestive gastric volumetric stimulus produced by ingested food accumulating in the cuff-closed stomach. Note that gastric stimuli are serving as the unconditioned stimulus for this association. This learned inhibitory control extinguished over three consecutive sham-feeding tests with the gastric cannula open.

In addition to testing the predictions of the Deutsch hypothesis, these experiments revealed new information about the ability of gastric stimuli to control meal size. When the cycle of two real-feeding tests and one sham-feeding test was repeated seven times, intake increased progressively in both types of tests (for details, see Fig. 4.6, Chapter 4). This was apparently due to the partial extinction of the acquired control of intake by pregastric stimuli.

The progressive increase of intake across the cycle of real-feeding and sham-feeding tests in this experiment was not observed in the prior experiments of identical design (Davis and Smith, 1990). The critical difference between the two experiments is that food ingested during real feeding could empty from the stomach in the first experiment, but not in the second experiment. This suggests that under these conditions postpyloric satiating stimuli are necessary and gastric satiating stimuli are not sufficient to maintain meal size. This was the first situation in which the combination of pregastric and gastric stimuli did not maintain meal size. In noting this, the authors suggested that the inadequacy of pregastric and gastric stimuli to control normal meal size may have been caused by the interposition of a sham-feeding test between each pair of real-feeding tests. This appears to be the case, because, when rats with pyloric cuffs closed underwent 14 consecutive real-feeding tests with 0.8 M sucrose, meal size did not change (Rauhofer et al., 1994).

Mechanisms of Gastric Satiation

The afferent fibers of gastric vagal nerves have been suggested as the major pathway by which the effects of gastric stimuli are carried to the hindbrain to be processed as negative-feedback information for the control of meal size. This suggestion is reasonable because gastric vagal afferents are sensitive to glucose, mechanical, thermal, osmotic, acid, alkali, and peptide stimuli (Schwartz et al., 1991, 1993, 1994; Grundy and Scratcherd, 1989). Total abdominal vagotomy that spared the hepatic branch, however, did not change meal size when rats ate and pyloric cuffs were closed (Gonzalez and Deutsch, 1981; Kraly and Gibbs, 1980).

In contrast to the chronic loss of vagal afferent and efferent innervation of the stomach produced by vagotomy, Rauhofer et al. (1996) investigated the effect of acute peripheral muscarinic blockade with atropine methylnitrate on meal size of 0.8 M sucrose in cuff-closed rats. This treatment blocked all cholinergic control of gastric smooth muscle contraction and, thus, should change the gastric contractile response to volume. The results were the same — meal size did not change.

Currently there are only three results consistent with vagal mediation of gastric, negative-feedback effects on intake. One is the blockade by hepatic-spared vagotomy of the inhibitory effect of a gastric infusion on intake in rats in which intake was large (Gonzalez and Deutsch, 1981). In the same experiments, vagotomy did not change meal size or the increased intake produced by withdrawal of gastric contents during ingestion. Thus, the vagal innervation of the stomach was not necessary for the gastric contribution to the control of meal size or the response to withdrawal of gastric contents under these conditions.

The other two results came from Lorenz and his colleagues. Lorenz (1983) showed that acute abdominal vagotomy increased intake with the pyloric cuff closed during 2 hours of suckling in rat pups on postnatal days 6–12. Acute vagotomy also attenuated the inhibitory effect of a preload under these conditions.

When diet was infused intermittently through an intraoral catheter for 60 minutes in rats of the same age, acute vagotomy also increased intake (Lorenz et al., 1982). Unfortunately, the pylorus was not closed in these experiments, so it is not clear that the critical effect of vagotomy was to block negative feedback from the stomach.

The paucity of evidence for vagal mediation of satiating effects from the stomach is surprising. Vagal afferent fibers do not appear to be necessary for the control of meal size in adult rats when they eat and the pylorus is closed. This conclusion is based on only two experiments, and both of them used peripheral vagotomy procedures that lesioned afferent and efferent fibers chronically and spared hepatic and duodenal branches of the vagus. It is possible that more refined lesions involving only gastric afferent fibers would yield different results.

The lack of evidence for vagal mediation of gastric satiating effects raises the possibility that spinal visceral afferents in the splanchnic nerves could be important. There are two relevant reports.

Lorenz (1982) observed in 6–12-day-old rat pups that when spinal visceral afferent information from the stomach (and all other abdominal organs) was removed by acute section of the cervical spinal cord and this lesion was combined with vagotomy, a much larger intake occurred during intraoral infusions than after vagotomy alone. Cord section alone, however, had no significant effect on intake.

In adult rats, Deutsch and Ahn (1986) used bilateral splanchnicotomy to remove visceral afferents to the stomach (and small intestine). Although splanchnicotomy failed to affect meal size or the inhibitory effect of gastric infusions in rats, splanchnicotomy did block the increase of intake when gastric contents were removed during ingestion of 50% corn oil. It should be noted that the pylorus was not occluded in these experiments. This interesting result must be viewed cautiously, however, because Deutsch (1990; 169) failed to replicate it.

Formulation

The most important fact to emerge from this review is that the combination of pregastric and gastric stimuli of ingested food can control meal size within the normal range. This is a robust phenomenon because it has been demonstrated after 0.5–19 hours of overnight food deprivation and with various test diets. Furthermore, the satiating effect of the combination of pregastric and gastric stimuli has been observed in rat pups less than 2 weeks old (Phifer and Hall, 1988; Phifer et al., 1986). Note that it is the *sequential* combination of pregastric and gastric stimuli that occurs during eating that is sufficient to control meal size within the normal range.

Pregastric stimuli alone are not adequate. Although pregastric stimuli terminate eating after a short deprivation, meal size is about twice normal. The satiating effect of pregastric stimuli is conditioned because most of it extinguishes over three or more consecutive sham-feeding tests (Davis et al., 1993; Davis and Smith, 1990).

The satiating potency of pregastric stimuli is markedly reduced by overnight food deprivation. Under these conditions, sham feeding persists for hours and the behavioral sequence of satiety does not occur. The satiating potency is also less in the dark than in the light.

In contrast to the information we have about the satiating potency of pregastric stimuli acting alone, we know nothing about the satiating potency of gastric stimuli acting alone. This ignorance is a constraint of measurement. The satiating potency of gastric stimuli is measured by a decrease of eating. Thus, all such experiments include an interaction between gastric and pregastric stimuli of unknown magnitude.

One way to deal with this problem is to subtract intake during pregastric and gastric stimulation from the intake during sham feeding. The difference is an estimate of the satiating potency of the gastric stimuli under these conditions. In the only case where this strategy was used, Davis et al. (1993) found that gastric stimuli contributed about 50% of the control of meal size. This measurement strategy assumes that the interaction between the satiating effects of pregastric and gastric stimuli is additive. The quantitative, compensatory, ingestive response to gastric withdrawal supports that assumption. But the relative insensitivity of the inhibition of intake produced by gastric volumetric infusions is not easily explained by an additive interaction. Identification of the various modes of interactions between pregastric and gastric stimuli is an important experimental task with fundamental implications for the central processing and integration of satiating stimuli from pregastric and gastric sites.

Despite the uncertainty about the measurement of the potency of gastric stimuli acting alone, recent work has revealed four important aspects of gastric satiating stimuli.

First, gastric volumetric stimuli decrease meal size, and there is one report that abdominal vagotomy blocked this effect (Gonzalez and Deutsch, 1981). It remains to be demonstrated that the effect of vagotomy was due to the removal of afferent fibers sensitive to gastric volume.

Second, gastric volumetric stimuli are synergistic with the satiating effect of glucose in the small intestine and with exogenously administered CCK (see Chapter 5).

Third, gastric nutrient infusions do not reduce intake more than what can be accounted for by their volume when the infusions are confined to the stomach by closing the pyloric cuffs. It is possible that gastric nutrient stimuli have satiating effects as Deutsch and his colleagues have claimed, but new evidence is required to support that hypothesis.

Fourth, gastric stimuli act as unconditioned stimuli for the formation of conditioned, inhibitory control of intake by pregastric stimuli (Davis et al., 1993).

In conclusion, the importance of pregastric and gastric satiating stimuli for the control of meal size has been established in the past two decades. Central processing of pregastric and gastric stimuli produces synergistic interactions of their effects on eating. Pregastric and gastric stimuli interact synergistically with each other, and both are synergistic with intestinal satiating stimuli (see Chapter 3). It is interesting that the experimental analysis designed to evaluate the separate contributions of pregastric and gastric (and intestinal) receptor sites has invariably led to evidence of potent synergistic interactions. Such interactions are likely to occur normally during a meal as ingested food passes from the mouth to the small intestine. The mechanisms of these interactions are in the central neural networks that process these peripheral stimuli. Other than the inference from the synergistic interactions that the networks must be capable of spatial summation, our ignorance about them is total.

That leads to my final point. Recent progress has depended entirely on analysis of the effects on meal size of changing chemical or volumetric stimuli at pregastric and gastric sites. Further progress requires that this sensory and behavioral analysis be combined with neurological, molecular, and genetic techniques.

ACKNOWLEDGMENTS

I thank Mr. Berkeley Cooley and Ms. Terri Popiel for expert processing of this manuscript. Doctors John Davis, Nori Geary, and Timothy Moran made constructive criticisms of earlier versions of this chapter that helped me immensely. I am supported by Research Scientist Award MH00149.

REFERENCES

Antin, J., J. Gibbs, J. Holt, R. C. Young, and G. P. Smith. Cholecystokinin elicits the complete behavioral sequence of satiety in rats. *J. Comp. Physiol. Psychol.* 89:784–790, 1975.

Antin, J., J. Gibbs, and G. P. Smith. Intestinal satiety requires pregastric food stimulation. *Physiol. Behav.* 18:421–425, 1977.

Antin, J., J. Gibbs, and G. P. Smith. Cholecystokinin interacts with pregastric food stimulation to elicit satiety in the rat. *Physiol. Behav.* 20:67–70, 1978.

Berthoud, H.-R., and B. Jeanrenaud. Sham feeding-induced cephalic phase insulin release in the rat. *Am. J. Physiol.* 242:E280-E285, 1982.

Black, R. M., and H. P. Weingarten. A comparison of taste reactivity changes induced by ventromedial hypothalamic lesions and stria terminalis transections. *Physiol. Behav.* 44:699–708, 1988.

Bolles, R. C. Grooming behavior in the rat. *J. Comp. Physiol. Psychol.* 53:306–310, 1960.

Cox, J. E., and G. P. Smith. Sham feeding in rats after ventromedial hypothalamic lesions and vagotomy. *Behav. Neurosci.* 100:57–63, 1986.

Davis, J. D., and C. S. Campbell. Peripheral control of meal size in the rat: Effect of sham feeding on meal size and drinking rate. *J. Comp. Physiol. Psychol.* 83:379–387, 1973.

Davis, J. D., and G. P. Smith. Learning to sham feed: behavioral adjustments to loss of physiological postingestional stimuli. *Am. J. Physiol.* 259:R1228–R1235, 1990.

Davis, J. D., G. P. Smith, and J. Miesner. Postpyloric stimuli are necessary for the normal control of meal size in real feeding and sham feeding rats. *Am. J. Physiol.* 265:R888–R895, 1993.

Deutsch, J. A. Dietary control and the stomach. *Prog. Neurobiol.* 20:313–332, 1983.

Deutsch, J. A. *Food intake: gastric factors*. In Stricker, E. M. (ed.): *Handbook of Behavioral Neurobiology. Vol. 10*. New York: Plenum Press, 1990, p. 151–182.

Deutsch, J. A., and S. J. Ahn. The splanchnic nerve and food intake regulation. *Behav. Neural Biol.* 45:43–47, 1986.

Deutsch, J. A., and M. F. Gonzalez. Gastric nutrient content signals satiety. *Behav. Biol.* 30:113–116, 1980.

Deutsch, J. A., and M. F. Gonzalez. Gastric fat content and satiety. *Physiol. Behav.* 26:673–676, 1981.

Deutsch, J. A., W. G. Young, and T. J. Kalogeris. The stomach signals satiety. *Science* 201:165–167, 1978.

Deutsch, J. A., and O. O. Tabuena. Learning of gastrointestinal satiety signals. *Behav. Neural Biol.* 45:282–299, 1986.

Deutsch, J. A., M. F. Gonzalez, and W. G. Young. Two factors control meal size. *Brain Res. Bull.* 5:55–57, 1980.

Falasco, J. D., G. P. Smith, and J. Gibbs. Cholecystokinin suppresses sham feeding in the rhesus monkey. *Physiol. Behav.* 23:887–890, 1979.

Forsyth, P. A., H. P. Weingarten, and S. M. Collins. Role of oropharyngeal stimulation in cholecystokinin-induced satiety in the sham feeding rat. *Physiol. Behav.* 35:539–543, 1985.

Gibbs, J., and J. D. Falasco. Sham feeding in the rhesus monkey. *Physiol. Behav.* 20:245–249, 1978.

Gibbs, J., S. P. Maddison, and E. T. Rolls. Satiety role of the small intestine examined in sham-feeding rhesus monkeys. *J. Comp. Physiol. Psychol.* 95: 1003–1015, 1981.

Gonzalez, M. F., and J. A. Deutsch. Vagotomy abolishes cues of satiety produced by gastric distention. *Science* 212:1283–1284, 1981.

Gowans, H., and W. P. Weingarten. Elevations of plasma glucose do not support taste-to-postingestive consequence learning. *Am. J. Physiol.* 261:R1409–R1417, 1991.

Grill, H. J. Caudal brainstem contributions to the integrated neural control of energy homeostasis. In Ritter, R. C., S. Ritter, and C. D. Barnes (eds.): *Feeding Behavior: Neural and Humoral Controls*. Orlando: Academic Press, 1986, pp. 103–129.

Grill, H. J., K. C. Berridge, and D. J. Ganster. Oral glucose is the prime elicitor of preabsorptive insulin secretion. *Am. J. Physiol.* 246:R88–R95, 1984.

Grill, H. J., and J. M. Kaplan. Sham feeding in intact and chronic decerebrate rats. *Am. J. Physiol.* 262:R1070–R1074, 1992.

Grundy, D., and Scratcherd, T. *Sensory afferents from the gastrointestinal tract*. In Wood, J. D. (ed.): *Handbook of Physiology, Section 6: The Gastrointestinal System, Vol. 1, Motility and Circulation, Part 1*. New York: Oxford University Press, 1989, pp. 593–620.

Hall, W. G. A remote stomach clamp to evaluate oral and gastric controls of drinking in the rat. *Physiol. Behav.* 11:897–901, 1973.

Hetherington, M., B. J. Rolls, and V. J. Burley. The time course of sensory-specific satiety. *Appetite* 12:57–68, 1989.

Joyner, K., G. P. Smith, R. Shindledecker, and C. Pfaffmann. Stimulation of sham feeding in the rat by sucrose, maltose, glucose, and fructose. *Soc. Neurosci. Abstr.* 11:1223, 1985.

Kaplan, J. M., W. Siemers, and H. J. Grill. Ingestion, gastric fill, and gastric emptying before and after withdrawal of gastric contents. *Am. J. Physiol.* 267:R1257–R1265, 1994.

Kirkham, T. C., and S. J. Cooper. Attenuation of sham feeding by naloxone is stereospecific: Evidence for opioid mediation of orosensory reward. *Physiol. Behav.* 43:845–848, 1988.

Kraly, F. S., W. J. Carty, and G. P. Smith. Effect of pregastric food stimuli on meal size and intermeal interval in the rat. *Physiol. Behav.* 20:779–784, 1978.

Kraly, F. S., B. J. Cushin, and G. P. Smith. Nocturnal hyperphagia in the rat is characterized by decreased postprandial satiety. *J. Comp. Physiol. Psychol.* 94:375–387, 1980.

Kraly, F. S., and J. Gibbs. Vagotomy fails to block the satiating effect of food in the stomach. *Physiol. Behav.* 24:1007–1010, 1980.

Kraly, F. S., and G. P. Smith. Combined pregastric and gastric stimulation by food is sufficient for normal meal size. *Physiol. Behav.* 21:405–408, 1978.

Liebling, D. S., J. D. Eisner, J. Gibbs, and G. P. Smith. Intestinal satiety in rats. *J. Comp. Physiol. Psychol.* 89:955–965, 1975.

Lorenz, D. N. Effects of gastric fillng and vagotomy on ingestion, nipple attachment, and weight gain by suckling rats. *Devl. Psychobiol.* 16:469–483, 1983.

Lorenz, D. N., S. B. Ellis, and A. N. Epstein. Differential effects of upper gastrointestinal fill on milk ingestion and nipple attachment in the suckling rat. *Devl. Psychobiol.* 15:309–330, 1982.

McHugh, P. R. Aspects of the control of feeding: application of quantitation in psychobiology. *Johns Hopkins Med. J.* 144:147–155, 1979.

McHugh, P. R., and Moran, T. H. The stomach: A conception of its dynamic role in satiety. In Sprague, J. M., and A. N. Epstein (eds.): *Progress in Psychobiology and Physiological Psychology. Vol. II.* New York: Academic Press, 1985, pp. 197–232.

Mindell, S., G. P. Smith, and D. Greenberg. Corn oil and mineral oil stimulate sham feeding in rats. *Physiol. Behav.* 48:283–287, 1990.

Mook, D. G. Oral and postingestional determinants of the intake of various solutions in rats with esophageal fistulas. *J. Comp. Physiol. Psychol.* 56:645–659, 1963.

Mook, D. G. Satiety, specifications and stoprules: Feeding as voluntary action. In Epstein, A. N., and A. R. Morrison (eds.): *Progress in Psychobiology and Physiological Psychology. Vol. 14.* New York: Academic Press, 1990, pp. 1–65.

Mook, D. G., R. Culberson, R. J. Gelbart, and K. McDonald. Oropharyngeal control of ingestion in rats: Acquisition of sham-drinking patterns. *Behav. Neurosci.* 97:574–584, 1983.

Mook, D. G., and S. Wagner. Glucose sham-drinking in rat: Some effects of hydration. *Appetite* 10:71–87, 1988.

Nissenbaum, J. W., and A. Sclafani. Sham-feeding response of rats to polycose and sucrose. *Neurosci. Biobehav. Rev.* 11:215–222, 1987.

Pavlov, I. P. *The Work of the Digestive Glands. 2nd Ed.* London: Charles Griffin & Company, Ltd., 1910.

Phifer, C. B., and W. G. Hall. Ingestive behavior in preweanling rats: Emergence of postgastric controls. *Am. J. Physiol.* 255:R191–R199, 1988.

Phifer, C. B., C. R. Sikes, and W. G. Hall. Control of ingestion in 6-day-old rat pups: Termination of intake by gastric fill alone? *Am. J. Physiol.* 250:R807–R814, 1986.

Phillips, R. J., and T. L. Powley. Gastric volume rather than nutrient content inhibits food intake. *Am. J. Physiol.* 271:R766–R779, 1996.

Rauhofer, E. A., D. Greenberg, and G. P. Smith. Peripheral atropinization does not change meal size controlled by prepyloric mechanisms. *Physiol. Behav.* 59:237–240, 1996.

Rauhofer, E. A., G. P. Smith, and J. D. Davis. Oral and gastric stimuli are sufficient to control meal size throughout 14 consecutive tests in adult rat. *Soc. Neurosci. Abstr.* 20:5861994.

Rauhofer, E. A., G. P. Smith, and J. Gibbs. Acute blockade of gastric emptying and meal size in rats. *Physiol. Behav.* 54:881–884, 1993.

Richter, C. A behavioristic study of the activity of the rat. *Comp. Psychol. Monogr.* 1, No. 2, 1922.

Rolls, B. J. Sensory-specific satiety. *Nutr. Rev.* 44:93–101, 1986.

Schwartz, G. J., P. R. McHugh, and T. H. Moran. Integration of vagal afferent response to gastric loads and cholecystokinin in rats. *Am. J. Physiol.* 261:R64–R69, 1991.

Schwartz, G. J., P. R. McHugh, and T. H. Moran. Gastric loads and cholecystokinin synergistically stimulate rat gastric vagal afferents. *Am. J. Physiol.* 265:R872–R876, 1993.

Schwartz, G. J., P. R. McHugh, and T. H. Moran. Pharmacological dissociation of responses to CCK and gastric loads in rat mechanosensitive vagal afferents. *Am. J. Physiol.* 267:R303–R308, 1994.

Sclafani, A., and J. W. Nissenbaum. Is gastric sham-feeding really sham-feeding? *Am. J. Physiol.* 248:R387–R390, 1985.

Seeley, R. J., J. M. Kaplan, and H. J. Grill. Effect of occluding the pylorus on intraoral intake: A test of the gastric hypothesis of meal termination. *Physiol. Behav.* 58:245–249, 1995.

Siegel, A., K. Joyner, and G. P. Smith. Effect of bilateral ibotenic acid lesions in the basolateral amygdala on the sham feeding response to sucrose in the rat. *Physiol. Behav.* 42:231–235, 1988.

Smith, G. P. Dopamine and food reward. In Fluharty, S., A. R. Morrison, J. Sprague, and E. Stellar (eds.): *Progress in Psychobiology and Physiological Psychology. Vol. 16.* New York: Academic Press, 1995a, pp. 83–144.

Smith, G. P. Pavlov and appetite. *Int. Physiol. Behav. Sci.* 30:169–174, 1995b.

Smith, G. P. The direct and indirect controls of meal size. *Neurosci. Biobehav. Rev.* 20:41–46, 1996.

Smith, G. P., and J. Gibbs. *Postprandial satiety.* In Sprague, J., and A. N. Epstein (eds.): *Progress in Psychobiology and Physiological Psychology. Vol. 8.* New York: Academic Press, 1979, pp. 179–242.

Swithers, S. E., and W. G. Hall. Does oral experience terminate ingestion? *Appetite* 23:113–138, 1994.

Swithers-Mulvey, S. E., and W. G. Hall. Control of ingestion by oral habituation in rat pups. *Behav. Neurosci.* 106:710–717, 1992.

Weingarten, H. P. Diet palatability modules sham feeding in VMH-lesion and normal rats: Implications for finickiness and evalaution of sham-feeding data. *J. Comp. Physiol. Psychol.* 96:223–233, 1982.

Weingarten, H. P., and O. T. Kulikovsky. Taste-to-postingestive consequence conditioning: Is the rise in sham feeding with repeated experience a learning phenomenon? *Physiol. Behav.* 45:471–476, 1989.

Weingarten, H. P., and T. L. Powley. A new technique for the analysis of phasic gastric acid responses in the unanesthetized rat. *Lab. Anim. Sci.* 30:673–680, 1980.

Weingarten, H. P., and S. D. Watson. Sham feeding as a procedure for assessing the influence of diet palatability on food intake. *Physiol. Behav.* 28:401–407, 1982.

Wirth, J. B., and P. R. McHugh. Gastric distension and short-term satiety in the rhesus monkey. *Am. J. Physiol.* 245:R174–R180, 1983.

Young, R. C., J. Gibbs, J. Antin, J. Holt, and G. P. Smith. Absence of satiety during sham feeding in the rat. *J. Comp. Physiol. Psychol.* 87:795–800, 1974.

Young, W. G., and J. A. Deutsch. The construction and surgical implantation of gastric catheters and a pyloric cuff. *J. Neurosci. Methods* 3:377–384, 1981.

3

Intestinal Satiety

DANIELLE GREENBERG

The small intestine is one of the sites where food stimuli act to produce satiety. Early evidence for this assertion came from an 1858 report by W. Busch of a patient with a fistula of the upper third of the small intestine (Busch, 1858). Most of the food that this patient ate drained rapidly from the fistula. Busch noted that although the patient ate very large meals, she did not report feelings of satiation. When asked if eating such large meals was not at least momentarily satiating, this patient answered that her stomach was full but she still felt hungry. This is particularly telling evidence for the role of the small intestine in satiety, because in this case food stimuli acted normally in the mouth and esophagus, nearly normally in the stomach, but failed to reach the lower small intestine. The result of the absence of food stimuli in the lower two-thirds of the small intestine was a complete absence of satiety and constant feelings of hunger. It must be noted that in this case gastric stimuli could have been abnormal because gastric emptying may have been accelerated due to rapid drainage of intestinal contents, and this may also have contributed to the absence of satiety.

This observation of absence of satiety when food drained out of the upper small intestine has been confirmed in monkeys by Gibbs et al. (1981). They allowed ingested food to drain from open duodenal cannulas after passing through the stomach and upper duodenum and found that intakes increased markedly above normal. Drainage from the duodenal cannulas, however, resulted in decreased gastric distention as well as lack of accumulation of food in the small intestine as a result of the increased rate of gastric emptying that occurred in this experimental condition. Thus the increased levels of food ingestion seen when food does not accumulate in the small intestine could be the result of the failure to activate satiety signals from the small intestine or the failure to activate satiety signals from stomach distention because the rate of gastric emptying increased.

An alternative approach to this question is to administer nutrients directly into the small intestine. When food is loaded into the small intestine, feeding is inhibited. In 1952, Hill et al. first showed that preloads of hypertonic glucose or nonnutritive fiber decreased food intake in rats. The effects of duodenal loads on feeding were first demonstrated in humans in 1973 by Love and Wilson. They showed that duodenal infusions of liquid food just prior to a test meal inhibited food intake and elicited reports of satiety. It is important to note that there were no reports of gastrointestinal distress or of illness in this study.

These studies and many others show that the small intestine can decrease food intake. It is necessary, however, to determine whether the small intestine acts as a site for satiating stimuli during a normal meal. Studies of the time course of gastric emptying and transit time of food entering the duodenum indicate that the duodenum is stimulated very rapidly after ingestion or stomach intubation of liquid food in the rat, rhesus monkey, and human (Fujimoto et al., 1992; McHugh and Moran, 1978, 1979; Hunt and Stubbs, 1975; Balagura and Fibiger, 1968; Card, 1941; Gibbs et al., 1981). In the case of glucose solutions, Kaplan et al. (1992) demonstrated that 25%–40% of the volume consumed empties by the end of the meal.

Solid food eaten after 2 hours of deprivation also penetrates at least 10 cm into the intestine within 5–10 minutes after ingestion begins in the rat (Wiepkema et al., 1972). Furthermore, when ingested food was labeled with dye so that newly ingested food could be differentiated from food of previous meals that remained in the stomach, newly ingested food emptied from the stomach before previously ingested food (Wiepkema et al., 1972). This preferential passage of newly eaten food could permit intestinal stimuli to activate satiating mechanisms to terminate a meal.

In this chapter, I review the evidence concerning the sites, adequate stimuli, and neural and endocrine or paracrine components of small intestinal satiating mechanisms. I am concerned with the action of small intestinal stimuli and mechanisms that provide a negative-feedback control of intake within a single test meal. The relationship of this satiating effect in a single meal to the control of body weight is outside the scope of this review.

Methods for Studying Intestinal Satiety

Parabiotic Rat Studies

One line of evidence for the importance of the small intestine in satiety came from studies on parabiotic rats. When rats are joined parabiotically without manipulation of their gastrointestinal tracts, both members of the pair eat normally (Han et al., 1963). If the stomach of one rat is connected to the duodenum of the other, only one member of the pair will eat (Lepkovsky et al., 1971). To investigate this phenomenon, Lepkovsky and his colleagues looked at gastric emptying of parabiotic rats with crossed intestines. If food was presented to these parabiotic rats after food deprivation in such a way as to have both members of the pair sample food, once one of the rats began to eat, the other stopped. It was found that the stomach and small intestine of the rat that was eating were empty, while the stomach and small intestine of the noneater were full. Further investigation showed that the rat that was eating would stop once gastric emptying of the other rat's stomach began. Thus a full stomach and a full duodenum in the noneating rat led to the suppression of feeding, but an empty stomach and a full duodenum in the eating rat also led to suppression of feeding (Lepkovsky et al., 1971).

It is still an open question as to the relative contributions of the duodenum, jejunum, and ileum in mediating intestinal satiety. When parabiotic rats have the stomach and duodenum intact but have exchanged jejunal and ileal sections (Koopmans, 1975), both members of the parabiotic pair eat normally, even if one member has been fed first. This is in contrast to duodenally crossed rats (Lepkovsky et al., 1971) in which food from the eating member of the parabiotic pair suppresses the feeding of the other rat. If only one member of jejunally crossed, parabiotic rats is permitted to feed, then the feeding animal shows no increase in intake (Koopmans, 1978). Furthermore, these rats were still able to compensate with increased intake of a given meal when nutrients were diluted. Thus, these rats were not insensitive to nutrient constituents, and satiety was elicited normally in the feeding rat by passage of food through the stomach and upper small intestine. These results on parabiotic rats with crossed intestines are consistent with the idea that the upper small intestine, perhaps the duodenum, is a site for the satiating effect of ingested food.

Intestinal Infusions During Normal Feeding

Another way to test the involvement of the small intestine in satiation is to study feeding while nutrients are introduced directly into the small intestine. Intestinal infusions of nutrients and some non-nutritive substances reduce food intake. This has been shown in dogs (Hill et al., 1952), humans (Welch et al., 1985, 1988; Love and Wilson, 1973), rabbits (Novin et al., 1974), pigs (Houpt, 1982; Houpt et al., 1983), and rats (Yin and Tsai, 1973; Glick and Modan, 1977; Snowdon, 1975; Ehman et al., 1971).

In the rat, a typical meal lasts less than 15 minutes (Woods et al., 1981; Kraly and Smith, 1978; Davis and Campbell, 1973; Kraly et al., 1978; West et al., 1984). We have begun to investigate the effects of intestinal infusions of Intralipid on normal meal size (Greenberg et al., unpublished results). Intralipid is a sterile, highly emulsified mixture of long-chain triglycerides, glycerol, and egg yolk phospholipids. Intralipid consists of triglycerides with component fatty acids as follows: linoleic, 50%; oleic, 26 %; palmitic, 10%; linolenic, 9%; stearic, 3.5%; and others, less than 1%. It is commonly used for intravenous parenteral feeding and is entirely miscible in water or saline. Infusion of 2.5 Kcal of Intralipid 10 minutes before feeding begins results in a 40% decrease in meal size. The effects of intestinal infusions of nutrients during real feeding requires much more work.

In human studies, it has been found that ileal and jejunal infusions of fats will immediately reduce meal size and caloric intake with no reports of discomfort or nausea (Welch et al., 1985, 1988). Only jejunal infusions, however, were able to affect the rate of ingestion or reduce hunger ratings (Welch et al., 1988), which suggests that satiating mechanisms in the upper small intestine may be more important than those in the lower small intestine for the control of meal size in humans. In an earlier study in rats, however, both duodenal and ileal infusions decreased subsequent 2-hour food intake equivalently (Glick and Modan, 1977).

Interaction of the Intestine and Gastric Emptying

During normal feeding, satiating effects of small intestinal stimuli act in concert with pregastric and gastric satiating mechanisms (see Chapter 2). A factor that relates gastric and intestinal satiety is gastric emptying rate (Hunt and Stubbs, 1975; McHugh and Moran, 1979). Differences in the suppressive effects of nutrients seen during real feeding and sham feeding may be attributed to the effects of these nutrients on gastric emptying, because the emptying rate determines in part the degree of gastric distention (see Chapter 2). A number of studies in humans (Hunt and Stubbs, 1975), monkeys (Robinson et al., 1988; McHugh and Moran, 1979), and rats (Reidelberger et al., 1983; Kalogeris et al., 1983) have shown that gastric emptying time is determined by the caloric density of a meal. That is, the number of calories delivered to the duodenum is quite constant regardless of the nutrient composition or osmotic properties of a meal. Thus fats, which are more calorically dense than proteins or carbohydrates, are more likely to remain in the stomach for a longer period of time.

Different nutrients are known to remain in the stomach for different time periods during normal digestion. The entry of different nutrients into the intestine slows gastric emptying differentially (Malagelada and Azpiroz, 1989; McHugh and Moran, 1979; Hunt and Stubbs, 1975). In addition, for nutrient fats, gastric secretion and emptying are inhibited by the presence of fats in the small intestine rather than in the stomach (Farrell and Ivy, 1926). The early work of Quigley et al. (1934) showed that the delivery of intravenous fats had no effect on the inhibition of gastric secretion and motility, but intraduodenal administration of fats in the same amounts slowed these measures considerably. There are also differences in the potency of medium-chain and short-chain fatty acids to slow gastric emptying; medium-chain fatty acids are much more potent (Hunt and Knox, 1968; Card, 1941).

The ability of nutrients to slow gastric emptying does not necessarily correlate with their ability to inhibit food intake. Welch et al. found that ileal infusions of a corn oil emulsion in human subjects slowed gastric emptying, but did not decrease food intake significantly (Welch et al., 1988). We have found that intestinal infusions of fats slows gastric emptying in obese Zucker rats to a significantly greater extent than in lean Zucker rats (Greenberg et al., 1987b), while intestinal fat infusions inhibit food intake similarly in both genotypes (French et al., 1994; Greenberg et al., 1986).

Studies with transplanted intestinal pouches in dogs led to the suggestion that the inhibition of gastric secretion and motility was due to a humoral factor (Quigley and Meshcan, 1941; Farrell and Ivy, 1926). This active humoral factor for control of gastric emptying and secretion was called *enterogastrone* by Kosaka and Lim (1930), indicating that this peptide was released from the small intestine and acted on the stomach.

One candidate for this humoral factor is cholecystokinin (CCK), which slows gastric emptying in a manner similar to intraduodenally infused nutrients in humans (Debas et al., 1975). CCK decreases gastric emptying (Moran and McHugh, 1982, 1988), and it is possible that in the normally feeding animal inhibition of gastric emptying may contribute to the satiating effect of CCK by activating gastric satiating mechanisms sensitive to volume (see Chapters 5 and 6). The satiating effect of intestinal nutrient infusions in the sham-feeding preparation indicates, however, that gastric distention is not necessary for the inhibition of meal size.

There is some evidence that the small intestine may be more effective than the stomach for reducing food intake. For example, Yin and Tsai (1973) found that intraduodenal glucose was more effective in suppressing feeding of rats than was intragastric administration of the same substance. Snowdon (1975) confirmed this with rats that had both intragastric and intraduodenal catheters and found that small infusions of hypertonic glucose (in a different concentration than that used by Yin and Tsai), hypertonic urea, or a mixed diet led to greater suppression of subsequent intake when infused intraduodenally than when infused intragastrically.

In monkeys, however, Wirth and McHugh (1983) found that when similar amounts of glucose entered the intestine there were large differences in glucose intake depending solely on stomach distention. These results suggest that the stomach may be more important than the small intestine for reducing food intake in response to a meal in monkeys under these conditions (see Chapter 2).

Intestinal Infusions During Sham Feeding

In all studies of the effects of intraduodenal infusions on normal feeding, infused substances are mixed with any ingested food that empties from the stomach. Thus, the nature of the intraduodenal stimuli that inhibit food intake is not clear. In addition, the contribution to the reduction of food intake by changes in gastric emptying rate and gastric distention resulting from duodenal infusions is not usually measured. Studies of sham feeding overcame these problems and permitted the study of intestinal satiating mechanisms in the absence of gastric satiating mechanisms and any effect of slowed gastric emptying on gastric distention (see Chapter 2). Sham feeding minimizes the stimulation of gastric nutrient receptors, eliminates the stimulation of gastric distention receptors, eliminates the role of gastric emptying, and minimizes or eliminates the stimulation of intestinal and postabsorptive receptors (Sclafani and Nissenbaum, 1985; Young et al., 1974; Davis and Campbell, 1973). This preparation also gives a clear measure of the termination of eating and the onset of postprandial satiety produced by an experimental treatment, because after overnight food deprivation, animals sham feed almost continuously and do not spontaneously satiate (Young et al., 1974).

Liebling et al. (1975) were the first to show that duodenal infusions of liquid food inhibited sham feeding and elicited the complete behavioral sequence of satiety (Antin et al., 1977). Infusions of equivolumetric saline did not suppress sham feeding, demonstrating that the mechanical effects of intestinal infusions were not sufficient for the satiating effect. The duodenal infusions of liquid food also did not inhibit food intake by producing aversive stimuli, because they did not function as the unconditioned stimulus for a conditioned taste aversion. The authors concluded that the entrance of nutrient into the duodenum stimulated intestinal satiating mechanisms that were sufficient to elicit satiety in the absence of any gastric distention or any effect of the intestinal stimuli on gastric emptying. They also suggested that endogenous release of the peptide CCK was one of the mechanisms that might mediate this effect (Liebling et al., 1975).

Antin et al. (1977) showed that intestinal satiety only operated when sham feeding was contiguous with intestinal infusion. If intestinal infusions were given 6–12 minutes before sham feeding was initiated, no suppression of intake was observed. Maximal suppression of sham feeding was obtained when the duodenal infusion began 12 minutes after the onset of sham feeding. These results suggest a synergy between intestinal and oral stimulation in eliciting satiety (see Chapter 2).

The question of synergy between oral and intestinal satiating effects has been further explored. Foster et al. (1996) presented sham-feeding rats a liquid diet (0.5 Kcal/ml), 0.3 M sucrose (0.4 Kcal/ml), or a 0.1 M solution of glucose polymers (Polycose 0.4 Kcal/ml). Rats sham fed more Polycose than sucrose and more sucrose than liquid diet. Duodenal infusions of saline or of Intralipid in concentrations of 5%, 2.5%, and 1.25% were then administered 12 minutes after the start of sham feeding. Although Intralipid suppressed intake of all three solutions in a dose-dependent manner, infusions of the lower two concentrations produced a larger suppression of intake of Polycose and sucrose, the solutions that elicited the highest rate of sham feeding. These results suggest an important synergy between oral and intestinal satiating stimuli that needs further investigation.

Gibbs et al. (1981) demonstrated that sham feeding in rhesus monkeys was also inhibited by duodenal infusions of nutrients. The infusions elicited the complete behavioral sequence for satiety, and the reductions in sham feeding were related to the chemical concentration of the nutrient infusions. Equivolumetric infusions of isotonic saline did not suppress sham feeding and did not elicit behaviors typical of satiety. Thus, in the monkey as well as in the rat and humans, the mechanical stimuli produced by the volume of saline infusions did not produce a satiating effect.

In the rat, a mixed diet (BioServ or Vivonex) and individual nutrients decrease sham feeding when infused into the duodenum (Yox et al., 1989, 1991a,b, 1992; Greenberg et al., 1986, 1987a,b, 1988, 1990; Reidelberger et al., 1983; Lew et al., 1983). Thus, infusion of nutrients into the duodenum inhibits intake in the absence of gastric distention.

When fats are infused into the duodenum of sham-feeding rats, a significant reduction in sham feeding occurs within 10 minutes of the start of the intestinal infusions. This time course suggests a preabsorptive site of action for the inhibitory effect of the intestinal infusion on feeding because fats of the chain length that we have infused (Greenberg et al., 1990) are not typically absorbed until 45 minutes after they reach the small intestine (Clark et al., 1969; Senior, 1964; Holt and Bennett-Clark, 1975).

We have recently investigated the effects of infusions of fats into the ileum of sham-feeding rats. When caloric loads of Intralipid that are highly effective at reducing sham feeding when infused into the duodenum (Greenberg et al., 1990) are administered into the ileum, they fail to reduce sham feeding (French et al., 1995). These results suggest that in sham feeding the upper small intestine is of greater importance for the satiating effect of fat than the lower small intestine, at least in the rat.

Thus, stimulation of the upper small intestine by nutrients leads to an inhibition of feeding and in some cases behaviors indicative of satiety. This is true whether the intestine is stimulated by prior surgical manipulation as in the parabiotic rat or by infusions of nutrients during normal or sham feeding. In normal feeding the satiating effects of intestinal stimulation interact with the effects of intestinal stimulation on the slowing of gastric emptying to increase gastric distension. Intestinal stimuli also interact with oral stimuli. The mechanisms of these synergistic interactions remain to be determined.

Nature of the Stimuli for Intestinal Satiety

The adequate stimuli for intestinally mediated satiation are beginning to be defined. They include the volume, colligative, and chemical properties of nutrients in the small intestine.

Intestinal Load

Liebling et al. (1975) suggested that the adequate stimulus was intestinal load, that is, the concentration of infusate multiplied by its volume. Several findings led to this conclusion. First, 3 ml of dilute (25% v:v) liquid diet infused intraduodenally in sham-feeding rats led to significant inhibition of feeding. Second, larger volumes of nutritive infusates produced larger inhibitions of feedings. Third, isovolumetric infusions of saline did not suppress feeding. Small volumes of concentrated liquid diet did not, however, lead to a suppression of feeding even though the total chemical load was as large as for effective dilute infusions. Therefore, it was concluded that activation of intestinal satiating mechanisms involves a threshold for volume. The reason for this may be that receptors for gastrointestinal neural and hormonal reflexes (Makhlouf, 1974) are widely distributed in the intestinal mucosa. A larger volume would distribute the chemical stimulus over a larger area of the mucosa, and this would result in the activation of more receptors.

Colligative Stimuli

When a mixed diet, cellulose, hypertonic saline, hypertonic glucose, and isotonic saline were infused directly into the duodenum, both non-nutritive bulk and hypertonic nutrients reduced subsequent intake of food, but isotonic saline did not (Ehman et al., 1971). Both bulk and osmotic pressure were suggested as adequate stimuli for intestinal satiation, but this interpretation is questionable. All infusions into the duodenum were given in large boluses at rapid rates that greatly exceeded normal emptying rates of ingested nutrients, and no tests for induced illness or discomfort from the infusions were performed. Thus, it is unclear whether these investigators were observing reductions in intakes due to satiation or to illness.

Experiments using hypertonic glucose and xylose showed that both of these substances suppressed food intake when infused intraduodenally and intragastrically, while isotonic sugars did not (Yin and Tsai, 1973). This was true even though xylose is metabolized much less efficiently than glucose; this suggested that an osmotic mechanism common to both sugars, rather than any metabolic effect, was the stimulus for the suppression of feeding. Snowdon's experiments (1975) with intraduodenally administered hypertonic urea and glucose also suggested a role for osmotic pressure. Davis and Collins (1978) added mannitol, a nonabsorbable carbohydrate, to the diet of rats and found that this led to a reduction in meal size that correlated with the length of time that the small intestine continued to be full. All of these results could be interpreted as evidence for a small intestinal satiating mechanism stimulated by hypertonicity.

In rats, isotonic glucose or urea infused into the duodenum inhibit lick rates and intake (Campbell and Davis, 1974), and isotonic or slightly hypotonic duodenal infusions of glucose, crystalline amino acids, or peanut oil lead to significant suppression of food intake (Novin et al., 1979). In rabbits, isotonic and hypertonic glucose infusions suppressed intake (VanderWeele et al., 1974). Intraduodenal glucose administration did not act as an unconditioned stimulus in the formation of a conditioned taste aversion. This suggests that the inhibition of intake was a satiating effect, not an aversive effect (Novin et al., 1974). In the pig (Houpt, 1982), duodenal infusions of hypertonic or isotonic glucose and saline administered prior to feeding also decreased food intake in a dose-related manner.

It appears that hypertonicity is a sufficient stimulus for intestinal satiety, but it may not be necessary. In contrast, the evidence is convincing that the specific chemical properties of nutrients are both necessary and sufficient stimuli for intestinal satiety.

Chemical Stimuli

Specific chemical properties of nutrients in the small intestine are also adequate stimuli for satiation. Furthermore, there is evidence that different physiological mechanisms mediate the effect of different chemical stimuli.

FATS

Intraintestinal fats have a satiating effect during real or sham feeding. Novin et al. (1979) found that short-term, intraduodenal infusions of peanut oil led to a 54% suppression of subsequent intake in rats. Intraduodenal administration of the salt of oleic acid significantly suppressed real feeding in pigs (Anika et al., 1981). In several studies in pigs, Gregory and colleagues (Gregory et al., 1989; Gregory and Rayner, 1987) showed that duodenal infusions of monoglycerides, oleic acid, Intralipid, and glycerol would all reduce meal size. Intraduodenal Intralipid suppressed real feeding in rats (Greenberg et al. unpublished results), and intraduodenal corn oil suppressed feeding in humans (Drewe et al., 1992; Welch et al., 1985, 1988). Hypertonicity is not necessary for inhibition of intake by oils or fats, because isotonic or slightly hypotonic duodenal infusions of peanut oil reduce food intake (Novin et al., 1979).

In all of these studies on real feeding, ingested food mixed with the infused stimuli in the duodenum as gastric emptying proceeded; thus the effective duodenal stimulus was uncertain. In the sham-feeding preparation, the delivery of the nutritive stimulus to the duodenum is precisely controlled and does not mix with ingested food because ingested food drains from the open gastric fistula (see Chapter 2). Under sham-feeding conditions, equivolumetric delivery of different caloric loads of Intralipid decreased intake and elicited the complete behavioral sequence of satiety as a direct function of the concentration of lipid in the infusate (Greenberg et al., 1986, 1990; Reidelberger et al., 1983). These infusions of different concentrations of Intralipid had different satiating potencies, but all were isotonic.

We have also tested the effects of duodenal infusions of glycerol, which is used in the emulsification of Intralipid. When we administered glycerol in concentrations over four times as large as is in the most concentrated load of Intralipid, we found no effect on sham feeding (Greenberg et al., unpublished results). These results show that the vehicle for the emulsification of Intralipid does not produce a satiating effect in the sham-feeding rat at the level of the intestine. It should be noted, however, that in real-feeding pigs, Gregory and Rayner (1987) showed a suppression of meal size with duodenal infusions of 15% glycerol when those infusions were maintained throughout the meal.

To determine if oily texture was an adequate stimulus for fat-elicited satiety, we compared the satiating potency of intraduodenal infusions of corn oil and mineral oil. Infusions of 50% mineral oil had no effect on sham feeding, while infusions of 12.5% corn oil elicited over a 50% reduction in sham feeding (Greenberg and Lewis, 1993b). These data show that oily texture alone is an insufficient stimulus for satiety.

This reduction of food intake was obtained with duodenal infusions that were administered at a rate apparently within the normal gastric emptying rate for ingested Intralipid (Kalogeris et al., 1983). Recent work using a different method of measuring gastric emptying following anesthesia, however, suggests that under some circumstances the oil portion of Intralipid may empty at a much slower rate (Horn et al., 1996). When Intralipid is infused at this slower rate, it still decreases intake significantly during sham feeding (Greenberg and Strohmayer, 1997). Further work is necessary to clarify the relationship between rate of infusion, volume of infusate, and magnitude of the satiating effect.

Fats are available in the small intestine as triglycerides, diglycerides, monoglycerides, and free fatty acids. There is considerable evidence that differences in the form, the degree of saturation, and the chain length of fats influence their digestion and metabolic utilization (Friedman, 1990; Senior, 1964; Holt and Bennett-Clark, 1975). To investigate the possibility that such differences also affect the satiating potency of fats, we measured the satiating potency of triglycerides and of free fatty acids infused into the small intestine of sham-feeding rats (Greenberg et al., 1988). We found that the satiating potency of the 18-carbon, long-chain fatty acid oleic acid was at least 20 times that of its triglyceride triolein. With a shorter chain length lipid, namely, the 8-carbon long fatty acid caprylic acid, we found the opposite relationship. That is, the triglyceride tricaprylin was significantly more effective at reducing food intake than caprylic acid. It must be noted, however, that at high concentrations of tricaprylin (2.5 Kcal or more), rats showed signs of illness and distress. Thus, in this range of doses, the inhibition of intake may be due to aversive effects of the infusions (Greenberg et al., 1987a; Greenberg et al., unpublished results).

Little work has been done on the satiating characteristics of mono- or diglycerides in rats. In pigs, it has been found that a mixture of monoglycerides (linoleic 62%, oleic 25%, palmitic 8%, and stearic 5%) was slightly less effective at reducing food intake (64% reduction) than a mixture of triglycerides (Intralipid; linoleic 50%, oleic 26%, palmitic 10%, linolenic 9%, and stearic 3.5%), which reduced food intake by 76% (Gregory et al., 1989). Specific comparisons of the satiating potency of a given fatty acid and its mono- or diglyceride have not been made.

The satiating potency of long-chain fatty acids that differ solely in their degree of saturation has also been investigated. When we administered isocaloric intestinal loads of oleic, linoleic, and linolenic acids in sham-feeding rats, we found that linoleic acid was more than twice as potent for reducing food intake than either oleic or linoleic acid. For example, intestinal infusion of 0.2 Kcal of linoleic acid produced over an 80% suppression of sham feeding than when saline was infused, while isocaloric loads of linolenic acid produced only a 25% suppression and oleic acid produced less than a 10% suppression. There is other evidence for a particular physiological effect of linoleic acid: Linoleic acid is more potent than oleic acid for slowing gastric emptying in humans (Card, 1941).

We have also explored the satiating potency of duodenal infusions of fatty acids that differ in chain length. When we examined the effects of duodenal infusions of the medium-chain fatty acids caprylic (C-8), capric (C-10), and lauric (C-12) acids, we found that lauric and capric acids, but not caprylic acid, decreased sham feeding. In addition, the threshold caloric load for lauric acid was 0.325 Kcal, while it was 0.65 Kcal for capric acid

TABLE 3.1.
Relative Satiating Potency of Intestinally Infused Fats

	Chain Length	Saturation	Satiating Potency
Fatty Acids			
Oleic acid	18	Mono-unsaturated	1.0
Linoleic acid	18	Di-unsaturated	4.0
Linolenic acid	18	Tri-unsaturated	2.0
Lauric acid	12	Saturated	0.25
Capric acid	10	Saturated	0.5
Caprylic acid	8	Saturated	0
Butyric acid	4	Saturated	0.25
Triglycerides			
Intralipid	14–22	Variable	4.0
Triolein	18	Mono-unsaturated	0
MCTs	8–12	Saturated	3.0
Corn oil	16–18	Variable	3.0
Non-Nutritive			
Mineral Oil	Very long-chain	Variable	0

Shown is a summary of the approximate relative satiating potencies of various fat stimuli in the rat. Potency was measured as the percent inhibition of sham-fed intake during the first 24 minutes after the duodenal infusion began. The potency of oleic acid is the reference point of 1.0. These comparisons were synthesized from many experiments referred to in the text.

(McCaffery et al., 1994). Thus, satiating potency increases as the chain length of medium-chain fatty acids increases.

These results show that the satiating potency of a fatty acid is determined by its chain length and degree of saturation (Table 3.1). These are the same aspects of the chemical structure of fats that have previously been shown to determine the digestion and absorption of fats (Hunt and Knox, 1968; Clark et al., 1969; Holt and Bennett-Clark, 1975). It also is clear that mixtures of medium- or long-chain triglycerides are more potent than most fatty acids or triolein. Thus, specific lipids differ greatly in their inhibition of intakes. Much more investigation is needed to understand the relationship between the structure of a fat and its satiating potency.

Amino Acids and Carbohydrates

Specific amino acids have different satiating potencies when infused intraduodenally. Lew et al. (1983) demonstrated that in the sham-feeding rat intraduodenal infusions of L-phenylalanine (3%) induced a significant suppression of intake compared with saline infusions, but D-phenylalanine did not. Lew et al. (1983) suggested that this effect was mediated by the release of CCK from the intestine. The satiating potencies of other amino acids and proteins remain to be tested.

Carbohydrates also elicit satiety when infused at the level of the intestine, although they may be less potent than amino acids or fats (Yox et al., 1991a). Isocaloric intestinal loads of sodium oleate or L-phenylalanine suppressed sham feeding in rats by approximately 55%. Isocaloric loads of maltose suppressed sham feeding by less than 30% (Yox et al., 1991a).

Earlier work (Snowdon, 1975), however, showed that intraduodenal infusions of 1 M glucose were more potent for reducing meal size than were 0.5 M urea infusions. Yin and Tsai (1973) also showed that duodenal infusions of glucose decreased meal size in relation to the concentration of glucose infused, and they compared the effects of xylose (25%) to glucose (25%) and found similar reductions in meal size. Thus, the effect of these stimuli could be due to hypertonicity.

More recently duodenal infusions of 1.2 Kcal of fructose did not decrease real feeding, while the same caloric load of glucose or maltose reduced 30-minute intakes significantly (Walls et al., 1995). These are the only direct comparisons of intestinal infusions of isocaloric sugar loads that have been explored. Given the well known differences in the physiological responses to different sugars, a comparison of the satiating potency of intestinal infusions of different carbohydrates is an area worthy of further exploration.

Relevant Studies in Genetically Obese Animals

The genetically obese Zucker rat (*fa/fa*) is hyperphagic compared to lean controls (*Fa/?*) (Bray and York, 1979; Bray et al., 1989). In addition, this hyperphagia is especially pronounced for fats rather than other nutrients (Castonguay et al., 1982, 1984). One explanation for this hyperphagia is that there are deficits in satiation in the obese Zucker rat. One possible deficit is that the satiating potency of nutrients at the level of the small intestine is reduced. Maggio et al. (1983) gave equicaloric intragastric loads of fats, proteins, and carbohydrates. Equicaloric loads of the macronutrients depressed intake equally in lean rats, but fats were less satiating than proteins or carbohydrates in obese rats. The difference in satiation seen in response to intragastric loads could be due to gastric satiating mechanisms, gastric emptying rates, intestinal satiating mechanisms, or any combination of these. We have attempted to determine which of these possibilities explains the differential satiating potency of fats in obese and lean Zucker rats.

In one study we examined differences in the rate of gastric emptying of saline in response to intestinal corn oil infusions in lean and obese Zucker rats. We hypothesized that if fats emptied more rapidly in obese rats, then gastric satiating mechanisms would be decreased and, thus, fats would be less satiating in these rats. Using a phenol red marker technique for measuring emptying, we found that corn oil slowed gastric emptying significantly more in obese than in lean rats. This contradicted our hypothesis (Greenberg et al., 1987b).

We have also investigated the possibility that fats are less satiating for obese Zucker rats than for lean rats by administering intestinal infusions of Intralipid, oleic acid, and linoleic acid during sham feeding in lean and obese rats. We found no difference in the satiating potency of Intralipid or oleic acid between the genotypes (French et al., 1995; Greenberg et al., 1986). For linoleic acid, however, a larger caloric load was necessary to elicit a significant reduction of feeding in obese than lean rats, but this difference was small and seems unlikely by itself to account for excessive fat intake by obese rats (French et al., 1995). Although there may be deficiencies in intestinal satiating mechanisms for fat in genetically obese rats that contribute to their excessive ingestion of fat, such deficiencies have not yet been demonstrated.

Summary

The specific chemical properties of nutrients are adequate stimuli for intestinal satiating effects in normal and genetically obese rats. Volume or

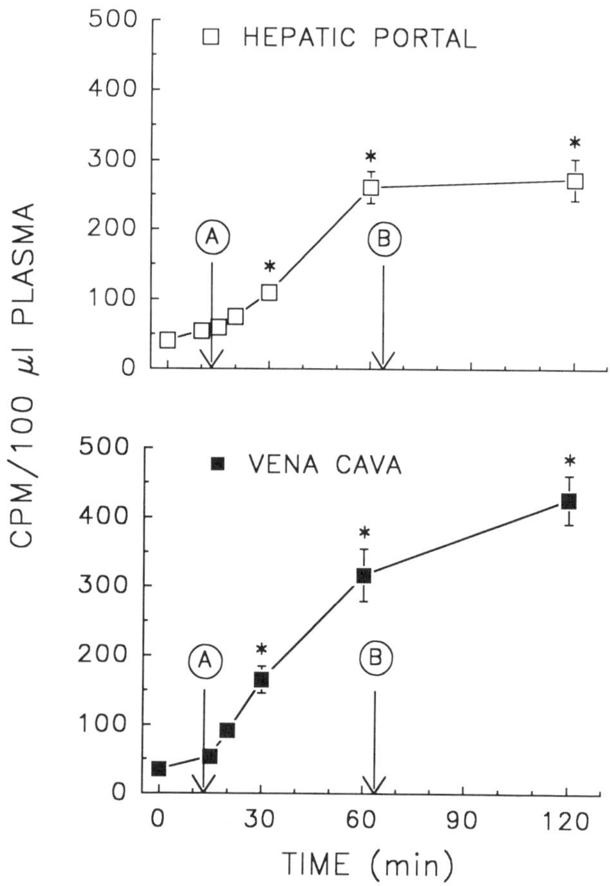

FIGURE 3.1. Mean ± SEM of the concentration of radioactivity in plasma sampled from the hepatic-portal vein (top) and from the inferior vena cava (bottom) just before (0) and at 10, 15, 20, 30, 60, and 120 minutes after a duodenal infusion of radioactive Intralipid. Radioactivity increased significantly at 30, 60, and 120 minutes in plasma sampled from both venous sites, *$p < 0.05$ compared with the plasma radioactivity at time 0. Previous experiments (Greenberg et al., 1990) showed that the satiating effect of duodenal infusions of Intralipid began within 10 minutes (A) and a second meal was initiated by 65 minutes (B). From Greenberg et al. (1995), with permission of publisher.

distention alone is not a sufficient stimulus, because intestinal infusions of saline that are isovolumetric with effective nutrient stimuli do not reduce food intake in a wide variety of experimental conditions.

Hypertonicity is a sufficient stimulus because hypertonic intestinal loads of various nutrients are able to reduce feeding, but hypertonicity is not a recessary stimulus because isotonic loads of different concentrations of fats have significantly different satiating potencies.

In each of the macronutrient classes, specific nutrients have different satiating potencies. Many more experiments need to be done to define more precisely the characteristics of the adequate stimuli for intestinal satiety. This is especially true for amino acids, proteins, and carbohydrates. Specific comparisons between the satiating potencies of intestinal stimuli by mixtures of nutrients as occurs in foods also remain to be done.

The Site of Action for Intestinal Satiation

Fats

There is considerable evidence that the inhibition of food intake produced by intraduodenally administered nutrients begins prior to the entry of digestive byproducts into the blood. In humans, ileal infusions of a corn oil emulsion reduced caloric intake and shortened the length of a subsequent meal, while intravenous infusions of Intralipid had no effect on food intake (Welch et al., 1985). In the sham-feeding rat, we have shown that duodenal infusions of total caloric loads of as little as 1.25 Kcal of Intralipid lead to significant reductions of intake, while total caloric loads of 20 Kcal delivered into either the hepatic-portal vein or the vena cava have no effect on sham feeding (Greenberg et al., 1990, 1993).

When plasma is sampled from either the hepatic-portal vein or the inferior vena cava subsequent to infusion of radioactively ^{14}C- labeled Intralipid into the duodenum at the same rates and volumes as in our feeding studies, there is no significant increase in plasma radioactivity until 30 minutes after the start of the duodenal infusions (Greenberg et al., 1995). This time course is dissociated from that of the suppression of feeding, because we find significant decreases in feeding 10 minutes after the start of duodenal infusions. Furthermore, at 60 minutes, when the plasma radioactivity is maximal, sham-feeding rats are initiating a second meal (Fig. 3.1). These results are strong evidence for the hypothesis that preabsorptive mechanisms mediate the intestinal satiating effect of fat stimuli.

Carbohydrates

The evidence of a preabsorptive site for the satiating effect of carbohydrates in the intestine is less clear. Yin and Tsai (1973) found that administration of xylose and glucose into either the mesentric or jugular vein did not suppress feeding in the rat, but similar amounts of xylose and glucose suppressed feeding when intraduodenally administered.

It is possible that intraduodenal infusions act postabsorptively in the liver in rats (Russek, 1971). Tordoff and Friedman (1986, 1988) showed that glucose infusions into the hepatic-portal vein suppressed feeding, but glucose infusions into the jugular vein did not. Furthermore, Campbell and Davis (1974) observed suppression of lick rates after infusions of glucose or urea into the portal vein that were similar to the suppressions induced by duodenal infusions.

In contrast to rats, the site of the satiating action of intraduodenal glucose infusions in rabbits appears to be preabsorptive (VanderWeele et al., 1974; Novin et al., 1974). In the pig, Houpt (1982) found that duodenal infusions of glucose decreased intake, but intravenous infusions that raised plasma glucose and osmotic pressure to values at least as high as those seen after duodenal infusions did not. It is clear that more research is needed to clarify the contribution of pre- and postabsorptive sites to the satiating effect of carbohydrates.

Summary

The likelihood that many intestinal nutrient stimuli act at preabsorptive sites is supported by two types of evidence. First, the ability of intestinal nutrient infusions to decrease intake rapidly and, in the case of fats, prior to their entry into the plasma suggests a preabsorptive site of action, but rapid, postabsorpitive effects from the liver have been described. Second, the minimal effect of systemic intravenous nutrient infusions on meal size supports this hypothesis as well.

Mechanisms of Intestinal Satiation

Fats

There is evidence that the arrival of fats in the duodenum is signaled to the central nervous system. Intraduodenal infusions of corn oil in the cat lead to increased blood flow in the superior mesenteric vein and to evoked spindles in EEG patterns that are typical of sedation. These events are also accompanied by behavioral events suggesting drowsiness (Fara et al., 1969).

One possible mechanism for mediating this effect and the satiating effect of fats at the level of the small intestine is stimulation of the vagus nerve. Two types of vagal afferent fibers are sensitive to lipids. One type responds preferentially to long-chain fats and the other type responds preferentially to short-chain fats and glycerol (Melone, 1986).

An early study of the vagal mediation of the suppressive effects of intraduodenal nutrient infusions showed that total subdiaphragmatic vagotomy in rats attenuated suppression of feeding elicited by intraduodenal glucose and amino acids, but did not have an effect on fats (Novin et al., 1979). This study was, however, carried out on normally feeding rats, and the confounding effects of ingested foods, gastric distention, and gastric emptying make these results difficult to interpret. This is especially true as vagotomy is known to affect the rate of gastric emptying (Snowdon, 1975; Waddell and Wang, 1953). More recently, selective celiac vagal deafferentation was shown to reverse suppression of intake produced by intestinal infusions of 1.4% oleic acid in real-feeding rats (Walls et al., 1995). This work suggests that in normal feeding the vagal innervation of the intestines is necessary for the satiating effect of fats.

Yox and colleagues (1991a) have shown that total subdiaphragmatic vagotomy abolished the suppression of sham feeding elicited by intestinal infusion of sodium oleate. We have replicated this finding and extended it to show that total abdominal vagotomy blocks the satiating effect of duodenally infused Intralipid, oleic acid, linoleic acid, or linolenic acid (Greenberg and Lewis, 1993a).

Yox et al. (1991b) also showed that pretreatment with capsaicin, a neurotoxin that selectively destroys small unmyelinated sensory fibers, into the fourth ventricle of rats attenuated the satiating effect of intestinal oleic acid during sham feeding. This implies that the effects of vagotomy are probably the result of lesioning afferent vagal fibers and is consistent with the results of Walls et al. (1995) in real feeding (see above).

To identify which branches of the abdominal vagus are most important for the satiating effect of fats in the sham-feeding preparation, we have begun to examine the effects of various selective abdominal vagotomies. At this time, we have completed only the study of the hepatic branch. Duodenally infused Intralipid and fatty acids retain their full satiating potency when the hepatic branch of the vagus is cut or when only the hepatic branch is left intact. Thus, the hepatic branch of the vagus is sufficient, but not necessary, for the mediation of fat-induced satiety in the sham-feeding rat (Greenberg and Lewis, 1993a).

Intestinal peptides also mediate intestinal satiating effects. The arrival of fats into the small intestine is a potent stimulus for the release of CCK in rats, dogs, pigs, and humans (Douglas et al., 1988, 1990; Green et al., 1989; Liddle et al., 1984, 1985; Lilja et al., 1984; Burhol et al., 1982; Makhlouf, 1974; Wang and Grossman, 1951). This is true not only for

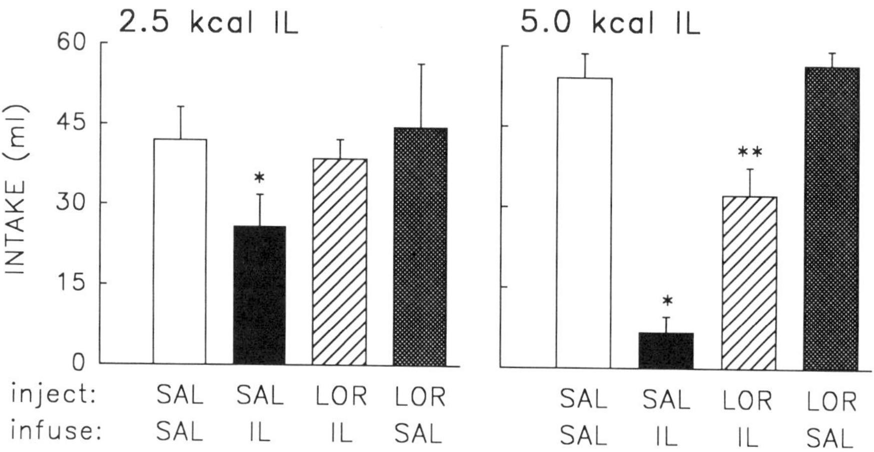

FIGURE 3.2. Data are mean ± SEM for sham-fed intakes of Sprague-Dawley rats. Rats were given intestinal infusions of either 0.15 M NaCl or 2.5 Kcal or 5 Kcal of Intralipid. Intestinal infusions of 10 ml were given at a rate of 0.44 ml/min. Ten minutes prior to the start of sham-feeding tests, rats were given intraperitoneal injections of either 0.15 M NaCl vehicle or of 600 μg/kg lorglumide, an A-type CCK antagonist. Lorglumide abolished the reduction in sham feeding elicited by intestinal infusion of 2.5 Kcal Intralipid, but only attenuated the reduction in sham feeding elicited by intestinal infusion of 5 Kcal of Intralipid. From Greenberg et al. (1989), with permission of the publisher.

orally ingested fats, but also when fats are infused intraduodenally. If CCK mediated the satiating effects of fats in the intestines, selective CCK antagonists, such as devazepide, loxiglumide, and lorglumide, should attenuate the inhibition of feeding by intraduodenal infusions of fats. In the rat, administration of either devazepide or lorglumide attenuated the suppression of sham feeding elicited by intraduodenal infusions of Intralipid (Greenberg et al., 1989) or oleic acid (Yox et al., 1992). In the pig, intestinal satiety elicited by fats was also attenuated by administration of devazepide (Ebenezer et al., 1990).

Differential release of CCK may also explain how adequate fat stimuli differ in satiating potency. For example, the differential satiating effect of fatty acids differing in saturation may be due to their differential release of endogenous CCK because more CCK is released by unsaturated fatty acids than by saturated fatty acids in humans (Beardshall et al., 1989), and this may also occur in the rat (Lewis and Williams, 1990).

The evidence for CCK mediation in the human is less clear. Drewe et al. (1992) administered loxiglumide before and during jejunal infusions of corn oil. When subjects were pretreated with intravenous administration of vehicle, jejunal fat infusions decreased the amount eaten in grams by 20% and the duration of a test meal by 16%, but, unlike the findings of Welch et al. (1988), they found no effect on kilocalories consumed. Based on their statistical analysis, the authors concluded that pretreatment with intravenous loxiglumide did not significantly attenuate the suppression of food intake, liquid intake, or meal duration elicited by intrajejunal corn oil. It should be noted, however, that loxiglumide did reduce the suppression of total food intake (in grams) from 20% to 12%, and, subsequent to loxiglumide pretreatment, intestinal fat infusion no longer significantly suppressed food intake (expressed as grams). Although Drewe et al. (1992) interpreted these results to mean that CCK does not mediate the satiating effect of intestinal corn oil in humans, this interpretation should be questioned for several reasons. First, loxiglumide had a tendency to attenuate the satiating effect of fat loads, although this was not significant. Second, the intestinal load of 41 g of fat was much larger than the fat load that would be found in a typical meal, and such a large load might induce satiating factors other than CCK. For example, we found that very high doses of lorglumide eliminated the inhibition of sham feeding elicited by a 2.5-Kcal duodenal infusion of Intralipid, but only attenuated the inhibition elicited by a 5.0-Kcal infusion (Greenberg et al., 1989) (Fig. 3.2).

Thus, it appears that CCK is involved in mediating the satiating effect of fat stimuli in animals. Whether endogenous CCK is also involved in mediating this effect in humans requires further study.

Another possible mediator of the satiating effect of fats is enterostatin, a pentapeptide (VAL-PRO-ASP-PRO-ARG) derived from the activation of pancreatic procolipase in the lumen of the upper small intestine by trypsin (Erlanson-Albertsson and Larsson, 1988).

Intraduodenal, intraperitoneal, intravenous, or intracerebroventricular administration of enterostatin decreases food intake (Erlanson-Albertsson and Larsson, 1988; Mei and Erlansson-Albertson, 1992; Mei and Erlansson-Albertson, 1996). Enterostatin inhibits the intake of high-fat diets more than the intake of high-carbohydrate diets (Okada et al., 1991; Erlanson-Albertsson et al., 1991). The satiating effect of intraduodenal enterostatin is reduced by pretreatment with intraduodenal tetracaine (Mei and Erlanson-Albertson, 1992). This suggests a local site of action.

The satiating effect of peripherally administered enterostatin is blocked by hepatic vagotomy (Tian et al., 1994). This suggests that the peripheral effect of enterostatin is tranduced into neural activity in hepatic vagal afferent neurons that project to the NTS in the hindbrain. To established that endogenous enterostatin participates in the satiation of eating by ingested fat, its site of action and cellular effect must be identified and a specific antagonist must increase meal size.

Other gastrointestinal peptides, such as gastroinhibitory polypeptide (Ohnedia et al., 1984; Falko et al., 1975) and neurotensin (Rosell and Rokaeus, 1979), are released by intraduodenal fat infusion (Thomas et al., 1976). Fats act preabsorptively to stimulate the release of neurotensin, because intravenous administration of fats had no effect on neurotensin levels (Rosell and Rokaeus, 1979). These peptides do not inhibit feeding when administered peripherally (Lorenz et al., 1979), but intracerebroventricular administration of neurotensin does (Luttinger et al., 1982).

Thus, these peptides are not as likely as CCK to be involved in mediating the satiating effect of intestinal fat. Further work is necessary to determine the role, if any, of these peptides in the inhibition of intake by fat in the intestine.

Finally, there is strong evidence for the hypothesis that apolipoprotein A-IV mediates at least part of the satiating effect of fat in the small intestine. Specifically, intragastric administration of 0.1 g triglyceride increased the concentration of A-IV in the plasma and lymph within 30 minutes to levels that were produced by doses of exogenous A-IV that decreased intake (Rodriguez et al., 1997; Fujimoto et al., 1992, 1993).

Amino Acids and Carbohydrates

The vagus nerve is also implicated in the mediation of the satiating effect of amino acids and carbohydrates in the intestine (Table 3.2). Yox et al. (1991a) have shown that total subdiaphragmatic vagotomy abolished the suppression of sham feeding elicited by intestinal infusion of maltose and attenuated the suppression of sham feeding elicited by L-phenylalanine. Recent work by Walls et al. (1995) showed that selective celiac deafferentation reversed the suppression of 30-minute intakes during real feeding elicited by duodenal infusions of 1.2 Kcal glucose or 1.2 Kcal maltose. This selective deafferentation also reversed the reduction of real feeding elicited by infusion of 12% L-phenylalanine (Walls et al., 1995), but only attenuated the suppression of intake elicited by 1.2 Kcal casein hydrolysate.

There is neurophysiological evidence that is consistent with stimulation of the abdominal vagus by carbohydrates and amino acids in the small intestine. For example, cells in the nodose ganglion respond to intraduodenal glucose in cats (Mei, 1978), and there is evidence for intestinal receptors that are sensitive to more than osmotic pressure (Hardcastle et al., 1978). In addition, intraduodenal glucose alters firing rates of neurons

TABLE 3.2.
Summary of the Effects of Three Experimental Interventions on the Satiating Potencies of Intestinal Nutrients

Stimulus	Abdominal Vagotomy	CCK Antagonists	Capsaicin Pretreatment
Fats	Reverses	Attenuates	Reverses
L-phenylalanine	Attenuates	Reverses	Attenuates
Casein	Not tested	Not tested	No effect
Maltose	Reverses	Reverses	Attenuates

in the dorsal vagal nucleus and the nucleus of the tractus solitarius in anesthetized rats (Ewart and Wingate, 1983). Some neurons had only excitatory responses to glucose infusions, while others had only inhibitory responses. Those neurons that responded to intraduodenal glucose were different from the neurons that responded to gastric distention produced by inflation of an intragastric balloon (Ewart and Wingate, 1983). Infusion of amino acids into the intestine also evokes responses in the dorsal vagal complex in the hindbrain of anesthetized cats (Jeanningros, 1982).

Although capsaicin administered into the fourth ventricle abolished the satiation elicited by intestinal infusion of maltose, capsaicin treatment did not affect satiation elicited by intestinal L-phenylalanine (Yox et al., 1991b). These results suggest that capsaicin-sensitive afferent fibers located near the fourth ventricle, some of which are likely to be vagal, mediate the satiating effect of maltose, but not of L-phenylalanine.

As is true for fats, CCK may also be involved in the satiating effect of amino acids and carbohydrates. L-phenylalanine, but not D-phenylalanine, is a potent releaser of CCK in the dog (Meyer and Grossman, 1972). It seemed possible that this differential release of CCK was responsible for the observed satiating effect of intestinal L-phenylalanine infusions in the rat (Lew et al., 1983) and of gastric preloads of L-phenylalanine, but not D-phenylalanine, in the rhesus monkey (Gibbs et al., 1976).

More recent work, however, contradicts this hypothesis. Although intraduodenal infusions of L-phenylalanine reduced sham feeding (Yox et al., 1992), plasma CCK was not elevated by these infusions when a bioassay of the increase in secretion of pancreatic acinar cells was used to measure CCK levels (Brenner et al., 1993). In addition, the suppression of sham feeding elicited by L-phenylalanine in rats was not reversed by administration of devazepide under the same conditions in which devazepide attenuated the suppression of sham feeding elicited by intraduodenal infusions of maltose (Yox et al., 1992).

Gastric inhibitory polypeptide (Ohnedia et al., 1984; Falko et al., 1975) and neurotensin (Rosell and Rokaeus, 1979) are released by intraduodenal amino acid infusion and could be involved in mediating satiety elicited by these stimuli (Thomas et al., 1976), but there are no relevant data.

There may also be interactions between neural and hormonal mediation of the satiating effects of nutrients in the small intestine. There are CCK receptors on vagal fibers (Corp et al., 1993; Garlicki et al., 1990; Moran et al., 1987; Davison and Clarke, 1988; Schwartz et al., 1992; Lin and Miller, 1992; Zarbin et al., 1981). Total abdominal or selective afferent vagotomy blocks the satiating effect of exogenous CCK (Smith et al., 1981, 1985; Lorenz and Goldman, 1982; South and Ritter, 1988). In addition, the satiating effect of exogenously administered CCK is attenuated by pretreatment with capsaicin (Ritter and Ladenheim, 1985). Thus, intestinal nutrients could directly activate vagal afferent fibers, or intestinal nutrients could release endogenous CCK that could then activate vagal afferent fibers.

Summary

It is clear that intestinal nutrient stimuli have satiating effects. The use of sham-feeding techniques derived from Pavlov's pioneering work (1910) has shown that nutrient stimulation of the small intestine in the absence of gastric distention is sufficient for satiety. Although this effect can be obtained in the absence of gastric distention, under normal conditions intestinal satiating mechanisms interact synergistically with gastric and oral satiating mechanisms (see Chapter 2).

There is strong evidence that intestinal infusions of fats act preabsorptively. First, intestinal fat infusions are potent stimuli for the inhibition of food intake; second, intravenous infusions of the same fats fail to inhibit food intake in many species, including humans: Third, the time course for the satiating effect of fats is much more rapid than their absorption into the blood. There is less evidence concerning amino acids, proteins, and carbohydrates, but, when intraduodenal and intravenous infusions of these nutrients are compared, intraduodenal infusions are usually more effective at reducing short-term intake.

The most salient adequate stimulus for intestinal satiety is the specific chemical conformation of fats, amino acids, or carbohydrates. Hypertonicity also appears to be a sufficient, although not a necessary, stimulus.

Both neural and peptide mediators of intestinal satiety have been demonstrated under certain experimental circumstances. The vagus nerve and

CCK are implicated in the mediation of intestinal satiating effects, but much more work is necessary to describe all of the adequate stimuli and mediating mechanisms for these effects.

The available data do not provide a clear answer to the question of whether intestinal infusions in experimental conditions reveal a physiological role for the satiating effect of nutrients entering the intestine during normal feeding. This also requires further experiments.

Finally, it remains to be determined whether deficiencies in the satiating effect of nutrients in the small intestine contribute to the abnormally large meal size observed in bulimia nervosa and some forms of obesity. Such information, when we have it, should provide insight into these disorders.

ACKNOWLEDGMENTS

This work was supported by NIH DK-38757 and The International Life Sciences Institute.

REFERENCES

Anika, S. M., T. R. Houpt, and K. A. Houpt. Cholecystokinin and satiety in pigs. *Am. J. Physiol.* 240:R310–R318, 1981.

Antin, J., J. Gibbs, and G. P. Smith. Intestinal satiety requires pregastric food stimulation. *Physiol. Behav.* 18:421–425, 1977.

Balagura, S., and H. C. Fibiger. Tube feeding: Intestinal factors in gastric feeding. *Psychosom. Sci.* 10:373–374, 1968.

Beardshall, K., G. Frost, Y. Morarji, J. Domin, S. R. Bloom, and J. Calam. Saturation of fat and cholecystokinin release: Implications for pancreatic carcinogenesis. *Lancet* II:1008–1010, 1989.

Bray, G. A., and D. A. York. Hypothalamic and genetic obesity in experimental animals: An autonomic and endocrine hypothosis. *Physiol. Rev.* 59:719–809, 1979.

Bray, G. A., D. A. York, and J. S. Fisler. Experimental obesity: A homeostatic failure due to defective nutrient stimulation of the sympathetic nervous system. *Vit. Horm.* 45:1–125, 1989.

Brenner, L., D. P. Yox, and R. C. Ritter. Suppression of sham feeding by intraintestinal nutrients is not correlated with plasma cholecystokinin elevation. *Am. J. Physiol.* 264:R972–R976, 1993.

Burhol, P. G., T. G. Jenssen, I. Lygren, T. B. Schulz, R. Jorde, and H. L. Waldum. Iodination with the iodo-gen and radioimmunoassay of cholecystokinin (CCK) in adidified plasma, CCK release, and molecular CCK composition in man. *Digestion* 23:156–158, 1982.

Busch, W. Contribution to the physiology of the digestive organs. *Arch. Pathol. Anat. Physiol. Klin. Med.* 14:140–186, 1858.

Campbell, C. S., and J. D. Davis. Licking rate of rats is reduced by intraduodenal and intraportal glucose infusion. *Physiol. Behav.* 12:357–365, 1974.

Card, W. I. A comparison of the inhibitory action of different fats and fatty acids introduced into the duodenum on gastric contractions. *J. Dig. Dis.* 8:47–53, 1941.

Castonguay, T. W., S. L. Burdick, M. A. Guzman, G. H. Collier, and J. S. Stern. Self-selection and the obese Zucker rat: The effect of dietary fat dilution. *Physiol. Behav.* 33:119–126, 1984.

Castonguay, T. W., W. J. Hartmann, E. A. Fitzpatrick, and J. A. Stern. Dietary self-selection and the Zucker rat. *J. Nutr.* 112:796–800, 1982.

Clark, S. B., B. Brause, and P. R. Holt. Lipolysis and absorption of fat in the rat stomach. *Gastroenterology.* 56:214–222, 1969.

Corp, E. S., J. McQuade, T. H. Moran, and G. P. Smith. Characterization of type A and type B CCK receptor binding sites in rat vagus nerve. *Brain Res.* 623:161–166, 1993.

Davis, J. D., and C. S. Campbell. Peripheral control of meal size in the rat: Effect of sham feeding on meal size and drinking rate. *J. Comp. Physiol. Psychol.* 83:379–387, 1973.

Davis, J. D., and B. J. Collins. Distention of the small intestine, satiety, and the control of food intake. *Am. J. Clin. Nutr.* 31:S255-S258, 1978.

Davison, J. S., and G. D. Clarke. Mechanical properties and sensitivity to CCK of vagal gastric slowly adapting mechanoreceptors. *Am. J. Physiol.* 255:G55-G61, 1988.

Debas, H. T., O. Farrooq, and M. I. Grossman. Inhibition of gastric emptying is a physiological action of cholecystokinin. *Gastroenterology* 68:1211–1217, 1975.

Douglas, B. R., J. B. M. J. Jansen, A. J. L. de Jong, and C. B. H. W. Lamers. Effect of various triglycerides on plasma cholecystokinin levels in rats. *J. Nutr.* 120:686–690, 1990.

Douglas, B. R., R. A. Woutersen, J. B. M. J. Jansen, A. J. L. de Jong, and C. B. H. W. Lamers. The influence of different nutrients on plasma cholecystokinin levels in the rat. *Experientia* 44:21–23, 1988.

Drewe, J., A. Gadient, L. C. Rovati, and C. Beglinger. Role of circulating cholecystokinin in control of fat-induced inhibition of food intake in humans. *Gastroenterology* 102:1654–1659, 1992.

Ebenezer, I. S., C. de la Riva, and B. A. Baldwin. Effects of the CCK receptor antagonist MK-329 on food intake in pigs. *Physiol. Behav.* 47:145–148, 1990.

Ehman, G. K., D. J. Albert, and J. L. Jamieson. Injection into the duodenum and the induction of satiety in the rat. *Can. J. Psychol.* 25:147–166, 1971.

Erlansson-Albertsson, C., M. Jie, S. Okada, D. A. York, and G. A. Bray. Pancreatic procolipase propeptide-enterostatin-specifically inhibits fat intake. *Physiol. Behav.* 49:1191–1194, 1991.

Erlansson-Albertsson, C., and A. Larsson. The activation peptide of pancreatic procolipase decreases food intake in rats. *Regul. Pept.* 22:325–331, 1988.

Ewart, W. R., and D. L. Wingate. Cholecystokinin octapeptide and gastric mechanoreceptor activity in rat brain. *Am. J. Physiol.* 244:G613-G617, 1983.

Falko, J. M., S. E. Crockett, S. Cataland, and E. L. Mazzaferri. Gastric inhibitory polypeptide (GIP) stimulated by fat ingestion in man. *J. Clin. Endocrinol. Metab.* 41:260–266, 1975.

Fara, J. W., E. H. Rubinstein, and R. R. Sonnenschein. Visceral and behavioral responses to intraduodenal fat. *Science* 166:110–111, 1969.

Farrell, J. I., and A. C. Ivy. Studies on the motility of the transplanted gastric pouch. *Am. J. Physiol.* 76:227–228, 1926.

Foster, L. A., K. Nakamura, R. Norgren, and D. Greenberg. Intestinal fat differentially suppresses sham feeding of preferred gustatory stimuli. *Am. J. Physiol.* 270:R1122–R1125, 1996.

French, S. J., J. McCaffery, and D. Greenberg. Lick rates during satiety elicited by duodenal linoleic acid infusions in Zucker rats. *Appetite* 23:393, 1994.

French, S. J., J. McCaffery, and D. Greenberg. The effect of ileal fat infusion on sham feeding in the rat. *Appetite* 24:381, 1995.

Friedman, M. I. Body fat and the metabolic control of food intake. *Int. J. Obesity* 14:53–67, 1990.

Fujimoto, K., J. A. Cardelli, and P. Tso. Increased apolipoprotein A-IV in rat mesenteric lymph after lipid meal acts as a physiological signal for satiation. *Am. J. Physiol.* 262:G1002-G1006, 1992.

Fujimoto, K., H. Machidori, R. Isakiri et al. Effect of intravenous administration of apolipoprotein A-IV on patterns of feeding, drinking and ambulatory activity of rats. *Brain Res.* 608:233–237, 1993.

Garlicki, J., P. K. Konturek, J. Majka, N. Kwiecien, and S. J. Konturek. Cholecystokinin receptors and vagal nerves in control of food intake in rats. *Am. J. Physiol.* 258:E40-E45, 1990.

Gibbs, J., J. D. Falasco, and P. R. McHugh. Cholecystokinin decreased food intake in rhesus monkeys. *Am. J. Physiol.* 230:15–18, 1976.

Gibbs, J., S. P. Maddison, and E. T. Rolls. Satiety role of the small intestine examined in sham-feeding rhesus monkeys. *J. Comp. Physiol. Psychol.* 95: 1003–1015, 1981.

Glick, Z., and M. Modan. Control of food intake during continuous injection of glucose into upper duodenum and upper ileum of rats. *Physiol. Psychol.* 5:7–10, 1977.

Green, G. M., S. Taguchi, J. Friestman, W. Y. Chey, and R. A. Liddle. Plasma secretin, CCK, and pancreatic secretion in response to dietary fat in the rat. *Am. J. Physiol.* 256:G1016-G1021, 1989.

Greenberg, D., P. A. Foelsch, P. M. Perez, and G. P. Smith. Intraduodenal medium-chain fats decrease food intake but intraportal medium-chain fats do not. *Soc. Neurosci. Abstr.* 13:462, 1987a.

Greenberg, D., J. Gibbs, and G. P. Smith. Intraduodenal infusions of fat inhibit sham feeding in Zucker rats. *Brain Res. Bull.* 20:779–784, 1986.

Greenberg, D., J. Gibbs, and G. P. Smith. Corn oil decreases gastric emptying more in obese than lean Zucker rats. *Eastern Psychol. Assoc. Abstr.* 58:48, 1987b.

Greenberg, D., R. A. Kava, D. R. Lewis, M. R. C. Greenwood, and G. P. Smith. Time course for entry of intestinally infused lipids into blood of rats. *Am. J. Physiol.* 269:R432–R436, 1995.

Greenberg, D., and D. R. Lewis. Total but not selective hepatic vagotomy attenuates fat-induced satiety. *Appetite* 21:180, 1993a.

Greenberg, D., and D. R. Lewis. Intraduodenal infusions of corn oil but not mineral oil decrease sham feeding. *Soc. Neurosci. Abstr.* 19:817, 1993b.

Greenberg, D., G. P. Smith, and J. Gibbs. Oleic acid inhibits sham feeding when duodenally infused while triolein does not. *Soc. Neurosci. Abstr.* 14:1196, 1988.

Greenberg, D., G. P. Smith, and J. Gibbs. Intraduodenal infusions of fats elicit satiety in the sham feeding rat. *Am. J. Physiol.* 259:R110–R118, 1990.

Greenberg, D., G. P. Smith, and J. Gibbs. Intravenous triglycerides fail to elicit satiety in sham feeding rats. *Am. J. Physiol.* 264:R409–R413, 1993.

Greenberg, D., and A. J. Strohmayer. Slow intestinal infusions of fat suppress sham feeding in the rat. *Soc. Neurosci. Abstr.* 23: (in press), 1997.

Greenberg, D., N. I. Torres, G. P. Smith, and J. Gibbs. The satiating effects of fats is attenuated by the cholecystokinin antagonist lorglumide. *Ann. N.Y. Acad. Sci.* 575:517–520, 1989.

Gregory, P. C., M. McFadyen, and D. V. Rayner. Duodenal infusion of fat, cholecystokinin secretion and satiety in the pig. *Physiol. Behav.* 45:1021–1024, 1989.

Gregory, P. C., and D. V. Rayner. The influence of gastrointestinal infusion of fats on regulation of food intake in pigs. *J. Physiol.* 385:471–481, 1987.

Han, P. W., J. Y. Mu, and S. Lepkovsky. Food intake of parabiotic rats. *Am. J. Physiol.* 205:1139–1143, 1963.

Hardcastle, J., P. T. Hardcastle, and P. A. Sandford. Effect of actively transported hexoses on afferent nerve discharge from rat small intestine. *J. Physiol. Lond.* 285:71–84, 1978.

Hill, R. G., E. C. Ison, W. W. Jones, and J. W. Archdeacon. The small intestine as a factor in the regulation of eating. *Am. J. Physiol.* 170:201–205, 1952.

Holt, P. R., and Bennett-Clark, S. Fat absorption. In Friedman, M. H. F. (ed.): *Function of the Stomach and Intestine.* Baltimore, MD: University Press, 1975, pp. 285–299.

Horn, C. C., M. G. Tordoff, and M. I. Friedman. Does ingested fat produce satiety? *Am. J. Physiol.* 270:R761–R765, 1996.

Houpt, K. A. Gastrointestinal factors in hunger and satiety. *Neurosci. Biobehav. Rev.* 6:145–164, 1982.

Houpt, T. R., B. A. Baldwin, and K. A. Houpt. Effects of duodenal osmotic loads on spontaneous meals in pigs. *Physiol. Behav.* 40:787–795, 1983.

Hunt, J. N., and M. T. Knox. A relation between the chain length of fatty acids and the slowing of gastric emptying. *J. Physiol.* 194:327–336, 1968.

Hunt, J. N., and D. F. Stubbs. The volume and energy content of meals as determinants of gastric emptying. *J. Physiol.* 245:209–225, 1975.

Jeanningros, R. Vagal unitary responses to intestinal amino acid infusions in the anesthetized cat: A putative signal for protein induced satiety. *Physiol. Behav.* 28:9–21, 1982.

Kalogeris, T. J., R. D. Reidelberger, and V. E. Mendel. Effects of nutrient density and composition of liquid meals on gastric emptying in feeding rats. *Am. J. Physiol.* 244:R865–R871, 1983.

Kaplan, J. M., A. C. Spector, and H. J. Grill. Dynamics of gastric emptying during and after stomach fill. *Am. J. Physiol.* 263:R813–R819, 1992.

Koopmans, H. S. Jejunal signals in hunger satiety. *Behav. Biol.* 14:309–314, 1975.

Koopmans, H. S. Regulation of food intake during long term loss of food from the intestines of the rat. *Proc. Soc. Exp. Biol. Med.* 157:430–434, 1978.

Kosaka, T., and R. K. S. Lim. Demonstration of the humoral agent in fat inhibition of gastric secretion. *Proc. Soc. Exper. Biol. Med.* 27:890–891, 1930.

Kraly, F. S., W. J. Carty, and G. P. Smith. Effect of pregastric food stimuli on meal size and intermeal interval in the rat. *Physiol. Behav.* 20:779–784, 1978.

Kraly, F. S., and G. P. Smith. Combined pregastric and gastric stimulation by food is sufficient for normal meal size. *Physiol. Behav.* 21:405–408, 1978.

Lepkovsky, S., P. Bortfeld, M. K. Dimik et al. Role of upper intestine in the regulation of food intake in parabiotic rats with their intestines "crossed" surgically. *Isr. J. Med. Sci.* 7:639–646, 1971.

Lew, M. F., J. Gibbs, and G. P. Smith. Intestinal satiety is elicited by L-but not by D-phenylalanine. *Soc. Neurosci. Abstr.* 9:900, 1983.

Lewis, L. D., and J. A. Williams. Regulation of cholecystokinin secretion by food, hormones, and neural pathways in the rat. *Am. J. Physiol.* 258:G512-G518, 1990.

Liddle, R. A., I. D. Goldfine, and J. A. Williams. Bioassay of plasma cholecystokinin in rats: Effects of food, trypsin inhibitor, and alcohol. *Gastroenterology* 87:542–549, 1984.

Liddle, R. A., I. D. Goldfine, M. S. Rosen, R. A. Taplitz, and J. A. Williams. Cholecystokinin bioactivity in human plasma: Molecular forms, reponses to feeding, and relationship to gallbladder contraction. *J. Clin. Invest.* 75:1144–1152, 1985.

Liebling, D. S., J. D. Eisner, J. Gibbs, and G. P. Smith. Intestinal satiety in rats. *J. Comp. Physiol. Psychol.* 89:955–965, 1975.

Lilja, P., I. Wiener, K. Inoue, G. M. Fried, G. H. Greely, Jr., and J. D. Thompson. Release of cholecystokinin in response to food and intraduodenal fat in pigs, dogs and man. *Surg. Gynecol. Obstet.* 159:557–561, 1984.

Lin, C. W., and T. R. Miller. Both CCK-A and CCK-B/gastrin receptors are present on rabbit vagus nerve. *Am. J. Physiol.* 263:R591–R595, 1992.

Lorenz, D. N., and S. A. Goldman. Vagal mediation of the cholecystokinin satiety effect in rats. *Physiol. Behav.* 29:599–604, 1982.

Lorenz, D. N., G. Kreielsheimer, and G. P. Smith. Effect of cholecystokinin, gastrin, secretin and GIP on sham feeding in the rat. *Physiol. Behav.* 23:1065–1072, 1979.

Love, A. H. G., and J. D. Wilson. Intestinal factors in the control of food intake in man. *Eur. Soc. Clin. Invest.* 174:139, 1973.

Luttinger, D., R. A. King, D. Sheppard, J. Strupp, C. B. Nemeroff, and A. J. J. Prange. The effect of neurotensin on food consumption in the rat. *Eur. J. Pharmacol.* 81:497–503, 1982.

Maggio, C. A., M. R. C. Greenwood, and J. R. Vasselli. The satiety effect of intragastric macronutrient infusion in fatty and lean Zucker rats. *Physiol. Behav.* 31:367–372, 1983.

Makhlouf, G. M. The neuroendocrine design of the gut. *Gastroenterology* 67:159–184, 1974.

Malagelada, J. R., and F. Azpiroz. Determinants of gastric emptying and transit in the small intestine. In Schultz, S. G., J. D. Wood, and B. B. Rauner (eds.): *Handbook of Physiology. 2nd Ed.* New York: American Physiological Society, 1989, pp. 909–937.

McCaffery, J., S. J. French, and D. Greenberg. Differential satiety effects after duodenal infusions of medium chain fatty acids in sham feeding rats. *Appetite* 23:415, 1994.

McHugh, P. R., and T. H. Moran. Accuracy of the regulation of caloric ingestion in the rhesus monkey. *Am. J. Physiol.* 235:R29–R34, 1978.

McHugh, P. R., and T. H. Moran. Calories and gastric emptying: A regulatory capacity with implications for feeding. *Am. J. Physiol.* 236:R254–R260, 1979.

Mei, J., and E. Erlanson-Albertsson. Effect of enterostatin given intravenously and intracerebroventricularly on high-fat feeding in rats. *Regul, Pept.* 41:209–218, 1992.

Mei, J., and C. Erlanson-Albertsson. Role of intraduodenally administered enterostatin in rats: Inhibition of food. *Obes. Res.* 4:161–165, 1996.

Mei, N. Vagal glucoreceptors in the small intestine of the cat. *J. Physiol.* 282:485–506, 1978.

Melone, J. Vagal receptors sensitive to lipids in the small intestine of the cat. *J. Auton. Nerv. Syst.* 17:331–341, 1986.

Meyer, J. H., and M. I. Grossman. Comparison of D- and L-phenylalanine as pancreatic stimulants. *Am. J. Physiol.* 222:1058–1063, 1972.

Moran, T. H., and P. R. McHugh. Cholecystokinin suppresses food intake by inhibiting gastric emptying. *Am. J. Physiol.* 242:R491–R497, 1982.

Moran, T. H., and P. R. McHugh. Gastric and nongastric mechanisms for satiety action of cholecystokinin. *Am. J. Physiol.* 254:R628–R632, 1988.

Moran, T. H., G. P. Smith, and A. M. Hostetler. Transport of cholecystokinin (CCK) binding sites in subdiaphragmatic vagal branches. *Brain Res.* 415:149–152, 1987.

Novin, D., J. D. Sanderson, and M. Gonzalez. Feeding after nutrient infusion: Effect of hypothalamic lesion and vagotomy. *Physiol. Behav.* 22:107–113, 1979.

Novin, D., J. D. Sanderson, and D. A. VanderWeele. The effect of isotonic glucose on eating as a function of feeding condition and infusion site. *Physiol. Behav.* 13:3–7, 1974.

Ohnedia, A., T. Kobayaski, and J. Nikei. Response of gastric inhibitory polypeptide to fat ingestion in normal dogs. *Regul. Pept.* 8:123–130, 1984.

Okada, S., D. A. York, G. A. Bray, and C. Erlanson-Albertsson. Enterostatin, (VAL-PRO-ASP-PRO-ARG), the activation peptide of procolipase selectively reduces fat intake. *Physiol. Behav.* 49:1185–1189, 1991.

Pavlov, I. P. *The Work of the Digestive Glands. 2nd Ed.* London: Charles Griffin & Company, Ltd., 1910.

Quigley, J. P., and I. Meshcan. Inhibition of the pyloric sphincter region by the digestion products of fat. *Am. J. Physiol.* 134:803–807, 1941.

Quigley, J. P., H. J. Zettleman, and A. C. Ivy. Analysis of the factors involved in gastric motor inhibition by fats. *Am. J. Physiol.* 108:643–651, 1934.

Reidelberger, R. D., T. J. Kalogeris, P. M. B. Leung, and V. E. Mendel. Postgastric satiety in the sham feeding rat. *Am. J. Physiol.* 244:R872–R881, 1983.

Ritter, R. C., and E. E. Ladenheim. Capsaicin pretreatment attenuates suppression of food intake by cholecystokinin. *Am. J. Physiol.* 248:R501–R504, 1985.

Robinson, P. H., T. H. Moran, and P. R. McHugh. Cholecystokinin inhibits independent ingestion in neonatal rats. *Am. J. Physiol.* 255:R14–R20, 1988.

Rodriguez, M. D., T. J. Kalogeris, X. L. Wang, R. Wolf, and P. Tso. Rapid synthesis and secretion of intestinal apolipoprotein A-IV after gastric fat loading in rats. *Am. J. Physiol.* 272:R1170–R1177, 1997.

Rosell, S., and A. Rokaeus. The effect of ingestion of amino acids, glucose and fat on circulating neurotensin-like immunoreactivity (NTLI) in man. *Acta Physiol. Scand.* 107:263–267, 1979.

Russek, M. Hepatic reception and the neurophysiological mechanisms controlling feeding behavior. In Ehrenpreis, S., and O. C. Solnitzky (eds.): *Neuroscience Research. Vol. 4.* New York: Academic Press, 1971, pp. 213–282.

Schwartz, G. J., P. R. McHugh, and T. H. Moran. Characterization of gastric vagal afferent mechanoreceptor responses to close arterial infusions of CCK in rats. *Soc. Neurosci. Abstr.* 18:1068, 1992.

Sclafani, A., and J. W. Nissenbaum. On the roles of the mouth and gut in the control of saccharin and sugar intake: A reexamination of the sham-feeding preparation. *Brain Res. Bull.* 14:569–576, 1985.

Senior, J. R. Intestinal absorption of fats. *J. Lipid Res.* 5:495–521, 1964.

Smith, G. P., C. Jerome, B. J. Cushin, R. Eterno, and K. J. Simansky. Abdominal vagotomy blocks the satiety effect of cholecystokinin in the rat. *Science* 213:1036–1037, 1981.

Smith, G. P., C. Jerome, and R. Norgren. Afferent axons in abdominal vagus mediate satiety effect of cholecystokinin in rats. *Am. J. Physiol.* 249:R638–R641, 1985.

Snowdon, C. T. Production of satiety with small introduodenal infusions in the rat. *J. Comp. Physiol. Psychol.* 88:231–238, 1975.

South, E. H., and R. C. Ritter. Capsaicin application to central or peripheral vagal fibers attenuates CCK satiety. *Peptides* 9:601–612, 1988.

Thomas, F. B., E. L. Mazzaferri, S. E. Crockett, H. S. Mekhjian, H. D. Grueman, and S. Cataland. Stimulation of secretion of gastric inhibitory polypeptide and insulin by intraduodenal amino acid perfusion. *Gastroenterology* 70:525–527, 1976.

Tian, Q., M. Nagase, D. A. York, and G. A. Bray. Vagal-central nervous system interactions modulate the feeding response to peripheral enterostatin. *Obes. Res.* 2:527–534, 1994.

Tordoff, M. G., and M. I. Friedman. Hepatic portal glucose infusions decrease food intake and increase food preference. *Am. J. Physiol.* 251:R192–R196, 1986.

Tordoff, M. G., and M. I. Friedman. Hepatic control of feeding: Effect of glucose, fructose, and mannitol infusion. *Am. J. Physiol.* 254:R969–R976, 1988.

VanderWeele, D. A., D. Novin, M. Rezek, and J. D. Sanderson. Duodenal or hepatic-portal glucose perfusion: Evidence for duodenally-based satiety. *Physiol. Behav.* 12:467–473, 1974.

Waddell, W. R., and C. C. Wang. Effect of vagotomy on gastric evacuation of high fat meals. *J. Appl. Physiol.* 5:705–711, 1953.

Walls, E. K., R. J. Phillips, F. B. Wang, M.-C. Holst, and T. L. Powley. Suppression of meal size by intestinal nutrients is eliminated by celiac vagal deafferentation. *Am. J. Physiol.* 269:R1410–R1419, 1995.

Wang, C. C., and M. I. Grossman. Physiological determination of release of secretin and pancreozymin from intestine of dogs with transplanted pancreas. *Am. J. Physiol.* 164:527–545, 1951.

Welch, I. M., K. Saunders, and N. W. Read. Effect of ileal and intravenous infusions of fat emulsions on feeding and satiety in human volunteers. *Gastroenterology* 89:1293–1297, 1985.

Welch, I. M., C. P. Sepple, and N. W. Read. Comparisons of the effects on satiety and eating behaviour of infusion of lipid into the different regions of the small intestine. *Gut* 29:306–311, 1988.

West, D. B., D. Fey, and S. C. Woods. Cholecystokinin persistently suppresses meal size but not food intake in free-feeding rats. *Am. J. Physiol.* 246:R776–R787, 1984.

Wiepkema, P. R., A. J. Alingh Prins, and A. B. Steffens. Gastrointestinal food transport in relation to meal occurrence in rats. *Physiol. Behav.* 9:759–763, 1972.

Wirth, J. B., and P. R. McHugh. Gastric distension and short-term satiety in the rhesus monkey. *Am. J. Physiol.* 245:R174–R180, 1983.

Woods, S. C., D. B. West, L. J. Stein et al. Peptides and the control of meal size. *Diabetologia* 20:305–313, 1981.

Yin, T., and C. Tsai. Effects of glucose on feeding in relation to routes of entry in rats. *J. Comp. Physiol. Psychol.* 85:258–264, 1973.

Young, R. C., J. Gibbs, J. Antin, J. Holt, and G. P. Smith. Absence of satiety during sham feeding in the rat. *J. Comp. Physiol. Psychol.* 87:795–800, 1974.

Yox, D. P., L. Brenner, and R. C. Ritter. CCK-receptor antagonists attenuate suppression of sham feeding by intestinal nutrients. *Am. J. Physiol.* 262:R554–R561, 1992.

Yox, D. P., Stokesberry, H., and Ritter, R. C. Suppression of sham feeding by intraintestinal oleate: Blockade by a CCK antagonist and reversal of blockade by exogenous CCK-8. In Hughes, J., G. Dockray, and G. Woodruff (eds.): *The Neuropeptide Cholecystokinin (CCK), Anatomy and Biochemistry, Receptors, Pharmacology and Physiology.* Chichester: Ellis Horwood, Ltd., 1989, pp. 218–222.

Yox, D. P., H. Stokesberry, and R. C. Ritter. Vagotomy attenuates suppression of sham feeding induced by intestinal nutrients. *Am. J. Physiol.* 260:R503–R508, 1991a.

Yox, D. P., H. Stokesberry, and R. C. Ritter. Fourth ventricular capsaicin attenuates suppression of sham feeding induced by intestinal nutrients. *Am. J. Physiol.* 260:R681–R687, 1991b.

Zarbin, M. A., J. K. Wamsley, R. B. Innis, and M. J. Kuhar. Cholecystokinin receptors: Presence and axonal flow in the rat vagus nerve. *Life Sci.* 29:697–705, 1981.

4

Measuring Satiety: From Meal Size to the Microstructure of Ingestive Behavior

JOHN D. DAVIS

Meal Size and Satiety

In thinking about the meaning of *satiety* and where the concept came from, it becomes readily apparent that the word has many meanings and that its origins predate any formal attempts to study it experimentally. Definitions of satiety range from those based on subjective feelings correlated with the tendency to eat or not, through those based on an assumed internal physiological or psychological state, to those based on some consequence of ingestive behavior, such as the amount of food ingested. It is a concept that originated in philosophical speculation about internal states of animals, not one that emerged from the results of experiments investigating the mechanisms for the control of ingestive behavior. For this reason, those interested in the problem of satiety have tended to adopt some definition or conceptualization of a state of satiety and then have chosen an appropriate objective measure to assess the magnitude of the assumed state under various experimental conditions.

Various methods have been developed in this way to study satiety in the laboratory rat (e.g., Booth, 1985; LeMagnen and Devos, 1980), but the most common one has been to measure the amount or volume of a test diet that is ingested by the subject during a fixed period of time. This is easy to do with almost any type of diet, and the interpretation of the measure seemed to be straightforward. Large meals seemed to imply that the diet was not very "satiating" and small meals that the diet was "satiating." As more was learned about meal size, however, it became apparent that at least two other variables, the flavor or type of the diet and the level of deprivation, could also affect it. Recognizing this, Smith and Gibbs (1979) argued that meal size could nevertheless be used as a measure of satiety if these variables were held constant.

On the face of it, measuring meal size under constant test stimulus and deprivation conditions to assess a state of satiety has much to recommend it. It is an unambiguous metric, it is a continuous variable so that variations in the degree of satiation can be assessed, and it is easy to use. For these reasons, it is probably the most commonly used measure of satiety in research conducted on human and nonhuman animals today.

This measure, however, is not without its problems, some of which are illustrated in Figure 4.1. It shows the results of four experiments my colleagues and I have conducted. Each of these experiments has something to do with the problem of satiety, and in each one the volume of a liquid diet ingested in a 30-minute period was measured. These experiments are discussed in detail later; here I will point out only what I see as a problem in interpreting the data shown in Figure 4.1.

In three of the four panels it is clear that there is a large difference in the volume ingested between the two groups. On the basis of the meal size definition of satiation, we would conclude, in each case, that members of one of the two groups were more satiated than the other. A common term, *satiation,* can be used to explain the smaller intake of one of the two groups in each of the three experiments. Does this mean that the state of satiety in the animals in the groups with the reduced intake is in any sense the same? Considering the fact that each of these experiments used a different procedure and was designed to answer a different question, I think few would be willing to agree with that conclusion. However, there is no information about the meal size variable with which to build or support a counterargument. We can say, since the experimental treatments were different, affecting different elements of the ingestive control system of the animals, that the "satieties" may have been different in each case. But that is all we can say. When the only information we have from the experiment is how much was ingested, there is no way to argue convincingly that the states of satiety, defined as reduced intake, in any two different experiments are qualitatively the same or different.

Another problem illustrated in Figure 4.1 is that meal size does not always reveal differences where a difference might be expected. Panel d, for example, shows the volume ingested of a low concentration (0.05 M, left bar) and a high concentration (0.8 M, right bar) of sucrose solution. Some might argue that a high concentration sucrose solution should satiate an animal more quickly than a low concentration diet and thus lead to a smaller meal. Yet it is clear by inspection that the volume ingested of the two different concentrations is the same. The test solutions were not the

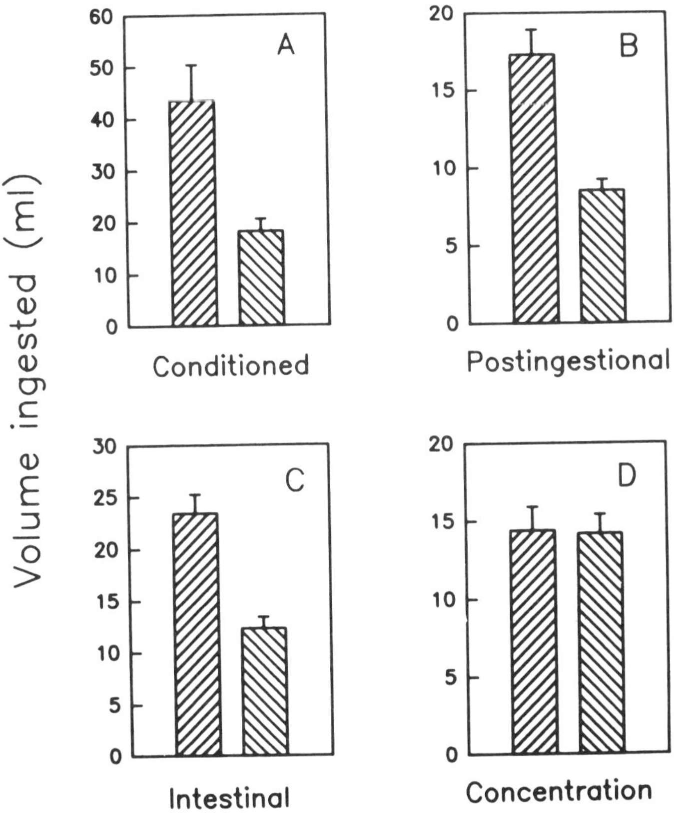

FIGURE 4.1. Volume ingested by two groups of animals in each of four experiments performed by the author and his collaborators. **(a)** Intake of milk on sham-feeding test for the first time (right bar) and on the fifth sham-feeding test following it (left bar). From Davis and Campbell (1975), with permission of the publisher. **(b)** The intake of 0.8 M sucrose on a real-feeding test (right bar) and on the first sham feeding test following it (left bar). From Davis and Smith (1990), with permission of the publisher. **(c)** The intake of a saccharin + glucose mixture (left bar) and the intake of the same solution with 0.3 M mannitol added to it (right bar). From Davis et al. (1975), with permission of the publisher. **(d)** shows the 30-minute intake of 0.05 M sucrose (left bar) and of 0.8 M sucrose (right bar) under real-feeding conditions.

same, so, in accordance with Smith and Gibbs (1979), we should not use these data to assess relative states of satiety. If we did, however, we would have to conclude that the two solutions are equivalent in their satiating potency. If we did not, then what could we do to measure the assumed differential satiating potency of these two solutions?

There are a number of solutions to these problems, one of which is relatively simple. Instead of just measuring the consequence of ingestive behavior, the amount ingested, we can record the ingestive behavior itself. The amount ingested in any fixed period of time depends on how much of that time was devoted to eating as opposed to the other activities that can occur, such as resting, grooming, rearing, exploring, and sleeping. Amount ingested is simply the sum of the number of ingestive acts times the amount consumed during each act. If we have a measure of the number, size, and distribution of the ingestive acts within the meal, we will have a fairly complete behavioral account of the meal, which should provide information about how the consumption of a particular amount was achieved. Knowing how the experimental treatment affected the specific components of ingestive behavior of the animal will not give us an answer to all of the interpretative questions we might ask, but it does gives us a richer data base with which to support an interpretation and can provide a deeper insight into the mechanisms involved. The emerging field of the microstructural analysis of ingestive behavior is beginning to provide us with these kinds of data.

Microstructural Analysis of Ingestive Behavior

During the past 20 years or so there have been a number of methodological and technological advances that now make it possible to go beyond meal size and examine the components of the ingestive behaviors of animals at a more detailed level. Videorecorders provide a means for recording and storing a record of the behavior of animals while they are eating or drinking. On-line computers provide a means for recording and analyzing the licking behavior of rats at millisecond resolution. These technological advances have led to the development of methods collectively called the *microstructural analysis of ingestive behavior,* since in effect they all apply a microscope focused on the behavior of the animal while it is engaged in ingesting food or fluid.

Several different methods of this kind have been developed. Weipkema (1971) introduced a method of observing and classifying the various gross motor acts that can occur during the time a rat is ingesting dry food. Grill and Norgren (1978) have videotaped and classified the orofacial responses of rats while a test solution is infused into the mouth. Following early work by Stellar and Hill (1952), Davis and Keehn (1959), Allison and Castellan (1970), and Knowler and Ukena (1973), among others, Davis and Smith (1992) developed a method for analyzing the licking behavior that rats exhibit when ingesting liquid diets. They reported that this activity is organized into bursts of licking separated by pauses of varying duration. The bursts of licking can be treated as a unit of the ingestive behavior that rats engage in when ingesting liquid diets. The size and distribution of the bursts vary throughout a meal and with differing experimental conditions and can reveal much about the ingestive behavior of the rat that is concealed when only the volume ingested in a test session is recorded (Davis and Smith, 1992; Davis and Perez, 1993; Asin et al., 1992).

Microstructural Analysis of Licking Behavior

The microstructural analysis used here has been described in detail elsewhere (Davis and Smith, 1992). The data recorded are the onset times of each lick made by the rat during the test session. These data are obtained from an electronic lickometer and are stored in a computer during the test session for later analysis. The size of the bursts of licking, which are runs of licking at a high constant rate, is analyzed by counting the number of licks that occur before a pause greater than 250 msec occurs. Because of the small variability in the distribution of the interlick intervals (ILIs) in this range (0–250 msec), they appear to reflect the operation of a motor pattern generator possibly located in the hypoglossal nucleus (Wiesenfeld et al., 1977).

Bursts of licking are separated by two types of pauses. One of them is a very brief pause, which generates an ILI in the range of 250–500 msec. These ILIs have a symmetrical distribution, with an average of about 330 msec. Bursts separated by these pauses, which we call *interburst intervals* (IBIs), are combined by adding them together to form clusters (of bursts) to describe the length of sustained bouts of licking with minimal interruption. The longer pauses have a very skewed distribution, ranging from 500 msec to 100 sec or more, with about half of them in the range of 0.5 to about 3 sec or so. These pauses are called *intercluster intervals* (ICIs). Figure 4.2 summarizes schematically these various components of the licking behavior of the rat.

The bursts and clusters define units of the ingestive act of licking. Their sizes tell us something about the lengths of the periods of persistent licking. Their distributions over time describe the rate at which they occur. This distribution is determined by summing the sizes of the bursts over successive time increments (usually 1 or 2 minutes) of the test period. These sums divided by the length of the increment gives the average rate of licking per unit time during that interval. This gives a measure of the probability of engaging in ingestive behavior at different time intervals during the test

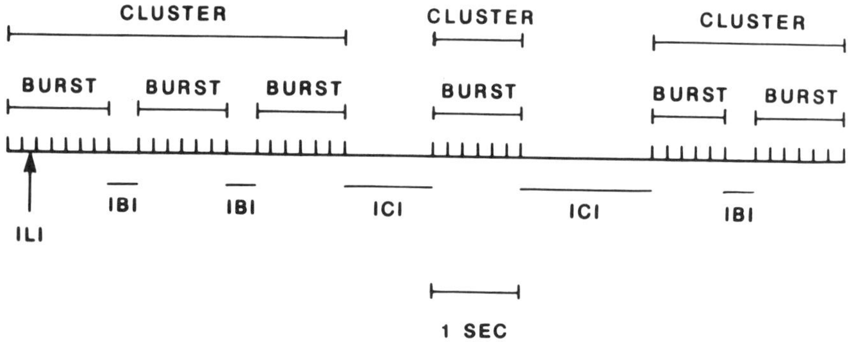

FIGURE 4.2. Schematic representation of a run of licks made by a rat ingesting a liquid diet to illustrate the terminology used in the text. The vertical lines represent individual licks and the spaces between them, interlick intervals (ILIs). ILIs are intervals separating licks by less than 0.26 second. Interburst intervals (IBIs) are ILIs in the range of 0.26 to 0.5 second. Intercluster intervals (ICIs) are ILIs greater than 0.5 second. From Davis and Smith (1992), with permission of the American Psychological Association.

session. This variable, the rate of licking at successive time intervals during a test session, is particularly valuable in the analysis of satiation because it describes how the probability of licking changes as the consequences of the act of ingesting fluid impact the motor system that controls the licking behavior. It provides a picture of the development of the negative-feedback signal(s) that participate in the control of ingestive behavior (Davis and Levine, 1977).

Applications of the Microstructural Analysis

Orophyrangeal and Postingestional Control

The clearest evidence for the existence of postingestional control of ingestive behavior comes from sham-feeding studies. It has been known for almost 100 years that if food is not allowed to accumulate in the gastrointestinal tract during a meal, intake increases dramatically (Pavlov, 1910). Figure 4.1b shows results from a recent sham-feeding study (Davis and Smith, 1990). In that study rats, with a gastric fistula drank an 0.8 M sucrose solution for 30 minutes. The bar on the right shows the volume drunk when the gastric fistula was closed and all of the ingested fluid remained in the gastrointestinal tract. The bar on the left shows the intake

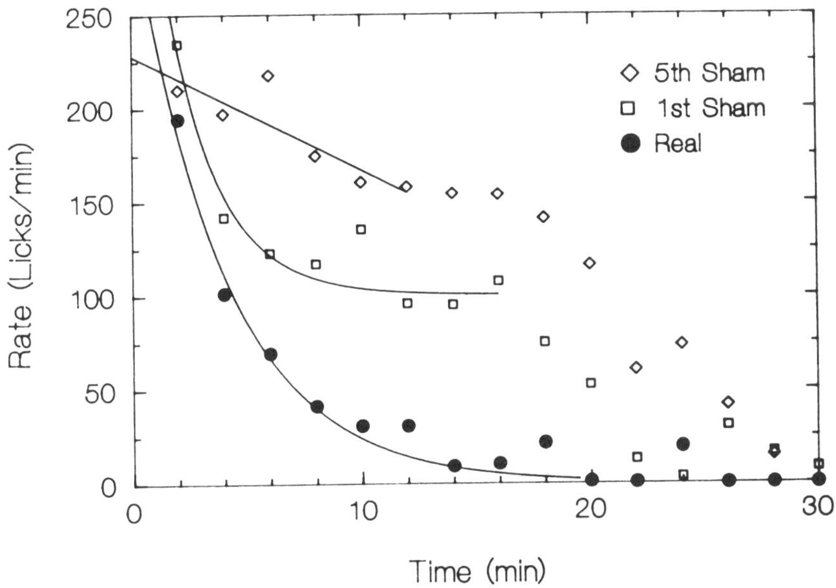

FIGURE 4.3. The average rate of licking calculated at 2-minute intervals for a real-feeding test (Real), the first sham-feeding test following it (1st Sham), and the fifth of five consecutive sham-feeding tests (5th Sham). The lines are least-squares curve fits to the data over the early portions of the meal to characterize the changes in the trends during this part of the meal. From Davis and Smith (1990), with permission of the American Physiological Society.

on the following day when the fistula was open and all of the ingested fluid drained out of the stomach. In the absence of gastrointestinal filling, intake increased significantly. The question that is not answered by the results shown in Figure 4.1b is what change in the behavior of the rat accounted for the increase. An examination of the rate of ingestion during the tests provides the answer.

Figure 4.3 shows the average rate of licking calculated at 2-minute intervals over the 30-minute test session for a real-feeding test (Real), a sham-feeding test on the following day (1st Sham), and the sham-feeding test after five consecutive sham-feeding tests (5th Sham). Looking at just the bottom two curves, two effects of sham feeding stand out. One is that sham feeding increased the duration of the meal. Thus, one of the effects of the accumulation of the ingested fluid in the gastrointestinal tract is to terminate the meal. A second effect is that it is only after the first 5 minutes of the meal that the rate of ingestion is appreciably greater with sham than with real feeding. The higher rate of licking in the sham-feeding relative to the real-feeding test from 5 minutes on is responsible for the increased intake that occurred when the gastric fistula was opened for the first time.

These results show that at least two different types of variables participate in limiting the size of a real meal. One exists at the very beginning of the meal and causes a progressive reduction in the rate of ingestion with time. The other is one that appears later in the meal. This second control variable has the effect of continuing the decay in the rate of ingestion until licking behavior ceases and the meal is terminated. Since both of these controls contribute to limiting the size of the meal, both of them can be considered to be satiety signals. The second control signal that coincides with the termination of a real meal is correlated with the accumulation of the ingested fluid in the stomach and intestine, and it is therefore reasonable to assume that it is generated by some direct physiological consequence of gastrointestinal stimulation or some postabsorptive effect of the meal.

The signal operating at the beginning of the meal begins to exert its influence on the rate of ingestion too early to be a direct unconditioned physiological consequence of the accumulation of the ingested fluid in the gastrointestinal tract. Furthermore, its effectiveness in reducing the rate of ingestion early in the meal declines with sham-feeding experience (Davis and Campbell, 1973); the result is an increase in intake. This can be seen in Figure 4.1a, which shows the intake during the first sham-feeding test (right bar) and during the fifth consecutive sham-feeding test (left bar). The extinction of control over the rate of licking early in the test can be seen in Figure 4.3 in the top curve labeled 5th Sham. This rate curve is the fifth of five consecutive sham-feeding tests (Davis et. al., 1993). The initial rapid decrease in the rate of licking in the early part of the test seen in the Real and 1st Sham curves is no longer apparent. This suggests that it is a conditioned signal based on an association between the orophyrangeal stimulation and some postingestional consequence of the ingested fluid. Weingarten and Kulikovsky (1989) have recently reported evidence supporting this interpretation.

If the increase in the rate of licking with consecutive sham-feeding experience is a consequence of the extinction of an acquired control over the rate of licking based on an association between orophyrangeal and postingestive stimulation, the increase should not occur if real-feeding tests are interspersed between sham-feeding tests. The real-feeding tests following the sham tests should recondition the inhibitory control that was extinguished in the sham tests, and the increase in the rate of licking early in the meal should not occur. Davis and Smith (1990) investigated this possibility in an experiment in which a group of rats was tested on four cycles of two real-feeding tests followed by a sham test. The result was that the progressive increase in intake seen when rats are given consecutive sham-feeding

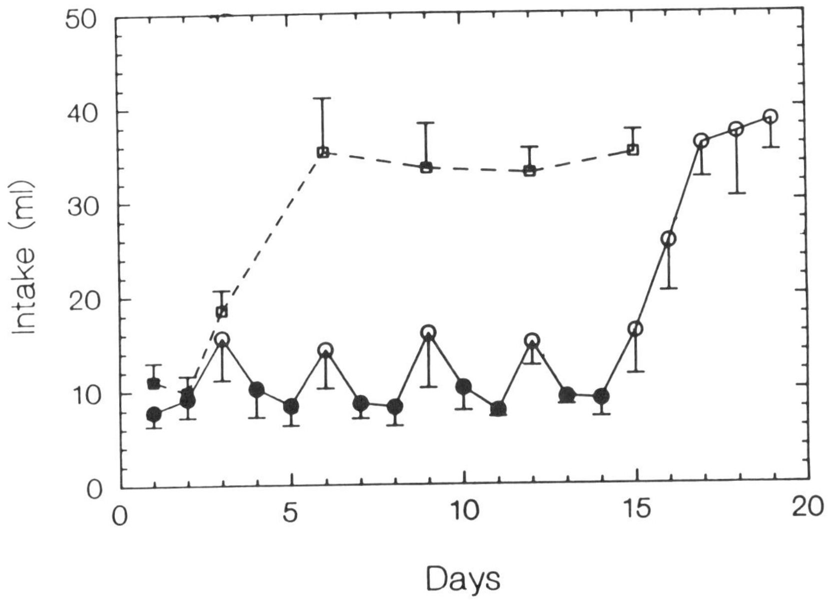

FIGURE 4.4. The volume ingested in a series of real- and sham-feeding tests. The closed circles represent real-feeding tests in which the gastric fistula was closed; the open circles, sham-feeding tests in which the gastric fistula was open. Open squares represent sham-feeding tests with no intervening real-feeding tests. From Davis and Smith (1990), with per-mission of the American Physiological Society.

tests did not occur. The intake in the sham tests was about twice that in the preceding real tests, but there was no progressive increase in the volume ingested in either the real or sham test over the series of 12 tests. When these animals were later tested in a series of five consecutive sham-feeding tests, intake increased progressively from an average of 16 ml to an average of 38 ml (Fig. 4.4).

The explanation for these effects on sham intake can be found in an examination of the rate of licking functions (Fig. 4.5). During the interrupted sham-feeding tests, the increase in the rate of licking during the first 4 minutes of the test that occurs in consecutive sham-feeding tests did not occur. The curves in Figure 4.5 labeled "First two" and "Last two" are the rate curves averaged over the first two and last two sham-feeding tests of the interspersed sham-feeding series. As can be seen, the increases in intake in the sham-feeding tests were due entirely to a continuation of licking from 6 minutes on. The two real-feeding tests intervening between the sham tests maintained the decrease in the initial rate of licking early in the meal.

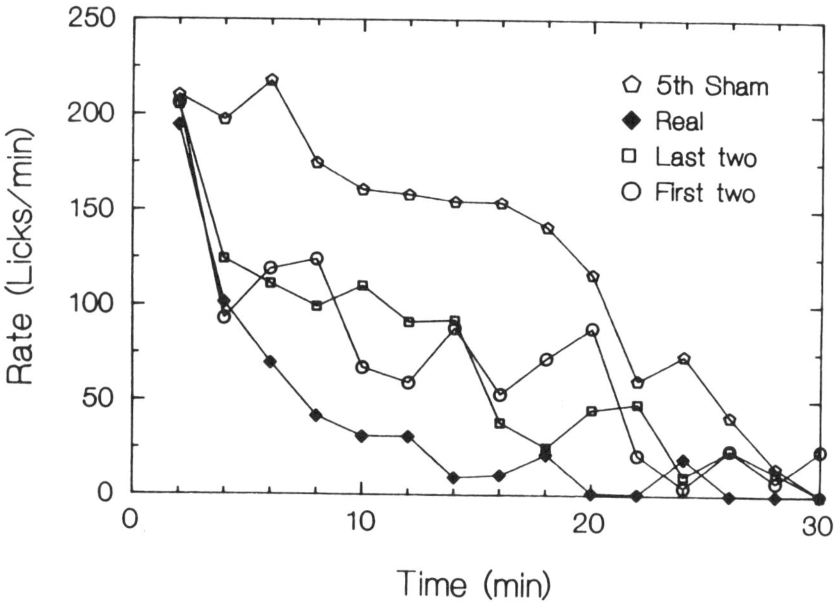

FIGURE 4.5. The rate of licking calculated at successive 2-minute intervals. Closed diamonds represent the real-feeding test; open symbols, sham-feeding tests. The open circles, "First two," are the average rates of the first two interrupted sham-feeding tests. The open squares, "Last two," are the average rates of the last two interrupted sham-feeding tests. The open diamonds, "5th Sham," is the last consecutive sham-feeding test. From Davis and Smith (1990) with permission of the American Physiological Society.

In the consecutive sham-feeding tests the rate of licking during the early part of the meal increased with each sham-feeding test, resulting in the progressive increase in intake with experience seen in Figure 4.4. By the fifth consecutive sham test there was no sign of the steep decline in the rate of ingestion during the first four minutes (Fig. 4.5, 5th Sham). Since interposing real-feeding tests between the sham tests prevented the increase in the rate of licking early in the meal, it must be assumed that these tests served to recondition the control over the rate of licking partially extinguished on the preceding sham test.

This type of control operating early in the meal could be called "early conditioned satiety" to distinguish it from the conditioned satiety proposed by Booth (1972), who has presented evidence for conditioned satiety that appears toward the end of a meal. A better term, however, might be "acquired control of the rate of ingestion," since this is what it is. This type of control of ingestion causes a reduction in meal size by decreasing the rate of licking at the beginning of the meal, not by decreasing the rate of licking at its end.

Postpyloric Control

The results of the study described above showed that control of the rate of ingestion of concentrated sucrose solutions early in the meal depended on an association between orophyrangeal stimulation by the test diet and its postingestional consequences. The postingestional stimulation serving as an unconditioned stimulus (UCS) for the conditioning could have originated in the gastric or in the intestinal compartment, or it could have had its origin in some postabsorptive event. There was no way to determine which of these possibilities is correct from the data obtained in that study.

Therefore, in an attempt to identify the site of origin of the UCS, we took advantage of the fact that postgastric and postabsorptive stimulation can be prevented during a meal by the use of a pyloric noose or cuff (Hall, 1973; Deutsch et al., 1978). Davis et al. (1993) repeated the experiment described above, but this time all of the animals were fitted with a pyloric cuff in

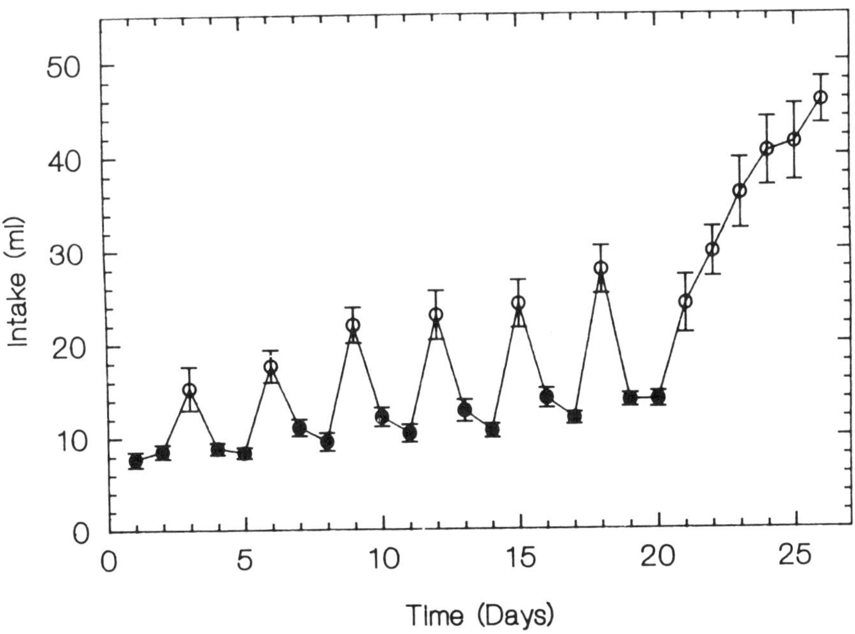

FIGURE 4.6. The volume ingested on successive real- and sham-feeding tests. Closed circles represent the intake in real-feeding tests; open circles, the intake in sham-feeding tests. From Davis et al. (1993), with permission of the American Physiological Society.

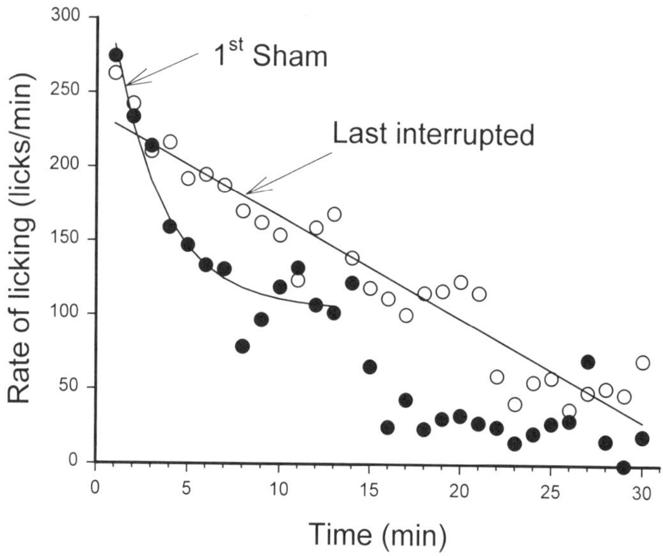

FIGURE 4.7. The rate of licking calculated at successive 1-minute intervals. The closed circles represent the data from the first sham-feeding test following a real-feeding test. The open circles represent the data from the seventh interrupted sham-feeding test. Redrawn from Davis et al. (1993), with permission of the American Physiological Society.

addition to the gastric fistula. In all the tests the cuff was inflated, confining all of the ingested fluid to the gastric compartment on the real-feeding tests. This device also served to eliminate any possibility of postgastric stimulation in the sham-feeding tests as well (Sclafani and Nissenbaum, 1985). Thus, a complete dissociation was obtained between orophyrangeal stimulation and postpyloric stimulation. An association was maintained, however, between orophyrangeal stimulation and gastric stimulation in the real-feeding tests, since all of the ingested fluid was retained there. This fluid was removed from the stomach at the end of these tests before the cuff was released.

The effect of this procedure on intake was that the volume ingested increased in both the real- and the sham-feeding tests as the animals gained experience ingesting the 0.8 M sucrose solution under these conditions (Fig. 4.6). Since this did not happen in the previous experiment, where postgastric stimulation occurred on all of the real-feeding tests, with perhaps a minimal amount occurring on the sham-feeding tests (Sclafani and Nissenbaum, 1985), the progressive increase in the volume ingested in the real- and sham-feeding tests must have been due to the experience of ingesting the fluid in the absence of postgastric stimulation.

The reason for the increase in intake in the interrupted sham-feeding

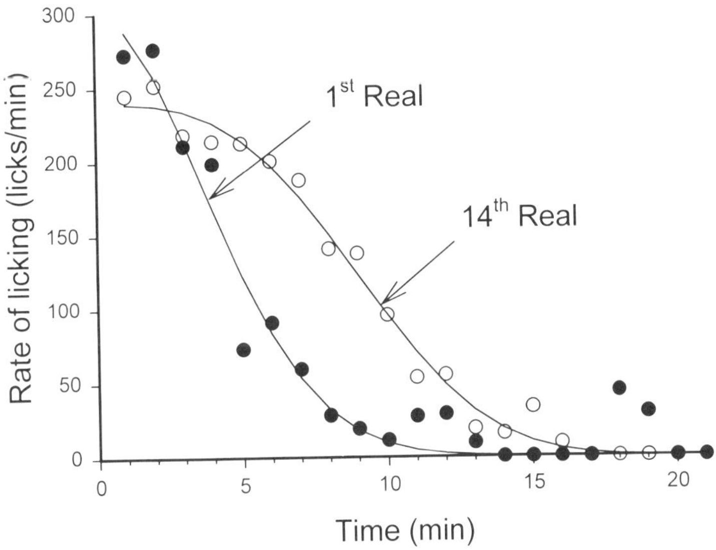

FIGURE 4.8. The rate of licking calculated at 1-minute intervals. The closed circles represent the data from the first real-feeding test in a series of interspersed real- and sham-feeding tests. The open circles represent the data from the last of a series of real-feeding tests interspersed between sham-feeding tests. The lines are the least-squares fit to the Weibull function. Redrawn from Davis et al. (1993), with permission of the American Physiological Society.

tests can be seen in Figure 4.7, which shows the rate of licking curves for the first sham test and the last interrupted sham tests. The rate of licking during the early part of the meal was greater in the seventh than in the first sham test. The same effect occurred in the real-feeding tests. Figure 4.8 shows the rate of licking curves in the first real-feeding test and on the 14th real-feeding test. It is clear that the increase in intake that occurred between the first and 14th real-feeding tests interspersed between the sham tests is a result of a decrease in the rate of decline in the rate of licking during the early part of the meal. Since this did not occur in our previous study in which postpyloric stimulation was permitted, it seems clear that postpyloric stimulation, either intestinal or postabsorptive, is necessary to maintain the initial decline in the rate of licking that occurs when rats ingest a concentrated sucrose solution.

Postorophyrangeal—Preabsorptive Control of Ingestive Behavior

The results of the study employing the pyloric cuff leave open the question of whether the UCS on which the conditioning of the initial rate of licking

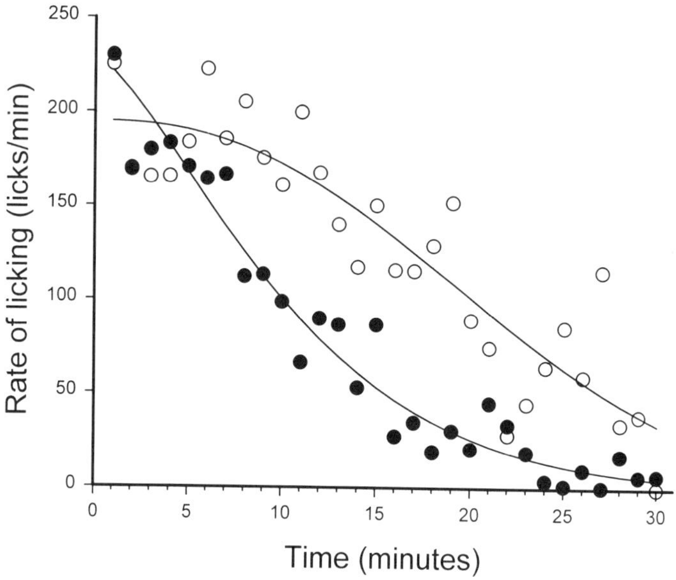

FIGURE 4.9. The rate of licking calculated at 1-minute intervals. The open circles represent data from a test in which the rats drank a saccharin + glucose solution. The closed circles represent the data from a test the first time that mannitol (0.3 M) was added to the test solution. The solid lines are from a least-squares fit to the Weibull function.

is based has a pre- or postabsorptive site of origin. There is evidence that preabsorptive intestinal stimulation can directly control meal size from a study by Davis et al. (1975), who reported that gastrointestinal filling alone can have a direct effect on the control of ingestive behavior. In their study rats ingested a saccharin–glucose mixture (6 mM saccharin + 0.1 M glucose) to which various concentrations of mannitol were added. When mannitol, a nonabsorbable, relatively tasteless sugar alcohol, is added to the mixture, fluid absorption in the intestine is retarded because of the osmotic gradient created by mannitol. The result is an accumulation of the ingested fluid in the intestine (Collins and Davis, 1978) and a reduction in intake in proportion to the concentration of mannitol in the solution (Davis et al., 1975). The left bar of Figure 4.1c shows the intake of the saccharin + glucose solution, and the right bar shows the intake when mannitol was added.

Figure 4.9, from a more recent study of the chronic effects of mannitol on ingestive behavior (Davis and Perez, 1993), shows the impact of mannitol on the ingestion of a saccharin–glucose mixture the first time it is added to the test diet. The curve represented by the open circles shows that the decline in the rate of licking the solution is relatively gradual when the solution contains no mannitol. The curve represented by the filled circles describes the rate of ingestion for the same group of rats ingesting the same solution but with 0.3 M mannitol added to it for the first time. Note that for the first 7 minutes or so there is not much difference between the two curves. However, from about 7 minutes on the curve describing the rate of licking for the animals ingesting the solution containing mannitol declines much more rapidly than does the one for the animals ingesting the mannitol-free solution. The impact of mannitol on the ingestive behavior is a delayed one, presumably because some time is required for the mannitol to accumulate in the gastrointestinal tract and trigger a feedback signal.

The rate of ingestion at any time during a meal depends on the size of the bursts or clusters of licking and the intervals between them. The decrease in the rate of licking caused by the addition of mannitol to the solution could have been caused by a reduction in the size of the bursts (SB) and clusters (SC) or by an increase in the intervals between the clusters, the ICI. To determine the relative contributions of these two variables to the decline in the rate of licking of both solutions, the clusters and ICIs for each animal in this test were divided into two groups at the midpoint of the meal. A cumulative probability distribution was then calculated for the ICIs and the SCs for the first and second half separately.

The cumulative probability distribution gives the probability that the variable in question (SC and ICI in this case) is greater than any particular value on the × axis. If this function has a simple exponential form, $y = e^{-at}$, then it is reasonable to infer that a random Poisson process is generating the variable (Wickens, 1982). The average of such a distribution is a^{-1}, and its variance is a^{-2}. If the function can be fit by the sum of two exponential functions, this suggests that two Poisson processes with different parameters may have contributed to the generation of the variable in question. In this case, the parameters a0 and a1 of the function

$$y = a_0 e^{-bt} + a_1 e^{-ct}, \ (a_0 + a_1 = 1)$$

give the probabilities of the two generating processes, i.e., the relative contribution of the two processes to the generation of the entire distribution, and the reciprocals of the parameters b and c give the average of the variables generated by the two processes. Davis, (1996) has described this type of analysis in more detail elsewhere.

Figure 4.10a shows the cumulative probability functions for the ICIs and Figure 4.10b the cumulative probability distribution for cluster sizes for the first and second halves of the clusters in the session. The numerical values of the functions were given by a least-squares fit to the data represented by the open symbols in Figure 4.10 and were used to generate the fit represented by the solid line. A three-parameter fitting function of the form

$$y = ae^{-bt} + (1 - a)e^{-ct}$$

was used to fit the ICI data, and the two-parameter Weibull function of the form

$$y = e^{-(an)}$$

was used to fit the SC distribution. The Weibull function was chosen as the fitting function for SC because it is a generalization of the simple exponential function and provides a better fit to data that are close to, but deviate somewhat from, the simple exponential form. Note that when the exponent b equals 1 the Weibull and the exponential functions are the same.

Figure 4.10a shows the cumulative probability distributions for the ICI distribution for the first and second halves of the test. It can be seen in both cases that the data are reasonably well fit by the sum of two exponentials. The logical implication of this fact is that these distributions describe two behavioral states of the animal that generate separate distributions of ICIs. One, described by the first of the two exponentials, describes the ICIs of the rat when it is in a state where its attention is focused on ingestion but it is not actually licking the tube. Observing a rat in a test session, one often sees the animal oriented in a position focused on the tube in a position to lick, but not actually doing so. In this state the animal does not move very far from the tube, so the time to initiate another burst of licking tends to be short. The second state, described by the second of the two exponentials, describes the state of the rat when its attention is focused on activities other than ingestion. In this state the behavior of grooming, rearing, exploring, and sleeping occur. I will call these two states State 1 and State 2 for the time to keep the designations neutral with respect to the nature of the underlying states that generate these two classes of ICIs.

The functions shown in Figure 4.10a show that the probabilities of being in State 1 and in State 2 are the same in the first and second halves of the session (0.6 in both cases). The averages of the ICIs for State 1 in the first and second halves were almost identical (1.3 and 1.4 for the first and second halves, respectively). The major difference is in the averages of the ICIs for State 2 in the first and last halves. The average of the ICIs for State 2 in the first half is 12.5 sec, and in the second half it is more than twice as long, 33.3 sec. Thus, although the proportion of long and short pauses

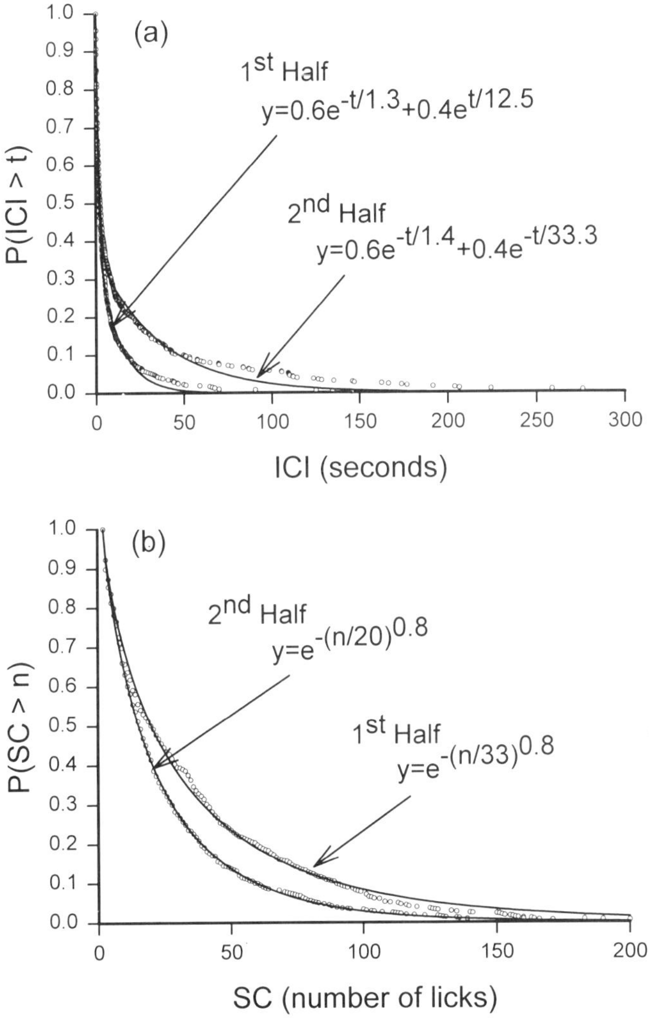

FIGURE 4.10. Cumulative probability functions for the intercluster intervals (ICI) and the size of clusters for the first and second halves in a test session in which the rats ingested a saccharin + glucose solution to which mannitol (0.3 M) was added for the first time. The solid lines are from a least-squares fit to the sum of two exponentials **(a)** and a Weibull function **(b).** Reciprocals of the rate parameters are shown in the functions to give the averages of the distributions.

remained approximately the same in the two halves of the test, when the rats were in the state generating the longer pauses (State 2), they took on average somewhat twice as long to return to licking in the second half of the test as they did in the first.

In Figure 10b it can be seen that a Weibull function,

$$y = e^{(-an)}, ^{b}$$

fits he cumulative probability distributions of the cluster sizes well; the coefficients of variations of the estimates were less than 2% in both cases. The shape parameter of the Weibull function was the same in both the first and second halves of the test (0.8), but the rate parameter was less in the first than in the second half of the test, indicating that the average size of the clusters in the first half was greater (33) than in the second half (20). In both cases the shape parameter of the Weibull function was less than 1 (0.8), indicating that there were slightly more small and slightly fewer large clusters than if the distribution were a true exponential. The average of the Weibull function is given by $a^{-1} - (1+1/b)$. Thus, the decline in the rate of licking during the last half of the ingestion interval, presumably when mannitol was beginning to influence the ingestive behavior, can be seen as reflecting both an increase in the duration of the ICIs generated by State 2 and a decrease in the size of the clusters.

Conditioning by Preabsorptive Stimulation

The intestinal control of the licking behavior described above is a direct unconditioned effect of gastrointestinal filling. The results from a recent study (Davis and Perez, 1993) indicate that gastrointestinal filling can serve as a UCS for a flavor-associated control of meal size as well.

In that study, one group of rats was trained to ingest a vanilla-flavored saccharin–glucose mixture (6 mM saccharin + 0.1 M glucose), and another was trained to ingest an orange-flavored saccharin–glucose mixture with 0.3 M mannitol added. Two additional groups had these flavor–mannitol associations reversed for the purposes of counter-balancing. After eight daily 30-minute exposures to these solutions, the mannitol was removed from the two flavored solutions containing it and was added to the two previously mannitol-free flavored solutions. The rats were then tested on these solutions for four additional daily tests.

The volume of fluid ingested in 30 minutes in the last training trial and the four following tests in which the flavor–mannitol associations were reversed are shown in Figure 4.11a. The data from the last day of training confirm those discussed above. Mannitol in the solution suppressed intake. Examination of the intake data in the first reversal test, when the flavor–mannitol association was switched, however, shows that the presence or absence of mannitol in the test solution no longer produced the differential

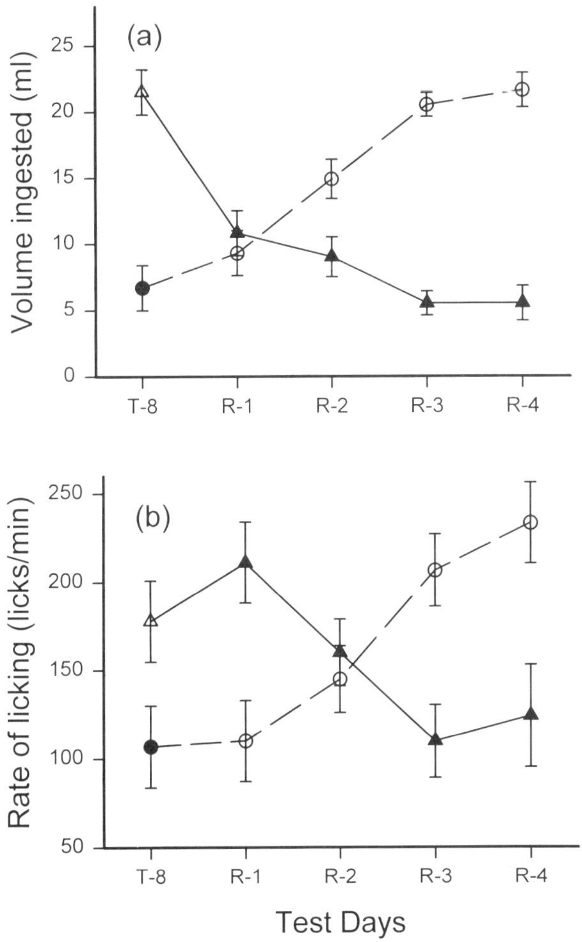

FIGURE 4.11. The volume ingested **(a)** and the rate of licking during the first 3 minutes of the test session **(b).** The data are from the last of eight training trials (T-8) and four following trials in which the flavor + mannitol association was switched. Closed symbols represent tests in which the test solution contained mannitol. Open symbols represent tests in which the test solution was mannitol-free. Circles and triangles differentiate groups with different initial training. Triangles represent groups initially trained on mannitol-free solutions; circles, groups trained on mannitol-containing solutions. Redrawn from Davis and Perez (1993), with permission of the publisher.

effect on the volume ingested. There was no significant difference in the volumes ingested among the groups. However, by the fourth reversal test the original differences established during training reappeared. The pattern of these results suggests that associative conditioning of ingestive behavior is involved in controlling meal size in this experiment, but the intake data provide no clues as to how the conditioning was expressed behaviorally.

An examination of the ingestive behavior of the animals provides the answer. The average rate of licking during the first 2 minutes of each of the test sessions is shown Figure 4.11b. On the last training trial (T-8) the average initial rate of licking was significantly greater among those animals ingesting the mannitol-free solution than among those ingesting the mannitol-containing solution. In the first test, in which the mannitol was removed from one test solution and added to the other (R-1), this difference was maintained, indicating that the flavor, not the mannitol content, of the solution controlled the initial rate of licking. With further experience with the new flavor–mannitol association the rates for the groups crossed, and by the fourth reversal test the initial rates were within the ranges established during the last training test with the original flavor–mannitol association.

These results show clearly that the postingestive effects of mannitol served as a UCS for the associative conditioning of a decrease in the rate of licking during the first 2 minutes of the test session and that this control could be reversed easily when a new flavor–mannitol association was arranged. They also provide an explanation for the lack of a difference in intake on the first reversal test. The intake of the animals that had previously associated the flavor with mannitol was controlled primarily by the low initial rate of licking acquired during training. The intake of the animals ingesting the solution with the familiar flavor, which now contained mannitol for the first time, was depressed, in spite of an initial high rate of licking, because of the direct effect of mannitol appearing later in the meal, which depressed the rate of licking. The net effect was essentially equal intake in both groups, but for very different reasons, one a conditioned effect and the other an unconditioned one. The volume measure by itself revealed none of this.

One might be tempted to call the low intake of the animals originally trained to ingest the mannitol-containing solutions when first tested on the mannitol-free solutions "conditioned satiety." With just the meal size data, however, it would be difficult to find an adjective to qualify "conditioned" and differentiate it from the other types of "conditioned control" of ingestion that have been discovered. Thus a better term might be the conditioned control of the initial rate of ingestion because microstructural analysis reveals that this is what it is.

Combined Orophyrangeal and Postingestional Control

It is sometimes assumed that satiation is proportional to the caloric density and/or concentration of the test diet, more concentrated solutions being

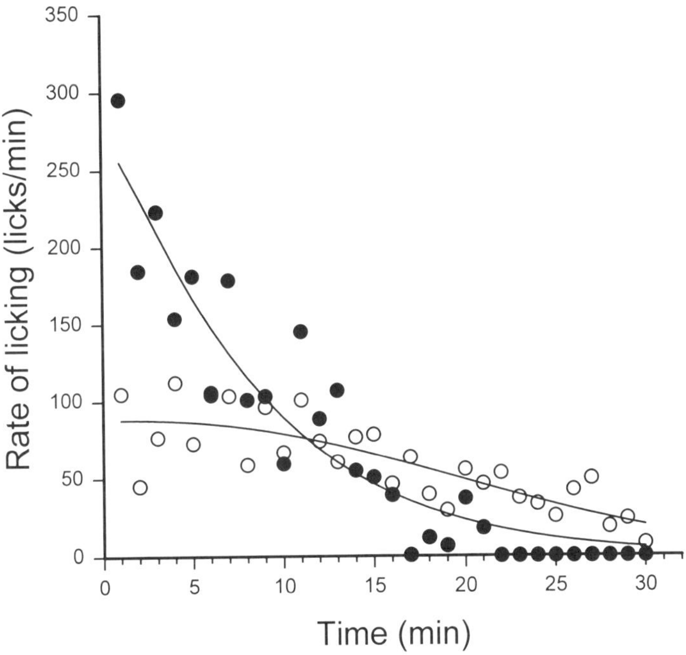

FIGURE 4.12. The rate of licking calculated at one minute intervals for a test session in which the test solution was 0.05 M sucrose (open circles) or 0.8 M sucrose (closed circles). The solid lines are from a least-squares fit by the Weibull function.

more satiating than less concentrated ones. If one uses variations in the amount ingested as a measure of the degree of satiation to support this argument, the evidence is not very convincing. Figure 4.1d provides an example. It shows the volume ingested of a 0.05 M (left bar) and an 0.8 M (right bar) sucrose solution by the same group of nonfood-deprived rats. The volumes ingested were not significantly different, but it is difficult to believe that an 0.8 M sucrose solution is not more satiating in the traditional sense than a 0.05 M solution. The problem is, as Smith and Gibbs (1979) have pointed out, that meal size is not an appropriate measure of satiety when either the type of the test solution or the deprivation conditions differ between the groups. However, by looking within the meal at the how these intakes were achieved clear differences due to concentration on ingestive behavior become apparent.

Figure 4.12 shows the rate of licking calculated at 1-minute intervals during the 30-minute test sessions for the volume data shown in Figure 4.1d. Integration of these two curves gives the total number of licks in the session, which were 1,978 + 311.5 licks (0.05 M) and 1,747 + 155 licks (0.8

M). The numbers of licks did not differ statistically, but their distribution throughout the test session did. The low concentration solution stimulated a low relatively constant rate of licking until the very end of the ingestion interval, whereas the high concentration solution stimulated a high rate of licking at the beginning of the interval followed by a rapid decline.

One interpretation of these results (Davis and Levine, 1977) is that the high concentration solution controlled the ingestive behavior by stimulating a high rate of ingestion at the beginning of the meal, leading to a rapid filling of the gastrointestinal tract. This activated a negative-feedback signal that terminated the behavior. The low concentration solution, by stimulating a low rate of licking at the beginning of the meal, did not fill the gastrointestinal tract rapidly, permitting clearance and absorption to keep pace with ingestion. Thus there was little accumulation of fluid in the gastrointestinal tract and therefore a weak negative-feedback signal to inhibit the ingestive behavior.

These two solutions differentially affected the ingestive behavior of the animals in other ways as well. The average length of the meal for the 0.05 M solution was almost twice that of the 0.8 M solution (25.8 + 2.1 minutes and 15.2 + 1.3 minutes, respectively). The rate of licking within the bursts was greater with the 0.05 M (6.89 licks sec^{-1}) than with the 0.8 M (6.41 licks sec^{-1}) solution, a small but statistically significant difference. The average size of the clusters of licking was larger (50 licks) for the 0.8 M than for the 0.05 M solution (36 licks).

The differences in the microstructures of the behaviors of the animals ingesting the two different solutions are brought into clear focus by an examination of the cumulative probability distribution functions for the ICIs and the size of the clusters for the two solutions (Fig. 4.13). Davis (1996) provides a fuller explanation of this type of analysis of the ICIs and SCs. The cumulative probability functions for the ICIs shows that when the animals were ingesting the 0.8 M sucrose solution State 1 dominated; 80% of the ICI belonged to this category in contrast to 50% when the rats were ingesting the 0.05 M sucrose solution. Furthermore, the average ICI for State 1 was 1.4 sec for 0.8 M sucrose and almost twice as long, 2.5 sec, when the rats were ingesting the 0.05 M sucrose solution. In short, when ingesting the more concentrated solution the animals were primarily in State 1, leaving the drinking spout for relatively short periods of time and returning sooner to resume licking than when ingesting the weaker solution.

The cumulative probability distribution of the size of the clusters (Fig. 4.13b) shows that the concentration of the solution affected this variable as well. The solid lines show the fits to the Weibull function. In both cases the

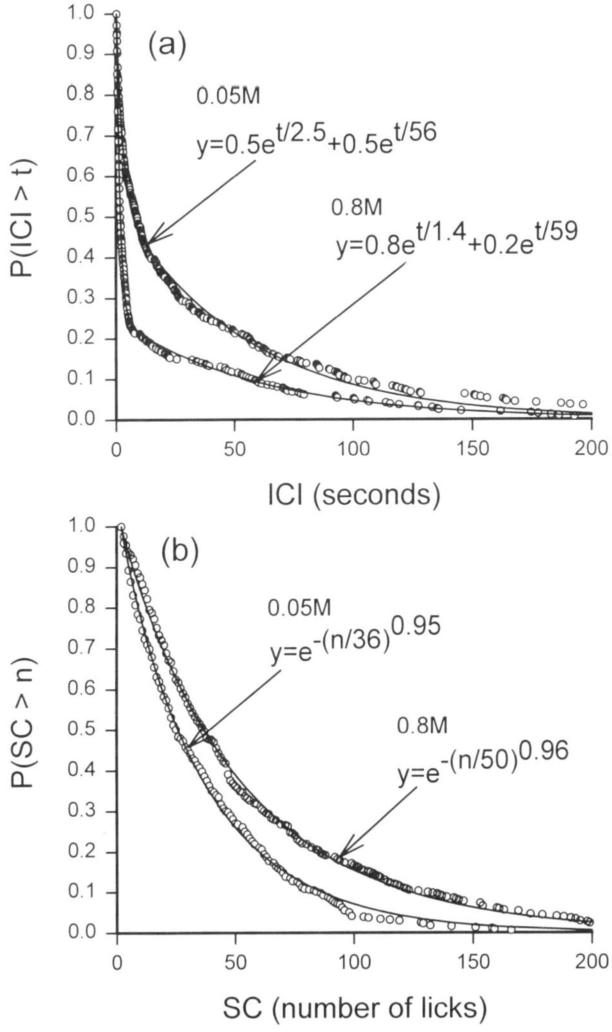

FIGURE 4.13. Cumulative probability functions for the intercluster interval **(a)** and the size of clusters **(b)** for a test session in which the test solution was 0.05 M sucrose and one in which it was 0.8 M sucrose. The solid lines are from a least-squares fit by the sum of two exponentials (a) and from a single Weibull function (b). Reciprocals of the rate parameters are shown in the functions to give the averages of the distributions.

Weibull exponents were very close to 1, indicating that the simple exponential, $y = e^{-an}$, describes the data well. The average of the size of the clusters for the 0.8 M sucrose solution was greater (50 licks) than for the 0.05 M sucrose solution (36 licks). This confirms an earlier report on the effect of sucrose concentration on the size of the clusters of licking (Davis and Smith, 1992).

When one compares the results of the microstructural analysis of the ingestive behavior of rats ingesting these two solutions to the results obtained when only the volume ingested is obtained, it becomes clear that the volume ingested in a fixed period of time is a very poor measure of anything. Not only does the measure, volume ingested, hide a rich complexity of control over the behavior of the animal by orophyrangeal and postingestional stimulation, but it gives no hint of the vast differences these two solutions have on the motor system that controls the licking behavior of the animal.

Summary

In this overview of some of the research my colleagues and I have conducted, I have attempted to show that with the technology now available it is possible to go beyond meal size as a measure of the effect of experimental manipulations on ingestive behavior and gain a better insight into the nature of the controls that determine the size of the meal. A relatively simple analysis of the distribution of the ingestive acts that make up a meal when a rat is tested with liquid diets reveals that meal size is determined in a complex way by the interaction of a variety of variables acting throughout the meal.

Oral stimulation acts as both an excitatory UCS to drive the motor system that controls the bursts of licking and as a conditioned stimulus to inhibit or modulate this excitatory input. This orophyrangeal control, acting throughout the meal, combines with direct unconditioned stimulation arising from the accumulated effects of the ingested substance to determine the total amount ingested. To understand the interplay of these variables over time on the determination of the volume of the test diet that will be ingested, it is necessary to measure directly the changes in the ingestive behavior as they occur throughout the meal. The microstructural analysis of the licking behavior of the rat provides one means of doing this.

An additional advantage of the microstructural approach is that it provides a means for measuring directly the output of the motor system responsible for the ingestion of liquid diets. This analysis and that of others (Grill and Norgren, 1978; Travers and Norgren, 1986; Wiesenfeld et al.; 1977) reveals that the brainstem motor control system that controls licking has the properties of a motor pattern generator that generates bursts of licking at a constant rate. By analyzing the size of these bursts and the intervals that separate them, we can now study directly the output of this motor system and thus are in a better position to understand the neurophysiological mechanisms that control these determinants of intake.

ACKNOWLEDGMENTS

The author expresses his appreciation to those people who have collaborated with him over the years in conducting the experiments discussed in this chapter and in developing many of the ideas presented here. They are Drs. Constance S. Campbell, Barbara J. Collins, James Gibbs, Michael Levine, Gerard P. Smith, and David Wirtshafter. I also express my appreciation to Dr. Paul A. S. Breslin for a critical reading of this and an earlier version of the manuscript, and for many helpful suggestions for improving its clarity.

I owe a special debt of gratitude to Gerard P. Smith and James Gibbs, who invited me to work at the Bourne Laboratory during a sabbatical leave in 1986 and during a number of summers thereafter. It was during that period that many of the ideas expressed in this chapter were developed and much of the research discussed here was carried out.

This work is supported by NIH DK 41563.

REFERENCES

Allison, J. N., and J. N. Castellan, Jr. Temporal characteristics of nutritive licking in rats and humans. *J. Comp. Physiol. Psychol.,* 70:116–125, 1970.

Asin, K., J. D. Davis, and L. Bednarz. Differential effects of serotonergic and catecholaminergic drugs on ingestive behavior. *Psychopharmacology* 109:415–421, 1992.

Booth, D. A. Conditioned satiety in the rat. J. Comp. Physiol. Psychol., 81:457–471, 1972.

Booth, D. A. Food-conditioned eating preferences and aversions with interoceptive elements: Conditioned appetites and satieties. In N. S. Braveman and P. Bronstein (eds): *Experimental Assessments and Clinical Applications of Conditioned Food Aversions. Vol. 443.* New York: New York Academy of Sciences, pp. 22–41, 1985.

Collins, B. A., and J. D. Davis. Long term inhibition of intake by mannitol. *Physiol. Behav.,* 21:957–965, 1978.

Davis, J. D. Deterministic and probabilistic control of the behavior of rats ingesting liquid diets. *Am. J. Physiol. Regul., Integ. Comp. Physiol.,* 270:R793–R800, 1996.

Davis, J. D., and C. S. Campbell. Peripheral control of meal size: Effect of sham feeding on meal size and drinking rate. *J. Comp. Physiol. Psychol.,* 83:379–384, 1973.

Davis, J. D. , B. J. Collins, and M. Levine. Peripheral control of meal size: Gastrointestinal filling as a negative feedback signal, a theoretical and experimental analysis. *J. Comp. Physiol. Psychol.* 89:985–1003, 1975.

Davis, J. D., and J. D. Keehn. Magnitude of reinforcement and consummatory behavior. *Science* 130:269–271, 1959.

Davis, J. D., and M. Levine. A model for the control of ingestion. *Psychol. Rev.* 89:379–412, 1977.

Davis, J. D., and M. C. Perez. The acquired control of ingestive behavior in the rat by flavor—associated postingestional stimulation. *Physiol. Behav.,* 54:1221–1226, 1993.

Davis, J. D., and G. P. Smith. Learning to sham feed: Behavioral adjustments to the absence of gastrointestinal stimulation. *Am. J. Physiol. Regul., Integ. Comp. Physiol.,* 28:R1228–R1235, 1990.

Davis, J. D., and G. P. Smith. Analysis of the microstructure of the rhythmic tongue movements of rats ingesting maltose and sucrose solutions. *Behav. Neurosci.,* 106:217–228, 1992.

Davis, J. D., G. P. Smith, and J. Miesner. Postpyloric stimuli are necessary for the normal control of meal size and sham feeding rats. *Am. J. Physiol., Regul. Integ. Comp. Physiol.,* 33:R888–R895, 1993.

Deutsch, J. A., W. G. Young, and T. J. Kalogeris. The stomach signals satiety. *Science* 201:165–167, 1978.

Grill, H. J., and R. Norgren. Taste reactivity as a measure of the neural control of gustatory stimuli in neurological normal animals. *Brain Res.,* 143:263–279, 1978.

Hall, W. G. A remote stomach clamp to evaluate oral and gastric controls of drinking in the rat. *Physiol. Behav.,* 11:897–901, 1973.

Knowler, W., and T. Ukena. The effects of chlorpromazine, pentobarbital, chlordiazepoxide and *d*-amphetamine on rates of licking in rats. *J. Pharmacol. Exp. Ther.* 184:385–397, 1973.

LeMagnen, J., and M Devos. Parameters of the meal pattern in rats. *Neurosci. Biobehav. Rev.* 4 (Suppl 1):1–11, 1980.

Pavlov, I. P. *The Work of the Digestive Glands.* W. H. Thompson, trans., 2nd English Ed. London: Charles Griffin, 1910.

Sclafani, A., and J. W. Nissenbaum. On the role of the mouth and the gut in the control of saccharin and sugar intake: A reexamination of the sham feeding preparation. *Brain Res. Bull.,* 14:569–576, 1985.

Smith, G. P., and J. Gibbs. Postprandial satiety. In J. M. Sprague and A. N. Epstein (eds): *Progress in Psychobiology and Physiological Psychology. Vol. 8.* New York: Academic Press, 1979, pp. 179–242.

Stellar, E., and J. H. Hill. The rat's rate of drinking as a function of water deprivation. *J. Comp. Physiol. Psychol.,* 45:96–102, 1952.

Travers, J. B., and R. Norgren. Electromyographic analysis of the ingestion and rejection of sapid stimuli in the rat. *Behav. Neurosci.,* 100:544–555, 1986.

Weingarten, H. P., and O. T. Kulikovsky. Taste-to-postingestive consequence conditioning: Is the rise in sham feeding with repeated experience a learning phenomena? *Physiol. Behav.,* 45:471–476, 1989.

Wickens, T. D. *Models for Behavior.* San Francisco: W. H. Freeman, 1982.

Wiepkema, P. R. Behavioral factors in the regulation of food intake. *Proc. Nutr. Soc.,* 30:142–149, 1971.

Wiesenfeld, Z., B. P. Halpern, and D. N. Tapper. Licking behavior: Evidence of hypoglossal oscillator. *Science* 196:1122–1124, 1977.

5

The Satiating Effects of Cholecystokinin and Bombesin-Like Peptides

GERARD P. SMITH and JAMES GIBBS

Cholecystokinin Inhibits Food Intake

In 1973, Gibbs, Young, and Smith reported that intraperitoneal administration of an impure extract of cholecystokinin (CCK) and the synthetic C-terminal octapeptide of cholecystokinin (CCK-8) produced inhibition of food intake in rats (Gibbs et al., 1973a). The effect was dose related and lasted 15–30 minutes. The effect was behaviorally specific because doses that inhibited food intake in food-deprived rats did not inhibit water intake in water-deprived rats. Furthermore, there were no signs of toxicity.

We proposed that this behavioral effect of exogenously administered CCK-8 was mimicking the function of the endogenous peptide that was known to be released from the upper small intestine by ingested food. We assumed that this behavioral effect of CCK-8 was a hormonal effect because CCK-8 (and larger forms of CCK) produced the classic visceral effects of contracting the gallbladder and stimulating pancreatic secretion by a hormonal mode of action. Subsequently, we referred to this as the "gut hormone hypothesis of postprandial satiety" (Smith, 1984).

In the 1973 paper, we proposed five criteria that a candidate peptide would have to satisfy to be considered a physiological, satiating mechanism for the inhibitory control of meal size (Table 5.1). The results obtained with CCK-8 satisfied the first four criteria, but not the fifth.

The paper was seminal for several reasons:

First, it raised the possibility that gut peptides were involved in the control of meal size. Because a large number of gut peptides are released during the ingestion of food, it was likely that other peptides could have a satiating effect. They did (see below and Chapters 6, 7, and 8).

TABLE 5.1.
Criteria for a Physiological Satiating Peptide (1973)

1. The peptide must be released during the ingestion of food
2. Administration of the peptide prior to a meal should decrease meal size, and the inhibitory effect should be dose related
3. The peptide should have a relatively rapid onset and brief duration of action in order to account for the termination of one meal and the recurrence of eating at the next meal
4. The inhibitory effect of the peptide should not be due to illness
5. The inhibitory effect should be obtained with doses of the peptide that produced circulating concentrations of the peptide within the range of the concentrations of the endogenous peptide observed during or at the termination of a meal

The statements are slightly modified from the original version (Gibbs et al., 1973a).

Second, the commercial availability of pure or synthetic gut peptides facilitated testing them in technically simple protocols.

Third, by 1973, investigators in the control of food intake were open to new ideas and approaches because they had found the glucostatic hypothesis (Mayer, 1953) to be an inadequate explanation for the control of meal size (Smith, 1976; Smith et al., 1972).

The inhibitory effect of CCK-8 was quickly confirmed and extended to a variety of species of animals as well as humans (for reviews, see Baldwin, 1992; Smith and Gibbs, 1992). Although the inhibitory effect was a robust phenomenon, its interpretation became controversial. This controversy is instructive because it reveals the experimental problems involved in deciding whether the inhibitory effect on meal size by a candidate peptide is a specific satiating effect.

Is the Inhibition a Satiating Effect?

BEHAVIORAL SPECIFICITY

A satiating effect is characterized by inhibiting meal size without producing abnormal or competing behaviors and by permitting or eliciting the normal behavioral sequence of satiety to appear (Antin et al., 1975). The C-terminal octapeptide of CCK (≤ 5 μg/kg, ip) inhibited meal size without producing competing behaviors, permitted the appearance of the satiety sequence when administered to rats prior to real feeding, and elicited the satiety sequence when administered prior to sham feeding (Antin et al., 1975; Gibbs et al., 1973a,b).

The same doses of CCK-8 decreased intake of liquid food in food-deprived rats, but did not decrease intake of water in water-deprived rats

(Gibbs et al., 1973a). Thus, CCK-8 did not produce a nonspecific effect on the oral movements of ingestion. Adding small amounts of sugar to water restored the inhibitory effect of CCK-8 (Waldbillig and O'Callaghan, 1980), presumably because the orosensory effects of sugar are associated with food.

These results showed that the inhibitory effect of CCK-8 was specific to the ingestion of food. It is logically possible, of course, that this apparent specificity is due to the fact that CCK interacts with some other consequence of food to produce a nonobservable, slight toxic effect similar to human nausea.

Conditioned Taste Aversion

The possibility that CCK-8 produced a toxic effect was tested in 1973 by investigating whether CCK-8 could serve as an unconditioned stimulus for the acquisition of a conditioned taste aversion (CTA). This is based on the assumption that the acquisition of a CTA only occurs when the unconditioned stimulus produces sickness. In our hands, CCK-8 failed to produce a CTA after one or repeated injections (Holt et al., 1974; Gibbs et al., 1973a). Although Deutsch and Hardy (1977) reported that CCK produced a CTA, subsequent work showed that doses of CCK sufficient to inhibit meal size by 30%–50% failed to produce a CTA. In a critical review of this issue, Kulkosky (1985) noted that CCK-8 is typically ineffective as an unconditioned stimulus in conditioned food aversion experiments. Kulkosky concluded that although aversions can be produced by relatively high doses of CCK-8 administered after completion of a consummatory session, when CCK was given prior to a meal and decreased meal size, no aversion was observed.

Conditioned Discrimination

Conditioned discrimination is another test of the specificity of the inhibitory effect of CCK-8. Animals are trained to respond to a stimulus associated with ingested or infused food. Rats and monkeys do not discriminate the effects of CCK-8 from the postingestive effects of food (Corwin and Schuster, 1993; Corwin et al., 1990) or from 1 hour of food deprivation (Davidson and Carretta, 1993). Under the same conditions, however, they discriminate anorectic drugs, such as *d*-amphetamine, fenfluramine, and lithium chloride, from the effects of food.

Taste Reactivity

A behavioral display consisting of oral gapes, chin rubbing, and forelimb flailing is a characteristic response to the orosensory properties of food stimuli that are aversive innately or by association with prior gastrointestinal toxicity (Grill and Norgren, 1978). Administration of CCK-8 in doses that decrease meal size significantly elicits some (Cabanac and Zhao, 1994) or no (Eckel and Ossenkopp, 1994; Grill, 1985) aversive reactions. This is characteristic of satiation (Grill, 1985) and negative alliesthesia (Cabanac and Zhao, 1994), but not aversion.

Microstructure of Licking

When CCK-8 decreased meal size significantly in a test situation in which licking was recorded continuously, CCK-8 did not change the initial rate of licking, but it did increase the rate of decay of licking (Davis et al., 1995). There was no evidence of impairment in the motor performance of rapid licking or other nonspecific effects. These changes in the microstructure of licking are characteristic of a satiating effect and are quite different from the changes in microstructure produced by an aversive agent, such as lithium chloride (J. D. Davis, unpublished observations).

Neuroendocrine Correlates

When these behavioral results failed to disclose the hypothetical aversive effect of CCK, Verbalis et al. (1986a,b) proposed a neuroendocrine approach to this problem. Their strategy was based on the observation that plasma arginine vasopressin (AVP) increased when human subjects were nauseated, but did not vomit, after apomorphine administration (Rowe et al., 1979).

This strategy was applied first to the rat, an animal that does not vomit. When agents that were known to produce toxic effects and CTAs in rats, such as apomorphine, lithium chloride, and copper sulfate, were administered, plasma AVP did not increase, but plasma oxytocin (OT) did in a dose-dependent manner (Verbalis et al., 1986b). The authors suggested that OT could be used to measure visceral illness despite the fact that many stressful stimuli unrelated to visceral illness were known to increase OT (Lang et al., 1983).

Verbalis et al. (1986a) showed that CCK-8 (0.1–100 μg/kg, iv or ip) produced a dose-related increase in OT in rats. They also observed that plasma OT increased significantly when rats ate large meals of liquid food. The maximal increase in OT produced by a meal was approximately 10 μU/ml (McCann et al., 1988, 1989; Flanagan et al., 1988). Large doses of CCK-8 (≥10 μg/kg) produced increases of OT larger than 10 μU/ml, but smaller doses in the range that were efficacious for the inhibition of meal size did not. For example, McCann et al. (1989) reported that CCK (5 μg/kg, ip) decreased meal size 64% and increased OT to only 8.9 μU/ml. Thus, the neuroendocrine strategy did not provide evidence for the hypothetical sickness in the rat.

Verbalis et al. (1987) did similar experiments in the rhesus monkey, an animal known to respond to emetic stimuli by retching and vomiting. Again, the results depended on dose: CCK (10 μg/kg, iv) produced a large increase in plasma AVP, and emesis occurred in five of eight tests. When 1 μg/kg (iv) was given, there was only a small increase of plasma AVP, and emesis did not occur. These results suggest that doses in the range of 1 μg/kg are not correlated with sickness. This is consistent with earlier reports that doses in this range decreased meal size significantly in rhesus monkeys without producing retching, emesis, or other behavioral signs of discomfort (Metzger and Hansen, 1983; Moran and McHugh, 1982; Falasco et al., 1979).

Having failed in rats and monkeys, Miaskiewicz et al. (1989) pursued the problem in humans. They found that an iv bolus injection of 0.05 μg/kg was the threshold dose of CCK-8 for increasing plasma AVP, for self-reports of nausea, and, in 1 of 14 subjects, vomiting. (Note that the design of this study was flawed: Subjects knew they were receiving CCK, CCK was not administered in the context of a meal, nausea was the only report requested, and the scale used to assess reports of nausea was not validated [Melzack et al., 1985; Morrow, 1984].)

Even with these limitations, the threshold dose for nausea was larger than doses that inhibited meal size in humans when CCK-8 was administered by slow intravenous infusion under double-blind conditions (Stacher et al., 1982; Pi-Sunyer et al., 1982; Kissileff et al., 1981). Thus, this series of neuroendocrine experiments also failed to produce evidence for the hypothesis that the inhibitory effect of CCK on food intake necessarily involved some kind of sickness.

Human Self-Reports

Kissileff and his colleagues (see Geary et al., 1992; Pi-Sunyer et al., 1982; Kissileff et al., 1981) used a battery of self-report measures, interviews, and behavioral observations in several experiments without finding evidence of significant sickness (Fig. 5.1). Furthermore, a recent meta-analysis of published experiments in lean and obese humans revealed that significant side effects during a slow, intravenous infusion of CCK in food intake tests were reported in only 2 of 160 subjects, a remarkably low incidence of 1.25% (Poeschla et al., 1997).

Eight of the 160 subjects (threee obese and five lean) reported slight side effects (Pi-Sunyer et al., 1982; Kissileff et al., 1981). These side effects were considered slight because no subject could identify the test in which CCK was infused compared with the test when vehicle was infused, no

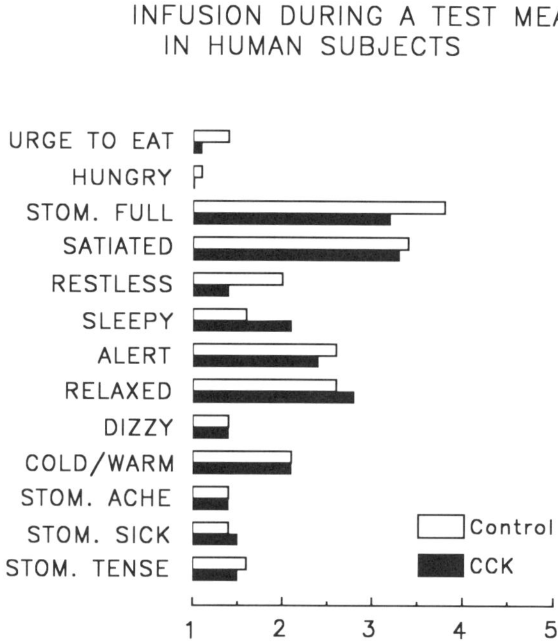

FIGURE 5.1. Intravenous CCK infusions produce a subjectivity-specific reduction in meal size. Data are category scale (1, not at all; to 5, extremely) estimates of the 13 internal perceptions indicated (STOM. = stomach) done immediately after meals in eight normal weight young men. Closed bars are mean ratings after meals during which 2 ng/kg/min CCK-8 was infused; open bars are ratings after control-infusion meals. CCK-8 reduced meal size from 856 to 587 g, standard error of the difference, 29 g; $p < 0.01$. There were no significant differences in postmeal perceptions. The standard errors of the differences were 0.1–0.3. Visual analogue scale ratings of hunger and satiety also failed to detect any nonspecific effect of CCK-8. Furthermore, during interviews after the last test, the men were unaware that they had eaten different amounts and could not distinguish test and control days. From Geary et al. (1992), with permission of the publisher.

subject dropped out of the experiments, and the occurrence of side effects did not correlate with the inhibition of food intake.

The conclusion is clear—the inhibitory effect of CCK-8 on meal size is a satiating effect of the peptide and is not due to a hypothetical, subclinical, nonspecific, toxic effect referred to as "malaise" in animals and "nausea" in humans (Table 5.2). There is no question that large doses of CCK-8, particularly when administered as a bolus injection intravenously, can cause toxic effects. This is true of most, if not all, peptides and is not relevant to the interpretation of the inhibitory effect of small doses on meal size as a satiating effect.

TABLE 5.2.
Tests of Behavioral Specificity of Cholecystokinin

Test	CCK
Water intake	Yes
Satiety sequence	
Real feeding	Yes
Sham feeding	Yes
Conditioned taste aversion	Yes
Conditioned discrimination	Yes
Taste reactivity	Yes
Microstructure of licking	Yes
Neuroendocrine correlates	Yes
Human self-reports	Yes

"Yes" means that at least one dose of the peptide that inhibited food intake significantly was tested and no nonspecific or toxic effect was obtained. See text for references.

Is the Satiating Effect Physiological?

Whether the satiating effect of exogenous CCK-8 was a physiological function of the endogenous peptide could not be answered until potent and specific antagonists of the CCK_A receptor were developed (see Chapter 6). When they became available, they were used to test the hypothesis that endogenous CCK released from the small intestine by ingested nutrients acted in a negative-feedback manner to control the size of the meal being eaten. The hypothesis predicted that administration of a CCK_A antagonist before a meal would increase the size of the meal because it would eliminate the usual satiating effect of endogenous CCK. This prediction has been fulfilled in more than 25 experiments under a variety of conditions in rats, mice, pigs, and monkeys (Smith and Gibbs, 1992).

A more rigorous test of the hypothesis came from experiments in which the inhibitory effect on meal size of stimuli known to release CCK from the small intestine, such as oleate (Yox et al., 1989), Intralipid (Greenberg et al., 1989), and soybean trypsin inhibitor (Weller et al., 1990), were blocked or attenuated by prior administration of a CCK_A antagonist.

Because the antagonists used in all of these experiments passed readily across the blood–brain barrier, it was possible that the site of action of the antagonist was not in the periphery as required by the hypothesis (see below), but was in the brain. This possibility appears to have been eliminated by recent experiments by Brenner and Ritter (1995). They showed that a peptide antagonist of CCK_A receptors that did not enter the brain increased meal size.

On the basis of the extensive evidence from these experiments, we concluded that the satiating effect of endogenous CCK was a physiological function of the peptide (Smith and Gibbs, 1992). Because most of this evidence was obtained in rats, it remains to be seen if this is also the case in other animals and humans. The limited results in humans are mixed so far. On the one hand, Lieverse et al. (1993b) have decreased meal size with infusions of CCK that produced plasma concentrations of CCK in the physiological range. On the other hand, four experiments with CCK_A antagonists failed to increase meal size (Lieverse et al., 1995; French et al., 1994; Drewe et al., 1992; Wolkowitz et al., 1990). The negative results with CCK_A antagonists are evidence against this hypothesis. More experiments are required to decide the issue.

Note that when a CCK_A antagonist increased meal size in animals, the maximal magnitude of the increase was about 25%. Thus, the antagonist did not prevent the termination of eating, it simply delayed it. This means that there are other satiating mechanisms operating during a meal. These include pregastric and gastric mechanisms (Chapter 2), small intestinal mechanisms (Chapter 3), other peptides (Chapters 7 and 8), and monoamines, especially serotonin (Chapter 9).

Having reviewed the evidence that CCK fulfills the five criteria proposed in 1973 to define a physiological, satiating function of a peptide (Table 5.1), we now review the evidence of how and where this peptide acts and how it interacts with other peripheral and central factors to control meal size.

Site and Mechanism of Action

The site of the satiating action of CCK-8 is in the abdomen. It involves CCK_A receptors of the low-affinity type. These receptors are in the circular muscle of the pyloric sphincter and on vagal afferent terminals. Capsaicin-sensitive vagal afferent fibers transmit the effect of CCK-8 on CCK_A receptors to the caudal portion of the nucleus of the tractus solitarius (NTS), where central processing begins (see below). The details of the experimental evidence on which this summary is based are given in Chapter 6.

Mode of Action

The mode of action depends on the species and the size of the CCK molecule being tested. In the rat, the mode of action is not hormonal because the satiating potency of CCK-8 is markedly reduced when CCK-8 is administered into the hepatic-portal vein, the natural site of entry of endogenous CCK-8 into the circulation, in comparison to when CCK-8 is administered intraperitoneally (Smith and Greenberg, 1988; Greenberg et al., 1987; Strubbe et al., 1989). From these results we suggested that the mode of action is paracrine, i.e., CCK-8 released from CCK-containing cells in the duodenal mucosa diffuses within the mucosa and submucosa to bind to CCK_A receptors on vagal afferent terminals in the region (Greenberg et al., 1987).

The reduction of potency after intraportal administration may be due to an uptake mechanism in the liver for CCK-8 (Gores et al., 1986), but when the first-pass of CCK-8 was avoided by doing an identical infusion into the inferior vena cava, CCK-8 also had no effect on food intake (Melville et al., 1993a). Longer, slow intravenous infusions of CCK-8, however, decreased intake (Cox et al., 1995; Reidelberger and Solomon, 1986; Lorenz et al., 1979), but the doses required were larger than doses that were efficacious after intraperitoneal administration.

Recent experiments with intraarterial injections are also consistent with a local action of CCK-8 in the upper duodenum or pyloric sphincter region. The evidence is that extremely low doses of CCK-8 decrease intake significantly after injection or infusion into or near the celiac artery (Cox et al., 1995; Calingasan et al., 1992), and the potency of CCK-8 is larger after such administration than in any other abdominal artery. Reidelberger et al. (1994) recently used passive immunization with a monoclonal antibody to CCK to support a nonendocrine, i.e., paracrine, mode of action for the satiating effect of endogenous CCK.

While accepting the nonhormonal mode of action, Ritter et al. (1994) have suggested that the critical pool of releasable CCK is secreted by enteric neurons in the upper small intestine rather than by mucosal cells. This difference of opinion points up the fact that not much is known about the details of the process within the wall of the upper small intestine by which CCK stimulates vagal afferent terminals.

When the experiments using different routes of administration were repeated with CCK-33, different results were obtained. Although CCK-33 was equipotent with CCK-8 after intraperitoneal administration (Melville et al., 1993b), it had significantly more satiating potency than CCK-8 after intraportal and intravenous administration (Smith et al., 1996). Furthermore, the satiating potency of CCK-33 was significantly higher after intraportal than intravenous administration. This suggests a hepatic site of action.

Thus, forms of CCK larger than CCK-8 may act locally in the mucosa and submucosa of the upper small intestine and hormonally in the liver. Although CCK-33 is not present in the rat, CCK-22 is, and when it becomes available, it should be tested by intraportal administration. The ability of CCK-33 , but not CCK-8, to act in the liver is consistent with an uptake process in the liver for peptides smaller than 10 amino acids (Strunz et al., 1978; Doyle et al., 1984).

If these results with CCK-33 are replicated, they raise the possibility that the variable satiating effect of individual fatty acids (see Chapter 3) could be due to the ratio of CCK-8 and larger forms released. When relatively more of the large form is released, it should be more potent because it acts locally and in the liver.

In contrast to the rat, the mode of action in monkeys and people may be hormonal because satiating doses of CCK administered intravenously appear to be within the physiological range (Lieverse et al., 1993b; Moran and McHugh, 1982).

Central Processing of the Peripheral Signal

Neurological Experiments

The brainstem caudal to a chronic supracollicular decerebration is sufficient to process the peripheral effect of CCK-8 into an inhibitory control of eating (Grill and Smith, 1988). In the intact brain, the forebrain influences the caudal hindbrain network and may carry on additional processing. For example, bilateral lesions of the paraventricular hypothalamic nucleus blocked the satiating effect of CCK-8 administered intraperitoneally (Crawley and Kiss, 1985), but this effect has not been replicated (Flanagan et al., 1992; Smith et al., 1981a). Clearly this issue deserves more experimental attention.

Pharmacological and Hormonal Experiments

There is suggestive evidence that serotonin and dopamine are involved in the central processing of CCK-8. Systemic administration of 8-hydroxy-2-(di-*n*-propylamino) tetralin (8-OH-DPAT) and of serotonergic antagonists that act preferentially at 5-HT$_{2C}$ receptors significantly attenuated the satiating potency of CCK-8 (Stallone et al., 1989; Poeschla et al., 1992, 1993). These results need to be followed up with experiments in which these drugs are administered centrally, preferably by local injection, in order to be sure that the effects are due to changes in central serotonin actions. (See Chapter 9 for further information about central serotonin and satiation.)

Södersten and his colleagues have provided evidence regarding central dopamine. They have shown that peripheral administration of CCK-8 is correlated with the release of dopamine from the forebrain and that dopaminergic antagonists attenuate the satiating effect of CCK-8 (see Bednar et al., 1991, 1992a,b).

Neurons that synthesize, store, and release CCK-8 are widely distributed in the brain, and there is considerable evidence that injection of CCK-8 into the lateral or third ventricles, anterior nucleus accumbens, or a variety of hypothalamic sites decreases food intake (Crawley and Corwin, 1994). It is not clear, however, if this is an anorectic, toxic, or satiating effect in the rat. The inhibitory effect of central CCK is mediated by CCK_A receptors (Baldwin, 1992; Schick et al., 1986, 1994). When CCK_A antagonists have been given intraventricularly to test the possibility that central CCK controls meal size, the results depend on the species tested: Meal size increased in pigs, sheep, fowl, and mice (Hirosue et al., 1993; Baldwin, 1992), but not in rats (Corp et al., 1997; Makovec et al., 1986). Although central administration of proglumide, a relatively nonspecific antagonist that binds to CCK_A and CCK_B receptors, prior to a meal increases meal size in rats (Corp et al., 1997; Lindén et al., 1989; Schwartz et al., 1988), the failure of more potent and selective antagonists given alone or together to produce the same effect makes it difficult to interpret the positive results with proglumide.

In addition to a possible independent role in the control of meal size, these central CCK neurons could be involved in processing the peripheral signal produced by CCK-8, but the attempts to demonstrate this have given mixed results. Lateral ventricular injection of proglumide reduced the satiating effect of peripheral CCK-8 (Corp et al., 1997; Lindén et al., 1989), but ventricular administration of more specific and potent CCK_A and CCK_B antagonists alone or in combination had no effect (Corp et al., 1997).

Two hormones secreted in the periphery appear to act centrally to affect the potency of peripheral CCK-8. Estrogen enhances the potency of CCK-8 after systemic administration (Geary et al., 1994; Lindén et al., 1990; Butera et al., 1993a) and after local injection into the paraventricular nucleus (Butera et al., 1993b). Insulin also enhances the potency of CCK-8 after intraventricular administration in baboons (Figlewicz et al., 1986b) and rats (Riedy et al., 1995).

It will be interesting to learn if the recently discovered Ob-protein (leptin) also enhances the potency of peripheral CCK-8 as part of its inhibition of food intake (Campfield et al., 1995; Zhang et al., 1994).

Summary of Central Processing

The current information about the sites and neuroendocrine and neurochemical mechanisms involved in the central processing of the satiating potency of peripheral CCK-8 is fragmentary. What little we know is accessible to further analysis with current techniques. The results from further work are relevant to a major problem in satiation: how the satiating potency of a mechanism of a short-term, direct control of meal size, such as peripheral CCK, is modulated by mechanisms, such as the central actions of insulin and estrogen, that exert long-term, indirect control of meal size and body weight (see Smith [1996] and Chapter 1 for the difference between direct and indirect controls).

Bombesin-Like Peptides Inhibit Food Intake

Bombesin, an amphibian tetradecapeptide (Anastasi et al., 1971), was the next peptide to be shown to inhibit food intake without toxic side effects. The work with bombesin (BN) grew out of the work with CCK. Because peripheral administration of BN released endogenous CCK from the small intestine (Erspamer et al., 1974), Gibbs began a series of experiments with BN in 1977 to test whether intestinal CCK had a satiating effect. Bombesin (ip) produced a dose-related inhibition of food intake, apparently confirming a satiating effect of endogenous CCK (Gibbs et al., 1979). When CCK-8 and BN were compared under the same conditions, however, an important difference was noted: BN decreased intake of liquid and solid food equally, while CCK-8 was two to five times more potent in decreasing intake of liquid food than of solid food (Gibbs et al., 1979). The difference led the authors to conclude that the inhibitory action of BN was not mediated solely by endogenous CCK, and subsequent work confirmed this conclusion. This new effect of BN was extended to other rodents (Taylor et al., 1985a; Miceli and Malsbury, 1985), cats (Bado et al., 1989), fowl (Denbow, 1989), baboons (Woods et al., 1983) and humans (Lieverse et al., 1993a; Muurahainen et al., 1993).

Gastrin-Releasing Peptide and Neuromedin B

The presence of bombesin-like immunoreactivity (BLI) in mammalian gut (Walsh et al., 1979) led Gibbs et al. (1979) to suggest that endogenous BN in the stomach was a putative satiety signal. This hypothesis was revised when it was discovered that BN was not present in mammals and that BLI was due to two families of mammalian homologues of BN, gastrin-releasing peptide ($GRP_{1\text{-}27}$ and $GRP_{18\text{-}27}$) and neuromedin B ($NMB_{1\text{-}10}$, $NMB_{1\text{-}30}$, and $NMB_{1\text{-}32}$). These two peptide families were shown to have genetically distinct receptors that bound GRP or NMB preferentially. Both receptors, however, bound BN with high affinity (see Chapter 6 for further details about these receptors).

When $GRP_{1\text{-}27}$, $GRP_{18\text{-}27}$, and $NMB_{1\text{-}10}$ were tested for an effect on food intake, they decreased intake significantly (DiPoala and Gibbs, 1985; Figlewicz et al., 1985; Stein and Woods, 1982). When both peptides were administered under the same experimental conditions, GRP was significantly more potent than NMB (Kirkham et al., 1995b; Ladenheim et al., 1996).

Although GRP was named for its release of gastrin, this effect does not mediate the inhibitory effect of GRP on food intake because large doses of gastrin do not change food intake (Lorenz et al., 1979). The potency of GRP appears to be a function of test diet. GRP decreased intake of a high fat diet more than intake of a low fat, high carbohydrate diet (Stratford et al., 1995b; Stein and Woods, 1982).

The potency of GRP and NMB is also changed by a recent meal (Kirkham et al., 1995b). For example, when a test meal occurred 30 minutes after the end of a 30-minute meal, the inhibitory potency of GRP or NMB was significantly larger than when the test meal occurred without the preceding meal.

The effect of prefeeding was a function of the doses of each peptide. Prefeeding increased the potency of GRP so that the maximal inhibition was achieved with the lowest dose tested (2.9 nmol/kg). The effect of prefeeding on the potency of NMB, however, was biphasic: Prefeeding increased the two largest doses (28.3 and 56.6 nmol/kg) significantly more than the two smaller doses (7.1 and 14.1 nmol/kg).

Although GRP is more potent than NMB, GRP was significantly less potent than molar equivalent doses of BN when inhibitory potency was measured by 30-minute intakes (Stratford et al., 1995b; Ladenheim et al., 1996). When potency was measured by 15-minute intakes, however, low molar-equivalent doses of BN and $GRP_{1\text{-}27}$ had equal potency (Stratford et al., 1995b). Larger doses of BN were significantly more potent primarily because they produced a longer duration of the suppression of intake than equally large doses of GRP (Stratford et al., 1995b).

Because BN activated GRP and NMB receptors, it was possible that the simultaneous activation of GRP and NMB receptors by BN, but activation of only GRP receptors by GRP, was the basis for the potency difference between the larger doses of BN and GRP. If this were so, then administration of a mixture of GRP and NMB that was equimolar with BN should produce inhibition of intake equivalent to that of BN (Stratford et al., 1995b; Stratford et al., 1994). Two experiments have tested this idea, and both obtained the predicted result (Stratford et al., 1996; Ladenheim et al., 1996) (see Chapter 6 for additional discussion).

Is the Inhibition a Satiating Effect?

The behavioral specificity of the inhibitory effect of BN was investigated in the original report (Gibbs et al., 1979). The dose of BN that produced approximately 60% suppression of liquid food intake had no significant effect on water intake. When BN decreased food intake, the normal satiety sequence appeared, but no abnormal behaviors were observed. Furthermore, when BN decreased sham feeding, it elicited the satiety sequence (Martin and Gibbs, 1980).

It was possible to produce a CTA when a large dose of BN was used as the UCS (Deutsch and Parsons, 1981), but in this range of doses BN loses its specificity because it inhibits water intake as well as food intake. When smaller doses of BN were used that inhibited food intake, but not water intake, BN was not efficacious as a UCS for the formation of a CTA (Kulkosky, 1985; Kulkosky et al., 1981).

Intraperitoneal administration of BN and GRP did not produce aversive behaviors in taste reactivity tests (Flynn, 1992). Bombesin changed the microstructure of licking in a manner consistent with a satiating, rather than a toxic, effect (Stratford et al., 1995b, 1996). Bombesin increased the rate of decay of licking without changing the initial rate of licking, and it decreased the number of bursts and clusters of licking without decreasing their size.

Doses of BN as large as 100 μg/kg did not increase plasma oxytocin in rats, but doses as small as 4 μg/kg increased plasma vasopressin in monkeys (Verbalis et al., 1989). The effect of GRP or NMB on the release of neurohyphophyseal hormones has not been reported.

Finally, in double-blind human experiments, slow intravenous infusion of BN or GRP decreased food intake or calorie intake significantly without producing self-reports of toxic side effects (Gutzwiller et al., 1994; Lieverse et al., 1993a; Muurahainen et al., 1993).

These results provide convergent evidence that the inhibitory effect of low doses of BN on food intake is a satiating effect and is not due to a hypothetical, subclinical sickness (Table 5.3). There is, however, one exception. Davidson and Carretta (1993) reported that rats could discriminate the effect of BN from 1 hour of food deprivation. This occurred in the same conditions in which rats could not discriminate CCK-8 from the 1 hour of food deprivation (see above).

The specificity of the weak, inhibitory effect of NMB on food intake has not been evaluated sufficiently to reach a conclusion, but the effect of GRP has. In every test of specificity that has been investigated, doses of GRP that decreased food intake did not have significant side effects (Table 5.3).

TABLE 5.3.
Tests of Behavioral Specificity of Bombesin and Gastrin-Releasing Peptide

Test	BN	GRP
Water intake	Yes	Yes
Satiety sequence		
Real feeding	Yes	Yes
Sham feeding	Yes	NA
Conditioned taste aversion	Yes	NT
Conditioned discrimination	No	NT
Taste reactivity	Yes	Yes
Microstructure of licking	Yes	Yes
Neuroendocrine correlates	Yes	NT
Human self-reports	Yes	Yes

"Yes" means that at least one dose of the peptide that inhibited food intake significantly was tested and no nonspecific or toxic effect was observed. "No" means that a nonspecific effect was observed. "NT" means that the peptide has not been tested. "NA" means that GRP did not decrease sham feeding, so the satiety sequence could not occur (Smith et al., 1994). See text for references.

Is the Satiating Effect Physiological?

Because BN is not present in mammals, its satiating effect cannot be a physiological effect. The satiating effect of peripherally administered BN can, however, be a model of the simultaneous stimulation of peripheral GRP and NMB receptors by gastrointestinal GRP and NMB if these peptides are released simultaneously during a meal. Unfortunately, the condition(s) under which such simultaneous release occurs has not been identified (see Mode of Action, below).

The best test of the physiological status of the satiating effect of peripheral GRP is to administer an antagonist that is selective for GRP receptors prior to a meal (see discussion of this issue in Cholecystokinin Inhibits Food Intake, above). If peripheral GRP provides an inhibitory control of meal size under these conditions, then meal size will increase after antagonist pretreatment. This experiment has been performed twice (Kirkham et al., 1994, 1995a), but meal size has not changed despite the fact that the antagonist blocked the inhibitory effect of exogenous BN and GRP under the same conditions. Thus, there is no evidence to support the hypothesis at this time.

Before the hypothesis can be considered to be refuted, however, further experiments should be done. If the mode of action of GRP is synaptic (see below), then effective antagonism of the effect of endogenous GRP will require larger doses of antagonist than those required to block the effect of local or circulating GRP produced by intraperitoneal administration. Such large doses will require potent and specific antagonists that lack partial agonist activity (Kirkham et al., 1994, 1995a).

Site and Mechanism of Action

When BN, GRP, and NMB are administered peripherally, they act in the periphery or in circumventricular organs because their size prevents them from crossing the blood–brain barrier. The major site of action of BN is in the upper abdominal organs perfused by the celiac artery because BN was significantly more potent after injections into the celiac artery than after injections into the superior mesenteric artery, hepatic-portal vein, inferior vena cava, or peritoneum (Yen et al., 1989; Miesner et al., 1989; Greenberg et al., 1988; Kirkham et al., 1991).

Bombesin differs from CCK-8, but not from CCK-33, by having a significant satiating potency after hepatic-portal administration (Greenberg et al., 1988). Presumably the larger size of BN accounts for its escape from hepatic inactivation by uptake or degradation, but this remains to be demonstrated. It is unfortunate that the effect of mode of administration on the satiating potency of GRP and NMB has not been investigated because they have molecular forms that differ in size, a characteristic that produced significant differences in the site of action of CCK-8 and CCK-33 (see above).

The satiating action of BN in the upper abdomen is transmitted to the brain over visceral afferent fibers of the abdominal vagus and of dorsal root ganglion cells up to the level of the third thoracic segment (Table 5.4) (Stuckey et al., 1985). Either the vagal or the spinal fibers are sufficient for normal satiating potency of BN; lesions of both groups of fibers are necessary to abolish the effect of BN (Stuckey et al., 1985; Smith et al., 1981b; Taylor et al., 1985b; Morley et al., 1982).

The level of the spinal afferent lesion is further evidence that the site of action is the upper abdomen. Cord section at T6 combined with total abdominal vagotomy did not change the effect of BN (Stuckey, unpublished observation). When bilateral dorsal rhizotomies of T3, T4, and T5 were added to total abdominal vagotomy and cord section at T6, however, the satiating effect of BN was abolished (Table 5.4) (Stuckey et al., 1985). There is little information about the afferent pathways required for the effects of GRP and NMB other than the preliminary report that total

TABLE 5.4.
Effect of cord section and dorsal rhizotomy, abdominal vagotomy, or the combination of cord section and dorsal rhizotomy plus abdominal vagotomy on inhibition of first meal size by BN

Group	n	0	2	4	8
Control	14	29.7 ± 1.2	23.3 ± 1.3 (−21 ± 4)	20.7 ± 1.9 (−31 ± 5)	16.8 ± 1.8 (−45 ± 4)
Cord section and dorsal rhizotomy	5	31.3 ± 4.1	21.0 ± 4.4 (−34 ± 8)	17.8 ± 4.8 (−44 ± 10)	19.6 ± 6.1 (−37 ± 13)
Cord section and dorsal rhizotomy and vagotomy	4	16.4 ± 0.8	18.0 ± 3.3 (+7 ± 13)*	18.0 ± 3.8 (+7 ± 12)*	15.2 ± 2.9 (−8 ± 19)*
Vagotomy	7	15.9 ± 1.0	12.3 ± 1.6 (−22 ± 11)	11.4 ± 1.2 (−29 ± 6)	7.4 ± 1.2 (−53 ± 8)

Values are mean intakes (in ml ± SEM). Values in parentheses are percent changes from saline injection (0 dose) ± SEM. Values for 0 dose were compiled from days on which rats received saline injections. There were seven controls for cord section and dorsal rhizotomy alone and seven controls for cord section, dorsal rhizotomy, and vagotomy; since there was no difference in intakes for these two control groups, their data were pooled. From Stuckey et al. (1985), with permission of the publisher.

*Significantly different from other groups, $p < 0.01$.

abdominal vagotomy did not change the satiating potency of GRP (Perez et al., 1994).

Mode of Action

GRP and NMB are synthesized, stored, and released by enteric neurons in the stomach, small intestine, and colon. Thus, their mode of action is either synaptic or local (neurocrine). It is very unlikely that endogenous GRP or NMB has hormonal actions except in the case of small-cell lung carcinomas that secrete GRP (Holst et al., 1989). This nonhormonal mode of action is consistent with the significantly larger potency of BN after celiac artery injection than after hepatic-portal or inferior vena caval administration (see above and Chapter 6).

Central Processing of the Peripheral Signal

Both vagal and spinal visceral afferent pathways project to the caudal brainstem. Lesions of this region that include the area postrema and the caudal portion of the NTS block the satiating effect of peripherally administered BN (Ladenheim and Ritter, 1989).

Flynn (1992) has shown that the caudal brainstem of the chronic decerebrate rat is sufficient to mediate the satiating effect of BN and GRP administered intraperitoneally. If the forebrain is necessary for processing the stimulus produced by peripheral BN, the sites have not been identified yet. In the three studies performed thus far, bilateral lesions of the ventromedial (West et al., 1982; Geary et al., 1986) and dorsomedial (Bellinger and Bernardis, 1984) hypothalamic nuclei did not change the effect of BN.

Merali and colleagues (1988) have implicated the extensive central network of neurons containing GRP or NMB in the processing of peripheral BN. They showed that administration of a specific antiserum to BN or a potent, GRP-preferring antagonist into the third ventricle attenuated the effect of peripheral BN. These investigators also have preliminary evidence for a role of corticotropin-releasing factor in central processing because pretreatment with a corticotropin-releasing factor antagonist reduced the effect of peripheral BN (Merali et al., 1994).

Central Bombesin-Like Peptides and Meal Size

Intraventricular Administration

Although initial experiments using intraventricular administration of BN and GRP demonstrated that BN and GRP were significantly more potent for inhibiting food intake after intraventricular than after peripheral administration, the central effect appeared to be nonspecific because water intake was also inhibited, grooming increased, and resting decreased (Morley and Levine, 1981; Kulkosky et al., 1982a,b). Nonspecific effects were also observed after central administration in hamsters and pigs (Parrott and Baldwin, 1982; Miceli and Malsbury, 1985), but apparently not in baboons (Figlewicz et al., 1986a). All of these effects were a result of BN because equivalent doses of the biologically inactive analogue of BN (D-Trp^8BN) had no effect on behavior (Kulkosky et al., 1982b).

The effects of BN on food intake and grooming were blocked by intraventricular administration of a specific antiserum to BN and by a GRP-preferring antagonist (Merali et al., 1988). When a GRP antagonist was injected prior to a meal, meal size increased (Merali et al., 1993). This suggests that central GRP receptors in the region of the third ventricle are involved in the control of meal size.

The area postrema and underlying NTS are necessary for the inhibitory effect of BN on food intake after intraventricular administration because lesion of these structures abolished the effect (Ladenheim and Ritter, 1989; LaCour et al., 1986). Recall that the same lesion also reduced the inhibitory effect of peripheral BN (Ladenheim and Ritter, 1989).

Effects in Forebrain

The lack of behavioral specificity after intraventricular administration was overcome when BN was microinjected into specific sites, such as the lateral hypothalamus (Stuckey and Gibbs, 1982), the hypothalamic paraventricular nucleus (PVN) (Willis et al., 1984), the amygdala and periaqueductal gray (Kyrkouli et al., 1987), and the substantia nigra (Calisher and Avery, 1984).

The positive result with microinjection of BN into the PVN has been complemented recently by the observation that the release of a BN-like peptide, presumably GRP, from push–pull perfusion of the anterior hypothalamus including the PVN, decreased during periods of food intake and increased back to or above control levels during intervals of nonfeeding (Plamondon and Merali, 1994). The decreased release during periods of feeding is consistent with disinhibition from the effect of GRP. If this correlation is robust, this will be strong evidence for a physiological role of GRP in the PVN in the control of food intake.

Effects in Hindbrain

Injection of BN into the fourth ventricle decreased food intake in a behaviorally specific manner in neurologically intact and chronic decerebrate rats (Ladenheim and Ritter, 1988; Flynn and Robillard, 1992; Flynn, 1989). Microinjection of BN into the NTS also inhibits food intake (Johnson and Merali, 1988; deBeaurepaire and Suaudeau, 1988).

To determine if endogenous GRP and GRP receptors in this region are involved in the control of meal size, GRP antagonists have been injected into the fourth ventricle prior to a meal in two experiments (Stratford et al., 1995a; Flynn, 1993). Meal size increased in both studies.

Summary of Central Bombesin-Like Peptides

The various results are evidence that central BN-like peptides are involved in the physiological control of meal size. The demonstration of increased meal size after treatment with GRP-preferring antagonists is particularly impressive and is evidence that GRP is exerting an inhibitory influence in the central network that controls meal size.

Although the current evidence from central experiments is consistent with a role for GRP, it does not exclude central NMB. The physiological role of central NMB will be evaluated best when potent and specific NMB receptor antagonists become available (see Chapter 6).

Formulation

This review of the satiating effects of CCK and BN-like peptides demonstrates how the analysis of an inhibition of food intake produced by the peripheral administration of these peptides has led to different conclusions about the physiological function of the endogenous peptides. In the case of CCK, research demonstrated that a satiating effect of intestinal CCK was a physiological function of the peptide under a variety of conditions and in numerous animals, although not yet in humans. In contrast, the physiological status of peripheral BN-like peptides is still uncertain despite being tested under experimental conditions in which the satiating effect of intestinal CCK can be demonstrated.

Less extensive evidence concerning the role of these peptides in the brain for the control of meal size suggests the opposite conclusion. There is more evidence for a physiological role of central BN-like peptides than for CCK. But the central effects of both peptides need much more investigation before firm conclusions can be reached.

ACKNOWLEDGMENTS

We thank B. Cooley and T. Popiel for expert processing of the manuscript and Dr. Nori Geary for helpful criticism of earlier versions. Our research was supported by grants from the National Institutes of Health (MH 40010, MH 00149, and DK 33248).

REFERENCES

Anastasi, A., V. Erspamer, and M. Bucci. Isolation and structure of bombesin and alytesin, two analogous active peptides from the skin of the European amphibians *Bombina* and *Alytes*. *Experientia* 27:166–167, 1971.

Antin, J., J. Gibbs, J. Holt, R. C. Young, and G. P. Smith. Cholecystokinin elicits the complete behavioral sequence of satiety in rats. *J. Comp. Physiol. Psychol.* 89:784–790, 1975.

Bado, A., M. J. M. Lewin, and M. Dubrasquet. Effects of bombesin on food intake and gastric acid secretion in cats. *Am. J. Physiol.* 256:R181-R186, 1989.

Baldwin, B. A. *CCK as a putative satiety factor in farm animals.* In C. T. Dourish, S. J. Cooper, S. D. Iversen, and L. L. Iversen (eds.): *Multiple Cholecystokinin Receptors in the CNS.* Oxford: Oxford University Press, 1992, pp. 206–221.

Bednar, I., G. Forsberg, A. Lindén, G. A. Qureshi, and P. Södersten. Involvement of dopamine in inhibition of food intake by cholecystokinin octapeptide in male rats. *J. Neuroendocrinol.* 3:491–496, 1991.

Bednar, I., G. A. Qureshi, and P. Södersten. A comparison between the effect of cholecystokinin octapeptide and apormorphine on ingestion of intraorally administered sucrose in male rats. *J. Neuroendocrinol.* 4:727–734, 1992a.

Bednar, I., G. A. Qureshi, and P. Södersten. Evidence that release of dopamine in the brain is involved in the inhibitory effect of cholecystokinin octapeptide on ingestion of intraorally administered sucrose in male rats. *J. Neuroendocrinol.* 4:735–741, 1992b.

Bellinger, L. L., and L. L. Bernardis. Suppression of feeding by cholecystokinin but not bombesin is attenuated in dorsomedial hypothalamic nuclei lesioned rats. *Peptides* 5:547–552, 1984.

Brenner, L., and R. C. Ritter. Peptide cholecystokinin receptor antagonist increases food intake in rats. *Appetite* 24:1–9, 1995.

Butera, P. C., D. M. Bradway, and N. J. Cataldo. Modulation of the satiety effect of cholecystokinin by estradiol. *Physiol. Behav.* 53:1235–1238, 1993a.

Butera, P. C., M. Xiong, and R. Davis. Modulation of the satiety effect of CCK by central implants of dilute estradiol. *Soc. Neurosci. Abstr.* 19:1823, 1993b.

Cabanac, M., and C. Zhao. Postingestive alliesthesia produced by exogenous cholecystokinin and blocked by abdominal vagotomy. *Am. J. Physiol.* 266: R633–R637, 1994.

Calingasan, N., S. Ritter, R. Ritter, and L. Brenner. Low-dose near-celiac arterial cholecystokinin supresses food intake in rats. *Am. J. Physiol.* 263:R572–R577, 1992.

Calisher, S. B., and D. D. Avery. Injections of bombesin into the substantia nigra produce hypothermia and hypophagia in food-deprived rats. *Neuropharmacology* 23:1201–1206, 1984.

Campfield, L. A., F. J. Smith, Y. Guisez, R. Devos, and P. Burn. Recombinant mouse OB protein: Evidence for a peripheral signal linking adiposity and central neural networks. *Science* 269:546–549, 1995.

Corp, E., M. Curcio, J. Gibbs, and G. P. Smith. The effect of centrally administered CCK receptor antagonists on food intake in rats. *Physiol. Behav.* 61:823–827, 1997.

Corwin, R. L., and C. R. Schuster. Anorectic specificity as measured in a choice paradigm in rhesus monkeys. *Pharmacol. Biochem. Behav.* 45:131–141, 1993.

Corwin, R. L., W. L. Woolverton, and C. R. Schuster. Effects of cholecystokinin, *d*-amphetamine and fenfluramine in rats trained to discriminate 3 from 22 hr. of food deprivation. *J. Pharmacol. Exp. Ther.* 253:720–728, 1990.

Cox, J. E., G. S. Perdue, and W. J. Tyler. Suppression of sucrose intake by continuous near-celiac and intravenous cholecystokinin infusions in rats. *Am. J. Physiol.* 268:R150–R155, 1995.

Crawley, J. N., and R. L. Corwin. Biological actions of cholecystokinin. *Peptides* 15:731–755, 1994.

Crawley, J. N., and J. Z. Kiss. Paraventricular nucleus lesions abolish the inhibition of feeding induced by systemic cholecystokinin. *Peptides* 6:927–935, 1985.

Davidson, T. L., and J. C. Carretta. Cholecystokinin, but not bombesin, has interoceptive sensory consequences like 1-hour food deprivation. *Physiol. Behav.* 53:737–745, 1993.

Davis, J. D., G. P. Smith, and T. M. Kung. Cholecystokinin changes the duration but not the rate of licking in vagotomized rats. *Behav. Neurosci.* 109:991–996, 1995.

deBeaurepaire, R., and C. Suaudeau. Anorectic effect of calcitonin, neurotensin and bombesin infused in the area of the rostral part of the nucleus of the tractus solitarius in the rat. *Peptides* 9:729–733, 1988.

Denbow, D. M. Centrally and peripherally administered bombesin decreases food intake in turkeys. *Peptides* 10:275–279, 1989.

Deutsch, J. A., and W. T. Hardy. Cholecystokinin produces bait shyness in rats. *Nature* 266:196, 1977.

Deutsch, J. A., and S. L. Parsons. Bombesin produces taste aversion in rats. *Behav. Neural Biol.* 31:110–113, 1981.

DiPoala, J. A., and J. Gibbs. Neuromedin C inhibits food intake in rats. *Soc. Neurosci. Abstr.* 11:38, 1985.

Doyle, J. W., M. M. Wolfe, and J. E. McGuigan. Hepatic clearance of gastrin and cholecystokinin peptides. *Gastroenterology* 87:60–68, 1984.

Drewe, J., A. Gadient, L. C. Rovati, and C. Beglinger. Role of circulating cholecystokinin in control of fat-induced inhibition of food intake in humans. *Gastroenterology* 102:1654–1659, 1992.

Eckel, L. A., and K. Ossenkopp. Cholecystokinin reduces sucrose palatability in rats: Evidence in support of a satiety effect. *Am. J. Physiol.* 267:R1496–R1502, 1994.

Erspamer, V., P. Improta, P. Melchiorri, and N. Sopranzi. Evidence of cholecystokinin release by bombesin in the dog. *Br. J. Pharmacol.* 52:227–232, 1974.

Falasco, J. D., G. P. Smith, and J. Gibbs. Cholecystokinin suppresses sham feeding in the rhesus monkey. *Physiol. Behav.* 23:887–890, 1979.

Figlewicz, D. P., A. Sipols, D. Porte, Jr., and S. C. Woods. Intraventricular bombesin can decrease single meal size in the baboon. *Brain Res. Bull.* 17:535–537, 1986a.

Figlewicz, D. P., L. J. Stein, S. C. Woods, and D. Porte, Jr. Acute and chronic gastrin-releasing peptide decreases food intake in baboons. *Am. J. Physiol.* 248:R578–R583, 1985.

Figlewicz, D. P., D. B. West, L. J. Stein, S. C. Woods, and D. Porte, Jr. Insulin alters the sensitivity of baboons to CCK-induced single meal suppression. *Am. J. Physiol.* 250:R856–R860, 1986b.

Flanagan, L. M., J. Dohanics, J. G. Verbalis, and E. M. Stricker. Gastric motility and food intake in rats after lesions of hypothalamic paraventricular nucleus. *Am. J. Physiol.* 263:R39–R44, 1992.

Flanagan, L. M., J. G. Verbalis, and E. M. Stricker. Naloxone potentiation of effects of cholecystokinin and lithium chloride on oxytocin secretion, gastric motility and feeding. *Neuroendocrinology* 48:668–673, 1988.

Flynn, F. W. Fourth ventricle bombesin injection suppresses ingestive behaviors in rats. *Am. J. Physiol.* 256:R590–R596, 1989.

Flynn, F. W. Caudal brain stem systems mediate effects of bombesin-like peptides on intake in rats. *Am. J. Physiol.* 262:R39–R45, 1992.

Flynn, F. W. Fourth ventricular injection of selective bombesin receptor antagonists facilitates feeding in rats. *Am. J. Physiol.* 264:R218–R221, 1993.

Flynn, F. W., and L. Robillard. Inhibition of ingestive behavior following fourth ventricle bombesin injection in chronic decerebrate rats. *Behav. Neurosci.* 106:1011–1014, 1992.

French, S. J., A. Bergin, C. P. Sepple, N. W. Read, and L. Rovati. The effects of loxiglumide on food intake in normal weight volunteers. *Int. J. Obesity* 18:738–741, 1994.

Geary, N., H. R. Kissileff, F. X. Pi-Sunyer, and V. Hinton. Individual, but not simultaneous, glucagon and cholecystokinin infusions inhibit feeding in men. *Am. J. Physiol.* 262:R975–R980, 1992.

Geary, N., G. P. Smith, and J. Gibbs. Pancreatic glucagon and bombesin inhibit meal size in ventromedial hypothalamus-lesioned rats. *Regul. Pept.* 15:261–268, 1986.

Geary, N., D. Trace, B. McEwen, and G. P. Smith. Cyclic estradiol replacement increases the satiety effect of CCK-8 in ovariectomized rats. *Physiol. Behav.* 56:281–289, 1994.

Gibbs, J., D. J. Fauser, E. A. Rowe, E. T. Rolls, and S. P. Maddison. Bombesin suppresses feeding in rats. *Nature* 282:208–210, 1979.

Gibbs, J., R. C. Young, and G. P. Smith. Cholecystokinin decreases food intake in rats. *J. Comp. Physiol. Psychol.* 84:488–495, 1973a.

Gibbs, J., R. C. Young, and G. P. Smith. Cholecystokinin elicits satiety in rats with open gastric fistulas. *Nature* 245:323–325, 1973b.

Gores, G. J., N. F. LaRusso, and L. J. Miller. Hepatic processing of cholecystokinin peptides. I. Structural specificity and mechanism of hepatic extraction. *Am. J. Physiol.* 250:G344–G349, 1986.

Greenberg, D., P. A. Foelsch, P. M. Perez, G. P. Smith, and J. Gibbs. Hepatic-portal infusions of bombesin reduce food intake in rats. *Ann. N.Y. Acad. Sci.* 547:502–504, 1988.

Greenberg, D., G. P. Smith, and J. Gibbs. Infusion of CCK-8 into the hepatic-portal vein fails to reduce food intake in rats. *Am. J. Physiol.* 252:1015–1018, 1987.

Greenberg, D., N. I. Torres, G. P. Smith, and J. Gibbs. The satiating effects of fats is attenuated by the cholecystokinin antagonist lorglumide. *Ann. N.Y. Acad. Sci.* 575:517–520, 1989.

Grill, H. J. Introduction: Physiological mechanisms in conditioned taste aversions. *Ann. N.Y. Acad. Sci.* 443:67–88, 1985.

Grill, H. J., and R. Norgren. The taste reactivity test. I. Mimetic responses to gustatory stimuli in neurologically normal rats. *Brain Res.* 143:263–279, 1978.

Grill, H. J., and G. P. Smith. Cholecystokinin decreases sucrose intake in chronic decerebrate rats. *Am. J. Physiol.* 254:R853–R856, 1988.

Gutzwiller, J.-P., J. Drewe, P. Hildebrand, L. Rossi, J. Z. Lauper, and C. Beglinger. Effect of intravenous human gastrin-releasing peptide on food intake in humans. *Gastroenterology* 106:1168–1173, 1994.

Hirosue, Y., A. Inui, A. Teranishi et al. Cholecystokinin octapeptide analogues suppress food intake via central CCK-A receptors in mice. *Am. J. Physiol.* 265:R481–R486, 1993.

Holst, J. J., M. Hansen, E. Bork, and T. W. Schwartz. Elevated plasma concentrations of C-flanking gastrin-releasing peptide in small-cell lung cancer. *J. Clin. Oncol.* 7:1831–1838, 1989.

Holt, J., J. Antin, J. Gibbs, R. C. Young, and G. P. Smith. Cholecystokinin does not produce bait shyness in rats. *Physiol. Behav.* 12:497–498, 1974.

Johnson, S. A., and Z. Merali. Specific neuroanatomical and neurochemical correlates of grooming and satiety effects of bombesin. *Peptides* 1:233–244, 1988.

Kirkham, T. C., J. Gibbs, and G. P. Smith. The satiating effect of bombesin is mediated by receptors perfused by the coeliac artery. *Am. J. Physiol.* 261: R614–R618, 1991.

Kirkham, T. C., J. Gibbs, G. P. Smith, and N. Geary. Meal pattern analysis in rats reveals partial agonist activity of the bombesin receptor antagonist BW 2258U89. *Pharmacol. Biochem. Behav.* 52:101–106, 1995a.

Kirkham, T. C., S. Perez, and J. Gibbs. Prefeeding potentiates anorectic actions of neuromedin B and gastrin releasing peptide. *Physiol. Behav.* 58:1175–1179, 1995b.

Kirkham, T. C., C. A. Walsh, J. Gibbs, G. P. Smith, J. Leban, and J. McDermed. A novel bombesin receptor antagonist selectively blocks the satiety action of peripherally administered bombesin. *Pharmacol. Biochem. Behav.* 48:809–811, 1994.

Kissileff, H. R., F. X. Pi-Sunyer, J. Thornton, and G. P. Smith. Cholecystokinin-octapeptide (CCK-8) decreases food intake in man. *Am. J. Clin. Nutr.* 34:154–160, 1981.

Kulkosky, P. J. Conditioned food aversions and satiety signals. *Ann. N.Y. Acad. Sci.* 443:330–347, 1985.

Kulkosky, P. J., J. Gibbs, and G. P. Smith. Behavioral effects of bombesin administration in rats. *Physiol. Behav.* 28:505–512, 1982a.

Kulkosky, P. J., J. Gibbs, and G. P. Smith. Feeding supression and grooming repeatedly elicited by intraventricular bombesin. *Brain Res.* 242:194–196, 1982b.

Kulkosky, P. J., L. Gray, J. Gibbs, and G. P. Smith. Feeding and selection of saccharin after injections of bombesin, LiCl, and NaCl. *Peptides* 2:61–64, 1981.

Kyrkouli, S. E., B. G. Stanley, and S. F. Leibowitz. Bombesin-induced anorexia—sites of action in the rat brain. *Peptides* 8:237–241, 1987.

LaCour, F. J., J. N. Kott, N. J. Kenney, and S. C. Woods. Intraventricular bombesin (BBS) reduces food intake of normal but not area-postrema-lesion rats. *Appetite* 7:273–274, 1986.

Ladenheim, E. E., and R. C. Ritter. Low-dose fourth ventricular bombesin selectively suppresses food intake. *Am. J. Physiol.* 255:R988–R992, 1988.

Ladenheim, E. E., and R. C. Ritter. Caudal hindbrain participation in suppression of feeding by central and peripheral bombesin. *Soc. Neurosci. Abstr.* 15:964, 1989.

Ladenheim, E. E., K. Wirth, and T. H. Moran. Receptor subtype mediation of feeding suppression by bombesin-like peptides. *Pharmacol. Biochem. Behav.* 54:705–711, 1996.

Lang, R. E., J. W. E. Heil, D. Ganten, K. Herman, T. Unger, and W. Rascher. Oxytocin unlike vasopressin is a stress hormone in the rat. *Neuroendocrinology* 37:314–316, 1983.

Lieverse, R. J., J. B. M. J. Jansen, A. van de Zwan et al. Bombesin reduces food intake in lean man by a cholecystokinin-independent mechanism. *J. Clin. Endocrinol. Metab.* 76:1495–1498, 1993a.

Lieverse, R. J., J. B. M. J. Jansen, A. v. Zwan, L. Samson, A. A. M. Masclee, and C. B. H. W. Lamers. Effects of a physiological dose of cholecystokinin on food intake and postprandial satiation in man. *Regul. Pept.* 43:83–89, 1993b.

Lieverse, R. J., A. A. M. Masclee, J. B. M. J. Jansen, L. Rovati, and C. B. H. W. Lamers. Satiety effects of the type A CCK receptor antagonist loxiglumide in lean and obese women. *Biol. Psychiatr.* 37:331–335, 1995.

Lindén, A., K. Uvnäs-Moberg, G. Forsberg, I. Bednar, and P. Södersten. Involvement of cholecystokinin in food intake: III. Oestradiol potentiates the inhibitory effect of cholecystokinin octapeptide on food intake in ovariectomized rats. *J. Neuroendocrinol.* 2:797–801, 1990.

Lindén, A., K. Uvnäs-Moberg, G. Forsberg, and P. Södersten. Plasma concentrations of cholecystokinin octapeptide and food intake in male rats treated with cholecystokinin octapeptide. *J. Endocrinol.* 121:59–65, 1989.

Lorenz, D. N., G. Kreielsheimer, and G. P. Smith. Effect of cholecystokinin, gastrin, secretin and GIP on sham feeding in the rat. *Physiol. Behav.* 23:1065–1072, 1979.

Makovec, F., M. Bani, R. Chiste, L. Revel, L. C. Rovati, and L. Setnikar. Different peripheral and central antagonistic activity of new glutaramic acid derivatives on satiety induced by cholecystokinin in rats. *Regul. Pept.* 16:281–290, 1986.

Martin, C. F., and J. Gibbs. Bombesin elicits satiety in sham feeding rats. *Peptides* 1:131–134, 1980.

Mayer, J. Glucostatic mechanisms in regulation of food intake. *N. Engl. J. Med.* 249:13–16, 1953.

McCann, M., J. G. Verbalis, and E. M. Stricker. Capsaicin pretreatment attenuates multiple responses to cholecystokinin in rats. *J. Auton. Nerv. Syst.* 23:265–272, 1988.

McCann, M., J. G. Verbalis, and E. M. Stricker. LiCl and CCK inhibit gastric emptying and feeding and stimulate OT secretion in rats. *Am. J. Physiol.* 256:R463–R468, 1989.

Melville, L. D., D. Greenberg, G. P. Smith, J. Gibbs, and M. J. Russ. Intraperitoneally administered CCK-8 acts locally, not hormonally, to inhibit food intake. *Soc. Neurosci. Abstr.* 19:816, 1993a.

Melville, L. D., G. P. Smith, and J. Gibbs. Comparison of the satiating potencies of cholecystokinin-33 and cholecystokinin-8. *Pharmacol. Biochem. Behav.* 45: 85–87, 1993b.

Melzack, R., Z. Rosberger, M. L. Hollingsworth, and M. Therlwell. New approaches to measuring nausea. *Can. Med. Assoc. J.* 133:755–761, 1985.

Merali, Z., T. W. Moody, and D. Coy. Blockade of brain bombesin/GRP receptors increases food intake in satiated rats. *Am. J. Physiol.* 264:R1031–R1034, 1993.

Merali, Z., T. Moody, P. Kateb, and H. Piggins. Antagonism of satiety and grooming effects of bombesin by antiserum to bombesin and by [Tyr4,D-Phe12] bombesin: Central versus peripheral effects. *Ann. N.Y. Acad. Sci.* 547:489–492, 1988.

Merali, Z., H. Plamondon, P. Kent, and K. Banks. Corticotropin-releasing factor (CRF) antagonist blocks the autonomic endocrine, ingestive and behavioral effects of central bombesin (BN). *Soc. Neurosci. Abstr.* 20:1284, 1994.

Metzger, B. L., and B. C. Hansen. Cholecystokinin effects on feeding, glucose and pancreatic hormones in rhesus monkeys. *Physiol. Behav.* 30:509–518, 1983.

Miaskiewicz, S. L., E. M. Stricker, and J. G. Verbalis. Neurohypophyseal secretion in response to cholecystokinin but not meal-induced gastric distention in humans. *J. Clin. Endocrinol. Metab.* 68:837–843, 1989.

Miceli, M. O., and C. W. Malsbury. Effects of putative satiety peptides on feeding and drinking behavior in Golden Hamsters. *Behav. Neurosci.* 99:1192–1207, 1985.

Miesner, J. A., E. I. Yen, J. Gibbs, A. Weller, and G. P. Smith. Intravenous administration of cholecystokinin antagonist increases food intake in the rat. *Soc. Neurosci. Abstr.* 15:1280, 1989.

Moran, T. H., and P. R. McHugh. Cholecystokinin suppresses food intake by inhibiting gastric emptying. *Am. J. Physiol.* 242:R491–R497, 1982.

Morley, J., and A. Levine. Bombesin inhibits stress-induced eating. *Pharmacol. Biochem. Behav.* 14:149–151, 1981.

Morley, J. E., A. S. Levine, J. Kniep, and M. Grace. The effect of vagotomy on the satiety effects of neuropeptides and naloxone. *Life Sci.* 30:1943–1947, 1982.

Morrow, G. R. The assessment of nausea and vomiting; past problems, current issues, and suggestions for future research. *Cancer* 53:2267–2280, 1984.

Muurahainen, N. E., H. R. Kissileff, J. Thornton, and F. X. Pi-Sunyer. Intravenous infusion of bombesin reduces food intake in humans. *Am. J. Physiol.* 264: R350–R354, 1993.

Parrott, R. F., and B. A. Baldwin. Centrally administered bombesin produces effects unlike short-term satiety in operant feeding pigs. *Physiol. Behav.* 28: 521–524, 1982.

Perez, S., G. P. Smith, D. Greenberg, and J. Gibbs. Abdominal vagotomy does not block the inhibition of food intake by gastrin-releasing peptide in rats. *Soc. Neurosci. Abstr.* 20:588, 1994.

Pi-Sunyer, X., H. R. Kissileff, J. Thornton, and G. P. Smith. C-terminal octapeptide of cholecystokinin decreases food intake in obese men. *Physiol. Behav.* 29:627–630, 1982.

Plamondon, H., and Z. Merali. Push–pull perfusion reveals meal-dependent changes in the release of bombesin-like peptides in rat paraventricular nucleus. *Brain Res.* 668:54–61, 1994.

Poeschla, B., J. Gibbs, K. J. Simansky, and G. P. Smith. The 5-HT1A agonist 8-OH-DPAT attenuates the satiating action of cholecystokinin. *Pharmacol. Biochem. Behav.* 42:541–543, 1992.

Poeschla, B., J. Gibbs, and G. P. Smith. Cholecystokinin, human satiation, and eating disorders. Manuscript in preparation, 1997.

Poeschla, B. D., J. Gibbs, K. J. Simansky, D. Greenberg, and G. P. Smith. Cholecystokinin-induced satiety depends upon activation of 5-HT1C receptors. *Am. J. Physiol.* 264:R62–R64, 1993.

Reidelberger, R. D., and T. E. Solomon. Comparative effects of CCK-8 on feeding, sham feeding, and exocrine pancreatic secretion in rats. *Am. J. Physiol.* 251:R97–R105, 1986.

Reidelberger, R. D., G. Varga, R.-M. Liehr et al. Cholecystokinin suppresses food intake by a nonendocrine mechanism in rats. *Am. J. Physiol.* 267:R901–R904, 1994.

Riedy, C. A., M. Chavez, D. P. Figlewicz, and S. C. Woods. Central insulin enhances sensitivity to cholecystokinin. *Physiol. Behav.* 58:755–760, 1995.

Ritter, R. C., L. A. Brenner, and C. S. Tamura. Endogenous CCK and the peripheral neural substrates of intestinal satiety. *Ann. N.Y. Acad. Sci.* 713:255–267, 1994.

Rowe, J. W., R. L. Shelton, J. H. Helderman, R. E. Vistal, and G. L. Robertson. Influence of the emetic reflex on vasopressin release in man. *Kidney Int.* 16:729–735, 1979.

Schick, R. R., V. Schusdziarra, T. L. Yaksh, and V. L. W. Go. Brain regions where cholecystokinin exerts its effects on satiety. *Ann. N.Y. Acad. Sci.* 713:242–254, 1994.

Schick, R. R., T. L. Yaksh, and V. L. W. Go. Intracerebroventricular injections of cholecystokinin octapeptide suppress feeding in rats—Pharmacological characterization of this action. *Regul. Pept.* 14:277–291, 1986.

Schwartz, D. H., D. B. Dorfman, L. Hernandez, and B. G. Hoebel. Cholecystokinin: 1. CCK antagonists in the PVN induce feeding 2. Effects of CCK in the nucleus accumbens on extracellular dopamine turnover. In Wang, R. Y., and R. Schoenfeld (eds.): *Cholecystokinin Antagonists.* New York: Alan Rt. Liss, Inc., 1988, pp. 285–305.

Smith, G. P. Humoral hypotheses for the control of food intake. In Bray, G. (ed.): *Obesity in Perspective. Vol. 2. Pt. 2.* Bethesda: National Institute of Health, 1976, pp. 19–29.

Smith, G. P. Gut hormone hypothesis of postprandial satiety. In Stunkard, A. J., and E. Stellar (eds.): *Eating and Its Disorders.* New York: Raven Press, 1984, pp. 67–75.

Smith, G. P. The direct and indirect controls of meal size. *Neurosci. Biobehav. Rev.* 20:41–46, 1996.

Smith, G. P., D. Dorre, and L. Melville. CCK-33 inhibits food intake after intraportal and intravenous administration. *Soc. Neurosci. Abstr.* 22:17, 1996.

Smith, G. P., and Gibbs, J. The development and proof of the CCK hypothesis of satiety. In Dourish, C. T., S. J. Cooper, S. D. Iversen, and L. L. Iversen (eds.): *Multiple Cholecystokinin Receptors in the CNS.* Oxford: Oxford University Press, 1992, pp. 166–182.

Smith, G. P., J. Gibbs, A. J. Strohmayer, and P. E. Stokes. Threshold doses of 2-deoxy-D-glucose for hypoglycemia and feeding in rats and monkeys. *Am. J. Physiol.* 222:77–81, 1972.

Smith, G., and D. Greenberg. Hepatic-portal infusion reduces the satiating potency of CCK-8. *Physiol. Behav.* 44:535–538, 1988.

Smith, G. P., C. Jerome, B. J. Cushin, R. Eterno, and K. J. Simansky. Abdominal vagotomy blocks the satiety effect of cholecystokinin in the rat. *Science* 213:1036–1037, 1981a.

Smith, G. P., C. Jerome, and J. Gibbs. Abdominal vagotomy does not block the satiety effect of bombesin in the rat. *Peptides* 2:409–411, 1981b.

Smith, J., G. P. Smith, J. Gibbs, and M. Russ. Gastrin-releasing peptide fails to inhibit sham feeding in rats. *Soc. Neurosci. Abstr.* 20:587, 1994.

Stacher, G., H. Steinringer, G. Schmierer, C. Schneider, and S. Winklehner. Cholecystokinin octapeptide decreases intake of solid food in man. *Peptides* 3:133–136, 1982.

Stallone, D., S. Nicolaïdis, and J. Gibbs. Cholecystokinin-induced anorexia depends on serotoninergic function. *Am. J. Physiol.* 256:R1138–R1141, 1989.

Stein, L. J., and S. C. Woods. Gastrin releasing peptide reduces meal size in rats. *Peptides* 3:833–835, 1982.

Stratford, T. R., J. Gibbs, D. H. Coy, and G. P. Smith. Fourth ventricular injection of the bombesin receptor antagonist [D-Phe^6]Bombesin(6-13)methyl ester, but not BW2258U89, increases food intake in rats. *Pharmacol. Biochem. Behav.* 50:463–471, 1995a.

Stratford, T. R., J. Gibbs, and G. P. Smith. Combined administration of neuromedin B (NMB) and gastrin-releasing peptide (GRP) reproduce the full satiety effect of bombesin (BN) in rats. *Appetite* 23:321, 1994.

Stratford, T. R., J. Gibbs, and G. P. Smith. Microstructural analysis of licking behavior following peripheral administration of bombesin or gastrin-releasing peptide. *Peptides* 16:903–909, 1995b.

Stratford, T. R., J. Gibbs, and G. P. Smith. Simultaneous administration of neuromedin B-10 and gastrin-releasing peptide (1-27) reproduces the satiating and microstructural effects of bombesin. *Peptides* 17:107–110, 1996.

Strubbe, J. H., J. G. Wolsink, A. M. Schutte, and A. J. A. Prins. Hepatic-portal and cardiac infusion of CCK-8 and glucagon induce different effects on feeding. *Physiol. Behav.* 46:643–646, 1989.

Strunz, M. T., M. R. Thompson, J. Elashoff, and M. I. Grossman. Hepatic inactivation of gastrins of various chain lengths in dogs. *Gastroenterology* 74:550–553, 1978.

Stuckey, J. A., and J. Gibbs. Lateral hypothalamic injection of bombesin decreases food intake in rats. *Brain Res. Bull.* 8:617–621, 1982.

Stuckey, J. A., J. Gibbs, and G. P. Smith. Neural disconnection of gut from brain blocks bombesin-induced satiety. *Peptides* 6:1249–1252, 1985.

Taylor, I. L., J. Elashoff, and R. Garcia. Effects of pancreatic polypeptide, caerulein, and bombesin in obese mice. *Am. J. Physiol.* 248:G277–G280, 1985a.

Taylor, I. L., R. Garcia, and J. Elashoff. Effects of vagotomy on satiety induced by gastrointestinal hormones in the rat. *Physiol. Behav.* 34:957–961, 1985b.

Verbalis, J. G., M. J. McCann, C. M. McHale, and E. M. Stricker. Oxytocin secretion in response to cholecystokinin and food: Differentiation of nausea from satiety. *Science* 232:1417–1419, 1986a.

Verbalis, J. G., M. J. McCann, and E. M. Stricker. Species-specific effects of bombesin on gastric emptying and neurohypophyseal secretion. *Peptides* 9: 1289–1293, 1989.

Verbalis, J. G., C. M. McHale, T. W. Gardiner, and E. M. Stricker. Oxytocin and vasopressin secretion in response to stimuli producing learned taste aversions in rats. *Behav. Neurosci.* 100:466–475, 1986b.

Verbalis, J. G., D. W. Richardson, and E. M. Stricker. Vasopressin release in response to nausea-producing agents and cholecystokinin in monkeys. *Am. J. Physiol.* 252:R749–R753, 1987.

Waldbillig, R. J., and M. O'Callaghan. Hormones and hedonics, cholecystokinin and taste: A possible behavioral mechanism of action. *Physiol. Behav.* 25:25–30, 1980.

Walsh, J. H., C. Wong, and G. J. Dockray. Bombesin-like peptides in mammals. *Fed. Proc.* 38:2315–2319, 1979.

Weller, A., G. P. Smith, and J. Gibbs. Endogenous cholecystokinin reduces feeding in young rats. *Science* 247:1589–1591, 1990.

West, D. B., R. H. Williams, D. J. Braget, and S. C. Woods. Bombesin reduces food intake of normal and hypothalamically obese rats and lowers body weight when given chronically. *Peptides* 3:61–67, 1982.

Willis, G. L., J. Hansky, and G. C. Smith. Ventricular, paraventricular and circumventricular structures involved in peptide-induced satiety. *Regul. Pept.* 9:87–99, 1984.

Wolkowitz, O. M., B. Gertz, H. Weingartner, L. Beccaria, K. Thompson, and R. A. Liddle. Hunger in humans induced by MK-329, a specific peripheral-type cholecystokinin receptor antagonist. *Biol. Psychiatr.* 28:169–173, 1990.

Woods, S. C., L. J. Stein, D. P. Figlewicz, and D. Porte, Jr. Bombesin stimulates insulin secretion and reduces food intake in the baboon. *Peptides* 4:687–691, 1983.

Yen, E. I., J. Miesner, J. Gibbs, A. Weller, and G. P. Smith. Intravenous bombesin reduces food intake in the rat. *Soc. Neurosci. Abstr.* 15:1280, 1989.

Yox, D. P., H. Stokesberry, and R. C. Ritter. Suppression of sham feeding by intraintestinal oleate: Blockade by a CCK antagonist and reversal of blockade by exogenous CCK-8. In Hughes, J., G. Dockray, and G. Woodruff (eds.): *The Neuropeptide Cholecystokinin (CCK), Anatomy and Biochemistry, Receptors, Pharmacology and Physiology.* Chichester: Ellis Horwood, Ltd., 1989, pp. 218–222.

Zhang, Y., R. Porcina, M. Maffei, M. Barone, L. Leopold, and J. M. Friedman. Positional cloning of the mouse obese gene and its human homologue. *Nature* 372:425–432, 1994.

6

Identification of Receptor Populations Mediating the Satiating Actions of Brain and Gut Peptides

TIMOTHY H. MORAN and ELLEN E. LADENHEIM

Introduction

The ability of exogenously administered or endogenously released cholecystokinin (CCK) or bombesin (BN) to influence feeding depends on their interactions with specific receptor populations. Thus, identification of the specific receptor populations through which these feeding inhibitory actions are mediated is essential for understanding the mechanisms underlying the satiety actions of these peptides. This chapter focuses on the identification of discrete receptor subpopulations whose pharmacological specificity and location make them leading candidates for mediating the satiety actions of CCK and BN-like peptides.

The methods for identifying receptor populations are now well characterized. While a complete detailing of receptor-binding techniques and their underlying assumptions is beyond the scope of this chapter*, a few of the basic principles on which these techniques depend are presented as the ideas are introduced.

* For a comprehensive discussion of receptor-binding methods and principles, see "Receptor Binding Techniques," 1980 Short Course Syllabus, Society for Neuroscience; or L. E. Limbird's "Cell Surface Receptors: A Short Course in Theory and Methods," Martinus Nijhoff Publishing, Boston, 1986.

Cholecystokinin Receptors in Satiety

Identification and Distribution of Cholecystokinin Receptor Subtypes

CCK was originally identified and characterized as a gut peptide with a variety of actions on gallbladder, pancreas, and intestinal motility. Shortly after CCK's actions in food intake were recognized (Gibbs et al., 1973), Vanderhaegen and colleagues (1975) discovered the presence of gastrin-like immunoreactivity in the rat brain. This was subsequently identified as CCK (Dockray, 1976). The presence of CCK in the brain led to the suggestion that it may act as both a hormone and a neurotransmitter. CCK had been found to exist in several molecular forms, including CCK-33 and CCK-8 (Larsson and Rehfeld, 1978). Both of these forms express the full range of CCK biological activity.

Since CCK exerted biological activity in the periphery and a role as a central neurotransmitter had been proposed, the identification and characterization of CCK receptors at both central and peripheral sites were sought. The first step in such studies is the identification of an appropriate radioligand. CCK-8 did not lend itself to tritiation or iodination, as the biological activity of the radiolabeled forms was greatly reduced. However, iodination of CCK-33 using Bolton-Hunter ^{125}I resulted in a ligand with complete biological activity as determined by its ability to release amylase from pancreatic acini (Sankaran et al., 1980). Initial CCK receptor-binding studies utilizing ^{125}I-labeled Bolton Hunter CCK-33 demonstrated saturable and specific binding to tissue homogenates from both pancreas and brain (Innis and Snyder, 1980; Sankaran et al., 1980; Saito et al., 1980).

In these experiments, a number of the criteria for the determination of receptor status from binding data were assessed. Tissue from the brain or pancreas was homogenized and exposed to ^{125}I-labeled Bolton Hunter CCK-33 under a variety of conditions. As stated above, binding of radiolabeled CCK-33 to both pancreas and brain was demonstrated to be saturable and specific. These are two of the primary criteria used to demonstrate that binding is in fact occurring to the receptor of interest. Saturability is critical, since a biological tissue can only contain a finite number of receptors. If a ligand is attaching to a biologically relevant site, as the concentration of ligand increases, the amount of binding should reach an asymptote. This was assessed by incubating the tissue sample with increasing levels of ^{125}I-labeled Bolton Hunter CCK-33 in the presence and absence of an excess of nonlabeled CCK-33. Two phenomena were demonstrated. First, there was greater association of the radioligand with the tissues in the absence of the unlabeled CCK-33 than in its presence, and, second, the difference in binding between these conditions reached an asymptote as the concentration of the ^{125}I-labeled Bolton Hunter CCK-33 was increased. Binding in the absence of the unlabeled CCK-33 is referred to as *total binding*. Binding in the presence of the unlabeled CCK-33 is the nonspecific binding. This is the binding that is noncompetitive. That is, the labeled and unlabeled forms of CCK-33 do not compete for these sites. The difference between the total binding and the nonspecific binding is referred to as *specific binding*. In a hypothetically pure receptor preparation, there should be no nonspecific binding; all the binding should be specific. However, various aspects of a tissue homogenate can provide sites for a radioligand to stick or interact. These are not all receptors.

Determination of the saturability of binding also allows assessments of the concentration of receptors present and the affinity of the receptor for the ligand. These are referred to as the B_{max} and K_d, respectively. B_{max} is the maximal binding attained at saturation. K_d is the concentration at which 50% of maximal binding occurs. Rosenthal (Scatchard) analysis allows the calculation of both of these factors. This analysis plots the bound/free radioligand against the bound radioligand. In this relationship the y intercept is the B_{max}/K_d, the inverse of the slope is the K_d, and the x intercept is the B_{max}. In the initial experiments with ^{125}I-CCK-33, the B_{max} ranged from 7 to 25 fmol/mg of protein for the brain and the K_ds were around 1 nM. In the pancreas, similar K_ds were found but the B_{max} or receptor density was significantly higher (Saito et al., 1980; Innis and Snyder, 1980).

While the saturation experiments in brain and pancreas yielded similar results, analysis of the specificity of binding in the two tissues did not. Determination of binding specificity is another step in the assessment of binding sites as potential receptors. If the binding of a radioligand represents attachment to a biological receptor, then other compounds known to have activity at this site should be capable of competing for the binding site while compounds without activity should be significantly less potent (Enna, 1980). Although both brain and pancreas demonstrated specificity for CCK-8 and CCK-33 over other nonrelated peptides, the relative affinity of desulfated CCK and CCK-4 for the two sites was markedly different (Innis and Snyder, 1980; Saito et al., 1981; Steigerwalt and Williams, 1981) (Fig. 6.1). ^{125}I-CCK-33 binding to brain homogenates could be inhibited with high affinity by both sulfated and desulfated CCK-8 and shorter CCK fragments. In contrast, ^{125}I-CCK-33 binding to pancreatic homogenates was relatively insensitive to the presence of nonsulfated CCK-8 or CCK fragments. Thus, these initial homogenate-binding experiments provided evidence for CCK receptor subtypes, differentiated by location (brain or pancreas) and pharmacological specificity.

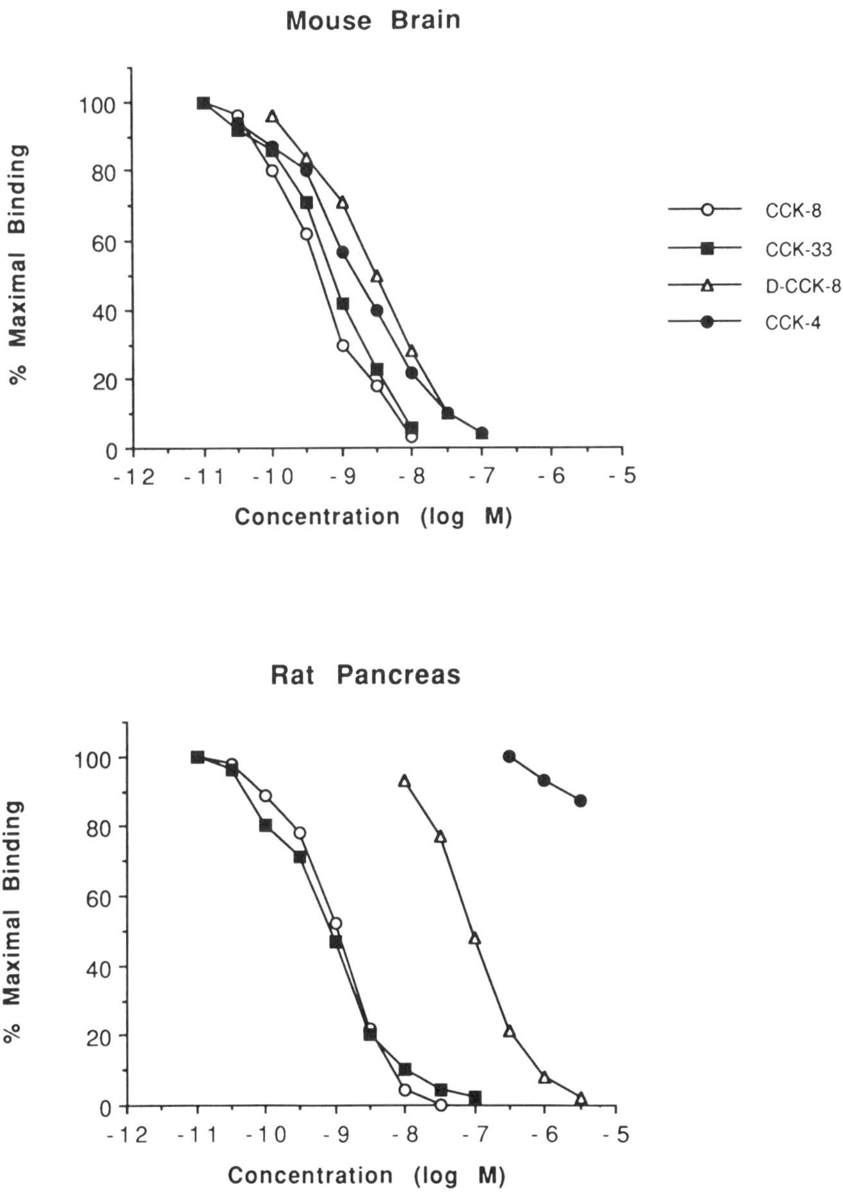

FIGURE 6.1. Pharmacological characterization of ^{125}I-CCK-33 binding to homogenates from mouse brain **(top)** and rat pancreas **(bottom).** While D-CCK-8 and CCK-4 inhibit binding in brain with high affinity, they have a lower affinity for pancreatic CCK-binding sites. (Adapted from Saito et al. (1981) and Steigerwalt and Williams (1981).

Pancreatic CCK-binding sites readily met criteria as functional receptors. Increasing concentrations of CCK added to pancreatic acinar cell preparations resulted in dose-related changes in amylase release, and the pharmacological characteristics of this response were the same as the pharmacological profile of binding to pancreatic homogenates (Jensen et al., 1980). In contrast, the functional relevance of brain CCK-binding sites was somewhat problematic. While homogenate-binding studies demonstrated that CCK brain receptors did not appear to differentiate between CCK, gastrin, or a variety of CCK analogues, electrophysiological and pharmacological studies had noted differential actions of CCK and its analogues at many brain sites (Crawley et al., 1985; White and Wang, 1984). For example, a variety of responses to local injections of CCK could be demonstrated to be specific to the sulfated CCK-8 analogue. That is, the responses were not stimulated or antagonized by nonsulfated CCK-8 or various CCK fragments even though these compounds appeared to have a high affinity for brain CCK-binding sites.

These apparently conflicting data suggested that either the CCK-binding sites identified and characterized in homogenate-binding experiments were not functional receptors or that CCK receptors in the brain were not a homogeneous population. The data available from these homogenate-binding experiments could not distinguish between these two possibilities.

Another method for studying receptors had been developed that focused on issues of cellular localization. Initially, this involved in vivo administration of a radioligand and the subsequent microscopic localization of the binding sites by autoradiographic techniques. Localization of receptors is of clear importance to understanding potential function, and these autoradiographic techniques maintained tissue integrity that was destroyed in the homogenization assays described above. These techniques, however, did not readily lend themselves to work with peptides. Peptides such as CCK did not cross the blood–brain barrier (Passaro et al., 1982), making a detailed analysis of the distribution of brain-binding sites impossible. Young and Kuhar (1979, 1980) adapted some of these in vivo techniques to use in an in vitro preparation. This in vitro autoradiography involved incubating slide-mounted tissue sections in the presence of a radioligand using the assay parameters established in homogenate studies. Following the incubation, the slides are rinsed, dried, and then opposed to radiation-sensitive film to produce an image of the binding in which the optical density is related to the amount of radiation or binding present at any site. Autoradiographic studies of the distribution of CCK-binding sites in the brain demonstrated that these binding sites were widely distributed and that the density of binding was quite variable—ranging from high levels in such structures as the olfactory tubercle and cingulate cortex to low densities in the thalamus (Zarbin et al., 1981a,1983).

Thus, two types of experiments characterizing CCK receptor-binding had been performed, each with a different goal. One, providing a characterization of the receptor binding, had depended on homogenate-binding studies in which large areas of the brain were homogenized to membrane fractions. These experiments had examined the ability of various CCK analogues to inhibit the binding of labeled CCK-33 in these preparations. The results demonstrated that brain CCK-binding sites demonstrated similar affinities for the various forms of CCK (Innis and Snyder, 1980; Saito et al., 1980). These homogenate-binding studies did not allow the detection of small localized populations of CCK receptors, which might have a different pharmacological profile than the majority present in the homogenate. In contrast, autoradiographic experiments had focused on the localization of binding sites (Zarbin et al., 1981a, 1983). That is, the specific location of receptor sites to a variety of brain regions had been identified, but the potential for variable pharmacological characteristics had not been assessed. Thus, homogenate-binding and autoradiographic experiments were carried out for different purposes.

When aspects of two approaches were combined, specific and differential distributions of CCK receptor subtypes were identified in the brain. Furthermore, autoradiographic experiments of this type have allowed the identification and pharmacological characterization of additional populations of CCK receptors in peripheral tissues other than the pancreas. Thus, by combining aspects of these techniques, we demonstrated two distinct patterns of CCK binding in the rat brain (Moran et al., 1986). One pattern was widely distributed, and binding to these sites was relatively nonspecific in that both sulfated CCK-8 and other CCK analogues such as desulfated CCK-8, CCK-4, and gastrin interacted with this receptor subpopulation with a high affinity (Fig. 6.2B). Binding at other locations, however, demonstrates the same kind of specificity that had been found in pancreatic homogenate-binding studies. That is, binding at these sites showed a much higher affinity for the sulfated form of CCK-8 than for desulfated CCK-8, gastrin, and CCK fragments (Fig. 6.2A). In our initial experiments, sites showing this pattern of binding were localized to the area postrema, the medial aspects of the nucleus tractus solitarius (NTS) (Fig. 6.3), the interpeduncular nuclei, and the dorsal medial aspects of the posterior hypothalamus. From this work with CCK agonists, we proposed the nomenclature of CCK_A and CCK_B for the two subtypes of CCK receptors (Moran et al., 1986). The designation CCK_A was based

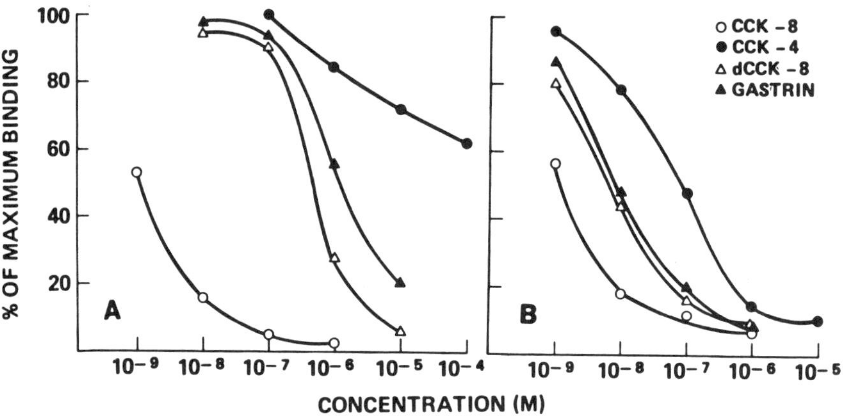

FIGURE 6.2. Pharmacological characterization of ^{125}I-CCK-33 binding to specific areas in rat brain. **(A)** Area postrema/nucleus tractus solitarius. **(B)** Caudate nucleus. From Moran et al. (1986), with permission of the publisher.

on *A* for the alimentary type as exemplified by the pancreatic CCK receptor—the receptor that showed a high preference for the sulfated form of CCK. The designation of CCK_B was for the CCK receptor that was widely distributed in the rat brain (*B*)—the receptor that was readily accessed by nonsulfated CCK-8, gastrin, and CCK fragments as well as by sulfated CCK-8. Subsequent experiments by Hill and colleagues (1987, 1988), using radiolabeled forms of specific CCK_A and CCK_B antagonists, confirmed the presence and distribution of these two receptor subtypes in the rat brain. They also demonstrated a variety of species differences in the relative distributions of CCK_A receptors in the central nervous system. For example, the primate central nervous system contains additional sites to which CCK_A receptors are localized (Hill et al., 1988b; Hill and Woodruff, 1990). These include such areas as the caudate nucleus and the spinal cord.

The existence of the two primary CCK receptor subtypes is now widely accepted. Recent work by Wank and colleagues (1992a,b) has resulted in the identification and cloning of the genes for the expression of both of these CCK receptors, and they clearly have different protein structures. Furthermore, the CCK_A receptors identified in brain in autoradiographic binding experiments have been demonstrated by in situ probes to be the same as pancreatic CCK receptors (Wank et al., 1992b). While there had been some controversy over whether CCK_B receptors and gastrin receptors were the same or different proteins, it has now been demonstrated that these represent a single gene product.

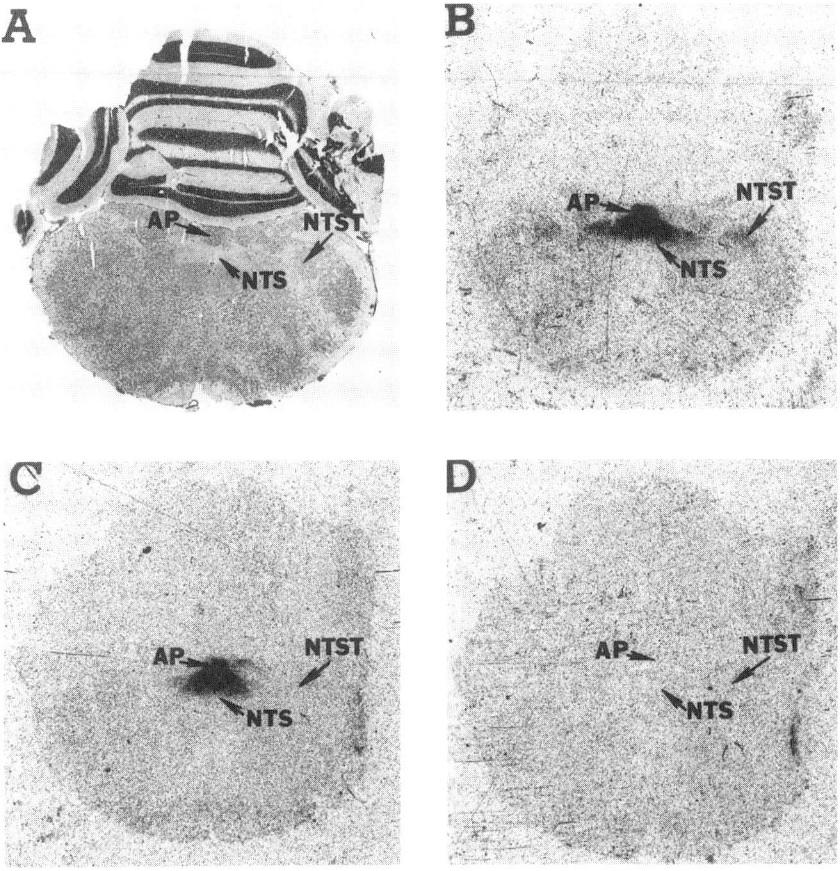

FIGURE 6.3. Photomicrograph of brainstem coronal section and corresponding autoradiographs showing total and competitively inhibited ^{125}I-CCK-33 binding. **(A)** Stained tissue section. **(B)** Autoradiograph of section in A showing total binding. **(C)** Autoradiograph of adjacent section showing CCK binding in the presence of 100 nM desulfated CCK-8 **(D)** Autoradiograph of adjacent section showing CCK binding in the presence of 100 nM sulfated CCK-8. AP, area postrema; NTS, nucleus tractus solitarius; NTST, nucleus of the spinal tract of the trigeminal nerve. Binding in the AP and medial NTS is inhibited in the presence of CCK-8, but not desulfated CCK-8. From Moran et al. (1986), with permission of the publisher.

Pharmacological Characterization of the Satiating Actions of CCK: Candidate Receptor Populations

Differentiation of CCK_A and CCK_B receptors facilitated the search for potential sites of action for CCK in food intake. The development and availability of specific antagonists at these sites has allowed a pharmacological characterization of this behavioral action of CCK. As outlined in more detail in Chapter 5, the feeding inhibitory actions of exogenously administered CCK can be clearly demonstrated to be mediated through the interaction of CCK with CCK_A receptors. This conclusion is based on a number of observations. Sulfated CCK-8 is effective in inhibiting food intake, but desulfated CCK-8 and gastrin are not (Gibbs et al., 1973; Lorenz et al., 1979) . Furthermore, prior administration of CCK_A, but not CCK_B antagonists, block the feeding inhibitory actions of exogenously administered CCK (Dourish et al., 1989a; Reidelberger and O'Rourke, 1989; Moran et al., 1992). On this point there does not appear to be any disagreement. However, the identification of the particular receptor subtype mediating the feeding inhibitory effects of endogenously released CCK has been somewhat controversial. While we and many others have demonstrated that prior administration of CCK_A antagonists can increase food intake in a variety of feeding paradigms (Hewson et al., 1988; Moran et al., 1992; Reidelberger and O'Rourke, 1989; Silver et al., 1989), Dourish et al. (1989b) demonstrated that administration of the CCK_B antagonist L365,260 to prefed rats was both more potent and more efficacious than the CCK_A antagonist devazepide at increasing food intake. However, we have not been able to replicate this effect with their paradigm (unpublished observation), and we, and others, have not been able to demonstrate a feeding stimulatory effect of CCK_B antagonists in other feeding paradigms where CCK_A antagonists can be demonstrated to increase intake (Brenner and Ritter, 1990; Moran et al., 1992; Reidelberger et al., 1990). Therefore, with this one caveat, the feeding inhibitory actions of both exogenously administered and endogenously released CCK appear to be mediated through the interaction of CCK with CCK_A receptors (see Chapter 5).

The satiating action of CCK is likely mediated at a site outside the blood–brain barrier. This conclusion is based on the size of the CCK peptide, its demonstrated inability to cross the blood–brain barrier (Passaro et al., 1982), and the lack of agreement on the ability of centrally administered CCK to inhibit food intake (Della-Fera and Baile, 1979; Crawley, 1985; Schick et al., 1986). While there may well be a feeding inhibitory action of centrally administered CCK, this appears to be a separate action of the peptide from that produced by peripherally administered CCK. Thus, the identification of CCK_A receptors in the periphery or at sites outside the blood–brain barrier would allow a direct assessment of the roles of these sites in the mediation of CCK satiety.

Area Postrema CCK Receptors

The only brain site that contains CCK_A receptors and meets the above criteria is the area postrema (Moran et al., 1986; Hill et al., 1988b*). The area postrema is a circumventricular organ open to circulating peptides. Van der Kooy (1984) demonstrated that electrolytic lesions that included the area postrema eliminated the feeding inhibitory actions of CCK. He concluded that this was the site of action for circulating CCK. Edwards et al. (1986), however, demonstrated that it was damage to the underlying NTS, the site of central termination of vagal afferents, rather than damage to the area postrema that produced this result. Thus, the CCK_A receptors localized to the area postrema do not appear to be involved in CCK satiety, and their function remains unknown.

Vagal CCK-Binding Sites

As well as being located in the pancreas and in specific brain sites, CCK_A-binding sites have been identified in a number of peripheral locations in the rat. Early work by Zarbin et al. (1981b) demonstrated the presence and axonal transport of CCK-binding sites in rat cervical vagal fibers. Receptor transport is a phenomenon that can be demonstrated for a variety of both peptide and nonpeptide receptors. It occurs through a mechanism of fast axonal transport and can be identified around a ligation as a time-dependent build up of binding sites proximal and often distal to the ligation site (Moran and McHugh, 1992). We extended this initial observation by demonstrating the presence and transport of CCK binding sites in all of the subdiaphragmatic vagal branches (Fig. 6.4) (Moran et al., 1987). Furthermore, we provided an initial characterization of the pharmacological specificity of these transported binding sites by comparing the abilities of sulfated and desulfated CCK-8 to inhibit ligand binding to these sites. The results of that experiment demonstrated that sulfated, but not desulfated, CCK-8 was a potent inhibitor of CCK binding (Moran et al., 1987). Therefore, we concluded that vagal CCK receptors were CCK_A receptors. Subsequent experiments have provided a more complete pharmacological characterization of vagal CCK-binding sites. Results from Mercer and his colleagues (1992) and from Corp et al. (1993), using specific CCK antagonists to compete for binding sites, have clearly demonstrated

* Recent work has suggested that the blood–brain barrier in some aspects of the NTS may not be complete (Gross et al., 1990). Thus, circulating or peripherally administered CCK may access some NTS sites containing CCK_A receptors. However, penetration of CCK to these sites has yet to be directly demonstrated.

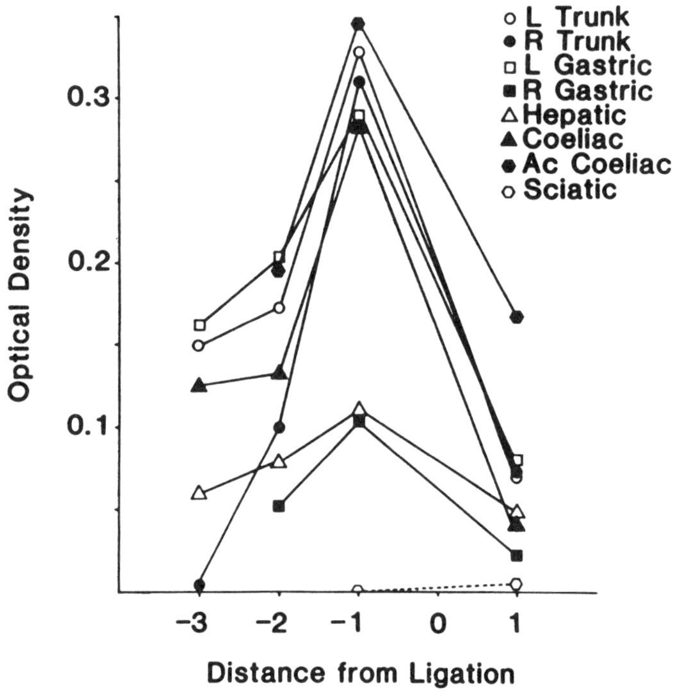

FIGURE 6.4. Mean CCK-binding site accumulation around a ligature in various vagal segments. Graph expresses specific optical densities of autoradiographs of ^{125}I-CCK-33 binding (total − nonspecific) to various vagal segments. Values along the abscissa indicate distances (mm) from a ligature site: Negative values correspond with areas proximal to the ligature, while positive values correspond with areas distal to the ligature. Increases in optical density correspond with increases in the density of binding. From Moran et al. (1987), with permission of the publisher.

that both CCK_A- and CCK_B-binding sites are present and transported in equivalent amounts in the vagus.

Transport of CCK-binding sites occurs in afferent, but not efferent, vagal fibers (Moran et al., 1990). Specific lesioning techniques were employed to study the potential central and peripheral transport of CCK-binding sites from the nodose ganglion (site of vagal afferent cell bodies) or the dorsal motor nucleus (site of vagal efferent cell bodies). Either supraganglionic vagotomies or specific transections of afferent rootlets as they enter the medulla resulted in the unilateral elimination of CCK-binding sites in the NTS. These results demonstrated that these CCK receptors were of vagal origin and were likely to be on vagal afferent terminals. This result was in agreement with that of Ladenheim et al. (1988), who demonstrated that, following nodose ganglionectomy, the density of binding in the NTS was significantly diminished.

We also demonstrated that the transport of CCK-binding sites to the periphery was occurring in afferent fibers (Moran et al., 1990). This was done by comparing the effects of supraganglionic and infraganglionic vagotomies on the transport of CCK-binding sites in the peripheral vagus. The rationale of this experiment was that if transport was occurring in efferent fibers both vagotomy procedures would disconnect the efferent cell bodies in the dorsal motor nucleus of the brainstem from the site of the ligation in the cervical vagus and both vagotomies would eliminate the build up of CCK-binding sites at the ligation. In contrast, if transport were occurring in afferent fibers, then only the infraganglionic vagotomy would disconnect the afferent cell bodies in the nodose ganglia from the ligation site in the cervical vagus. We found that infraganglionic, but not supraganglionic, vagotomy eliminated the transport of CCK-binding sites (Fig. 6.5).

Vagal CCK-binding sites are a leading candidate for the mediation of the satiating effect of CCK. There are a number of reasons for this conclusion. As discussed elsewhere in this book (see Chapter 5), the satiety effect of peripherally administered CCK is dependent on an intact vagus. That is, vagotomy eliminates or significantly attenuates the feeding inhibitory actions of exogenously administered CCK (Smith et al., 1981a), and this depends on afferent vagal fibers (Smith et al., 1985). Second, we and others demonstrated that vagal afferent activity can be stimulated by CCK (Davison and Clarke, 1988; Schwartz et al., 1991). Close arterial administration of CCK increases activity in vagal afferent fibers with a gastric receptive field that are sensitive to gastric distention (Fig. 6.6). Thus, CCK mimics the effect of gastric distention in these fibers. CCK also sensitizes these fibers to subsequent gastric distention (Fig. 6.6). Furthermore, there is a synergistic relationship between the effects of local administration of CCK and gastric distention on activity in these fibers (Schwartz 1993). The pharmacological profile of this response to CCK is consistent with mediation through CCK_A receptors in that the responses can be blocked with the CCK_A antagonist devazepide, but are not affected by the CCK_B antagonist L365,260 (Schwartz et al., 1994).

Recent results have also demonstrated similar responses to CCK and distention in vagal afferents with a duodenal receptive field (Schwartz et al., 1995). In addition, vagal fibers that apparently represent duodenal mucosal afferents have been shown to be activated by CCK (Blackshaw and Grundy, 1990). Thus, CCK-induced electrophysiological activity has been demonstrated in multiple vagal afferent populations.

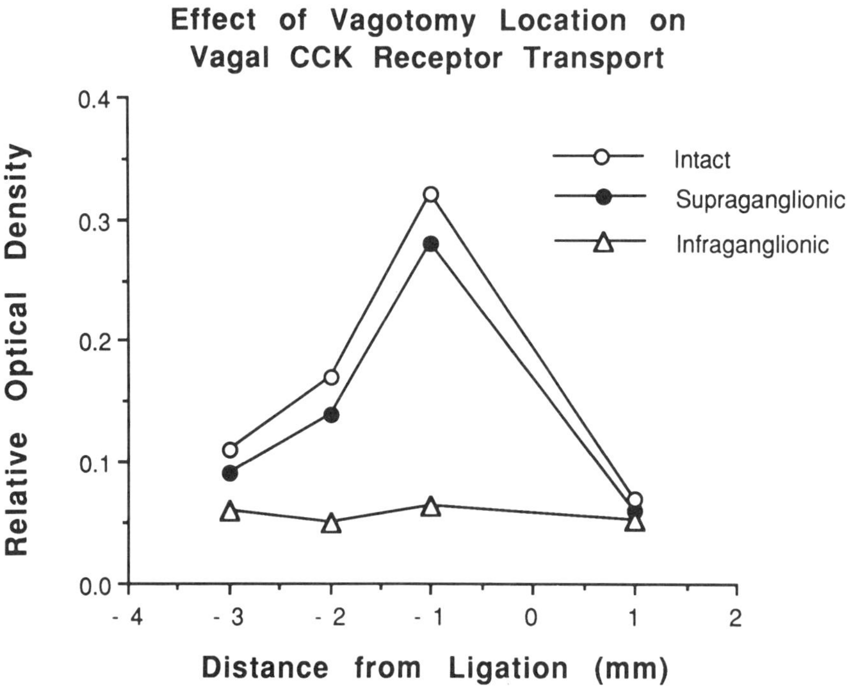

FIGURE 6.5. Mean CCK-binding site accumulation around cervical vagal ligations expressed as relative optical densities in intact rats and rats with either supra- or infranodose ganglionic vagotomies. Transections below the level of the nodose ganglion eliminated the accumulation of binding sites proximal to the ligature, while transection above the ganglion did not. From Moran and McHugh (1992), with permission of the publisher.

While we can demonstrate both the presence of CCK-binding sites on vagal afferent fibers and the ability of CCK-8 to activate vagal afferents, it is not necessary that the vagal afferent responses to CCK are directly mediated through the activation of these binding sites. It is conceivable that the effects of CCK that we are studying are mediated through CCK's actions at a nonvagal site, the result of which is activating vagal afferent fibers. Thus, we are still referring to these vagal afferent sites as "CCK-binding sites." We have no evidence that these are functional receptors and no evidence that the binding sites that we are studying while they are undergoing transport become functional receptors at vagal terminations. However, with this caveat in mind, we do believe that vagal CCK-binding sites are prime candidates for mediating at least some aspects of the satiating effect of CCK.

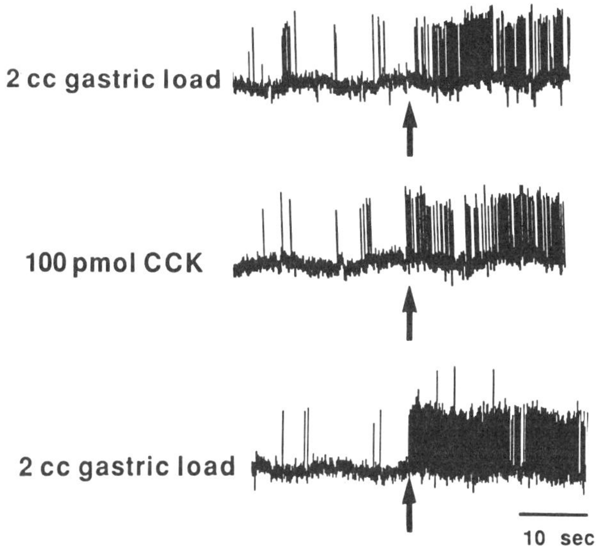

FIGURE 6.6. Electrophysiological responses of a gastric vagal afferent fiber to **(top)** a 2-ml intragastric saline load, **(middle)** a 100-pmol celiac artery infusion of CCK, and **(bottom)** a 2-ml intragastric saline load 10 minutes following CCK administration, when baseline activity had returned to pre-CCK levels. Arrows indicate the stimulus onset.

Pyloric CCK Receptors

We have also demonstrated the presence of CCK receptors in the circular muscle layer of the pyloric sphincter in the rat (Smith et al., 1984) and other species (Robinson et al., 1987; Strohmayer et al., 1987; Moran et al., 1992). Again, these results depended on receptor autoradiography. Homogenate-binding analysis of segments of the gastrointestinal tract did not identify any significant specific binding in any segment. Binding was found, but this binding was not inhibited in the presence of unlabeled CCK. Receptor autoradiography, however, revealed specific CCK binding to the circular muscle layer of the pyloric sphincter and significant nonspecific binding to the mucosal layer (Fig. 6.7). This nonspecific mucosal binding prevented the identification of the specific binding to the pyloric circular muscle in the homogenate of pyloric tissue. Pharmacological analysis of this binding demonstrated that binding was occurring to CCK_A sites (Moran and McHugh, 1987).

In vitro physiological experiments have demonstrated that these binding sites are likely to be functional receptors (Murphy et al., 1987; Margolis et al., 1989). In an isolated pyloric preparation, CCK has the ability to induce

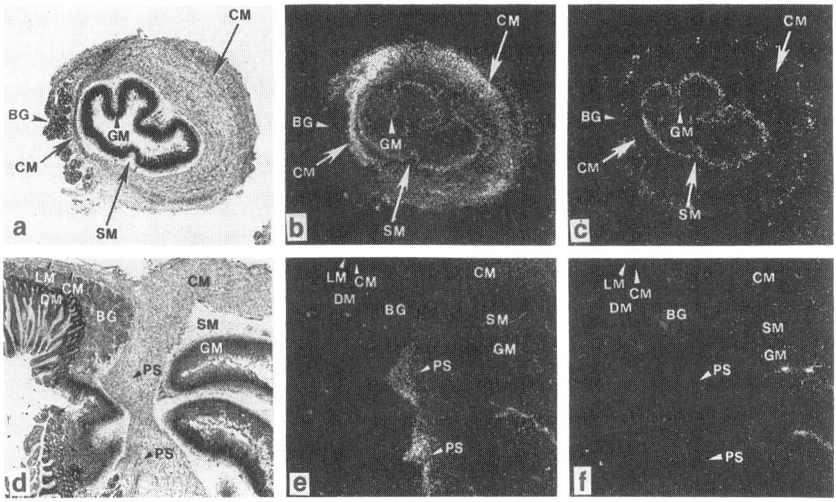

FIGURE 6.7. Photomicrographs of stained tissue section of rat pyloric sphincter (PS) and corresponding autoradiographs showing total and nonspecific ^{125}I-CCK 33 binding. **(a)** Cross section of distal pyloric sphincter. **(b)** Autoradiograph of section showing total binding. **(c)** Autoradiograph of adjacent section showing nonspecific binding. **(d)** Longitudinal section through the pyloric sphincter. **(e)** Autoradiograph showing total CCK binding. **(f)** Autoradiograph of adjacent section showing nonspecific CCK binding. LM, longitudinal muscle; CM, circular muscle; SM, submucosa; BG, Brunner's glands; GM, gastric mucosa; DM, douodenal mucosa. From Smith et al. (1984), with permission of the publisher.

contractions, and the pharmacology of this contractile response matches the pharmacological profile of receptor-binding experiments (Moran et al., 1991b). These CCK receptors appear to be localized directly on the circular muscle fibers in that contraction is not sensitive to either tetrodotoxin or atropine, and in an isolated muscle cell preparation CCK produces muscle contraction (Murphy et al., 1987, 1988).

A role for these receptors in the mediation of the satiety effect of CCK has been suggested. We have demonstrated that pylorectomy, the surgical removal of the area of the pylorus containing these receptors, results in a temporary attenuation of the ability of CCK to inhibit food intake (Moran et al., 1988b). The data demonstrated that, in the period shortly following pylorectomy, the CCK dose–response curve was truncated. The potency of low doses of CCK to inhibit food intake was normal, but the ability of higher doses to produce further inhibitions of intake was eliminated. However, following a 6-week or 2-month recovery period, the potencies of all doses of CCK were normal (Moran et al., 1988b; Smith et al., 1988). This recovery of the normal CCK dose–response curve appears to coincide with the appearance of functional CCK receptors at the new gastroduodenal junction. The presence of CCK-binding sites at the new gastroduodenal junction can be demonstrated autoradiographically, and these appear to be functional receptors since CCK produces contraction of the new gastroduodenal junction in in vitro experiments.

While the specific mechanisms by which pyloric CCK receptors play a role in mediating the satiety effect of CCK is unclear, the above data suggest that a portion of the feeding inhibitory action of exogenously administered CCK is mediated through the activation of these receptors.

High and Low Affinity CCK Receptors

Rosenthal (Scatchard) analysis of homogenate-binding data not only provides information about the dynamics of binding, it can also determine whether binding is occurring to a single or multiple sites (Bylund, 1980). A linear relationship between bound ligand and bound-over free ligand is consistent with binding occurring to a single site. That is, the affinity of the ligand for the receptor is constant over a range of ligand concentrations. If binding is occurring to two or more sites differentiated by their affinity for the ligand, the relationship between bound ligand and bound-over free ligand will not be linear, and the relationship will be the outcome of the ligand's interaction with the different sites at various concentrations.

Data generated from homogenate-binding experiments examining the dynamics of CCK binding to rat pancreatic receptors demonstrated a nonlinear Rosenthal relationship. Analysis of the curve indicated that the best fit was to a two-site model: one, a high-affinity, low-capacity CCK-binding site and the second, a low-affinity, high-capacity site (Sankaran et al., 1982). While the differentiation of high- and low-affinity binding sites depended on binding data, there was also a functional correlate to this distinction. The stimulation of pancreatic amylase secretion is a physiological action of CCK, and the relationship between amylase secretion and CCK dose is not linear. Low doses of CCK stimulate secretion in a nonmonotonic, yet dose-dependent fashion. Specifically, with increases in CCK dose, amylase secretion becomes maximal and then decreases with higher CCK doses. The dose-dependent increase in amylase secretion depends on the interaction of CCK with pancreatic high-affinity CCK-binding sites. With increasing CCK concentration and activation of the low-affinity site, pancreatic secretion is actively inhibited such that the dose–response curve for amylase secretion with increasing concentrations of CCK is an inverted U-shaped function. Work by Martinez and his colleagues with various CCK analogues demonstrated that one of the compounds they had synthesized

(JMV-180) acted as a functional agonist at the rat high-affinity site but as a functional antagonist at the rat low-affinity site (Galas et al., 1988; Matozaki et al., 1989; Stark et al., 1989). JMV-180 mimics the actions of CCK at the high-affinity site: It stimulates pancreatic secretion. It antagonizes the action of CCK at the low-affinity site in that the presence of JMV-180 blocks the inhibition of pancreatic secretion produced by high concentrations of CCK.

This compound has become a valuable tool in evaluating the receptor populations through which CCK's satiety effects are mediated. Recent data (Asin and Bednarz, 1992; Weatherford et al., 1993) have demonstrated that, in rats, the satiating effect of CCK seems to be mediated through activation of the low-affinity receptor site. Administration of JMV-180 alone had no effect on food intake, but JMV-180 blocked the inhibition of food intake by CCK. This result is consistent with data from Reidelberger and Solomon (1986) comparing the relative potencies of CCK for stimulating pancreatic secretion and inhibiting food intake. They demonstrated that the doses of CCK that inhibited food intake were the same as those that suppressed pancreatic secretion.

We have recently instituted further functional studies with JMV-180 to evaluate whether CCK-induced pyloric contraction or vagal afferent activity are mediated through high- or low-affinity CCK-binding sites. Close arterial administration of JMV-180 did not activate vagal afferent activity itself, but blocked the ability of CCK to activate vagal afferent fibers (Schwartz et al., 1994). In a similar fashion, JMV-180 inhibited the ability of CCK to contract the isolated pyloric sphincter and to inhibit gastric emptying (Moran et al., 1994). Thus, as with CCK satiety, these other actions of CCK are mediated through interactions with low-affinity CCK receptors. This similarity further supports the candidacy of these receptors in the mediation of CCK satiety.

Summary

Two subtypes of CCK receptors have been identified based on their pharmacological characteristics: CCK_A and CCK_B. These receptors are found both within the brain and in the periphery. Both high- and low-affinity CCK_A receptors have been identified, and these two affinity states can be functionally differentiated. Pharmacological characterizations of CCK's satiety actions have demonstrated that low-affinity CCK_A receptors located in the periphery mediate the satiating effects of exogenous and endogenous CCK (see Chapter 5). CCK receptors with these characteristics are found in the circular muscle layer of the pyloric sphincter. CCK_A-binding sites are localized to and undergoing transport in afferent vagal fibers, and the afferent vagal response to CCK is mediated through activation of low-affinity CCK_A receptors. Together these pyloric and vagal populations remain the leading candidates for the mediation of the satiating action of CCK.

Bombesin-Like Peptide Receptors in Satiety

We have employed similar strategies for understanding the receptor mechanisms underlying the satiating actions of BN-like peptides. That is, we attempted to identify distinct receptor subtypes, to map their distribution in the brain and the gastrointestinal tract, and to evaluate the potential roles of these receptor populations in the inhibition of food intake by BN-like peptides. While this line of research is not as advanced for BN-like peptides as it is for CCK, considerable progress has been made.

The overall view expressed from this work is that the BN family of peptides is comprised of a number of compounds that are structurally related to BN. Gastrin-releasing peptide (GRP) and neuromedin B (NMB) are two members of this family that have been isolated from mammalian tissue and are biologically active in the rat brain and in the periphery. A shown in Table 6.1, GRP and $GRP_{18\text{-}27}$ (neuromedin C [NMC]), the mammalian equivalents of the amphibian peptide BN, share most of the carboxyl-terminal peptide sequence with BN except for the substitution of a His residue in place of the Gln at position 8. Although NMB shares much structural homology with GRP and BN, it more closely resembles the amphibian peptides litorin and ranatensin in that it shares the identical tetrapeptide sequence with these compounds, which involves a phenyalanine substitution for the leucine at the penultimate position.

Identification and Distribution of Receptor Subtypes for Bombesin-Like Peptides

Initial homogenate-binding studies aimed at identifying bombesin receptors demonstrated that ^{125}I-Tyr_4 BN bound with high affinity to BN-binding sites in a variety of tissues. In the rat brain, high densities of binding were found in homogenates from the hypothalamus and hippocampus and moderate densities in the cortex, striatum, thalamus, midbrain, and hindbrain (Moody et al., 1978). Subsequent receptor autoradiographic studies identified the specific locations of these binding sites and demonstrated that their densities were variable and structure specific (Wolf and Moody, 1985;

TABLE 6.1.
Structure of Bombesin and Related Peptides

Peptide	Amino Acid Sequence													
Bombesin	pGlu	Gln	Arg	Leu	Gly	Asn	Gln	Trp	Ala	Val	Gly	His	Leu	Met	NH_2
$GRP_{14\text{-}27}$	Met	Tyr	Pro	Arg	Gly	Asn	His	Trp	Ala	Val	Gly	His	Leu	Met	NH_2
Neuromedin C [$GRP_{18\text{-}27}$]					Gly	Asn	His	Trp	Ala	Val	Gly	His	Leu	Met	NH_2
Litorin						pGlu	Gln	Trp	Ala	Val	Gly	His	Phe	Met	NH_2
Ranatensin				pGlu	Val	Pro	Gln	Trp	Ala	Val	Gly	His	Phe	Met	NH_2
Neuromedin B					Gly	Asn	Leu	Trp	Ala	Thr	Gly	His	Phe	Met	NH_2

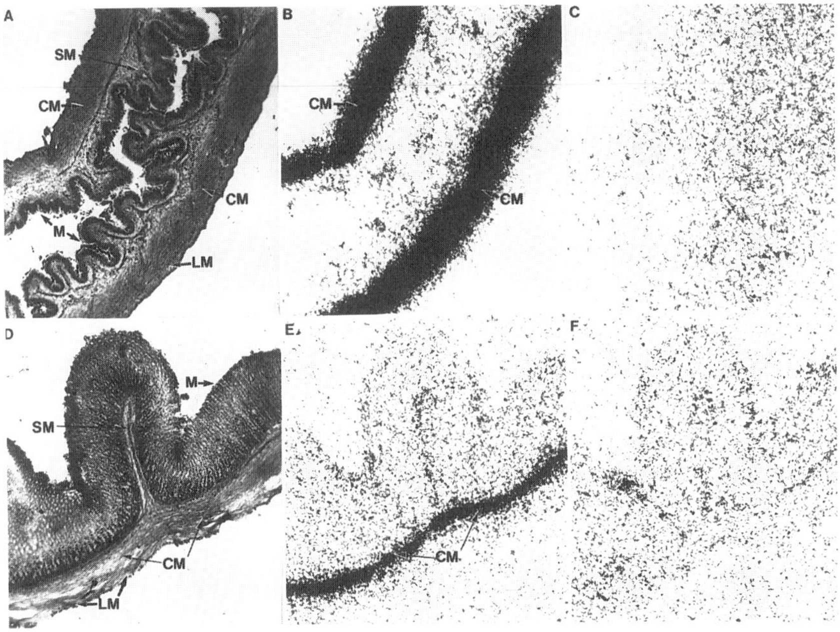

FIGURE 6.8. Photomicrographs of stained gastric sections and corresponding autoradiographs showing BN binding. **(A)** Cross section of gastric fundus. **(B)** Autoradiograph of section showing total ^{125}I-Tyr$_4$ BN binding. **(C)** Autoradiograph of adjacent section showing nonspecific binding. **(D)** Cross section from gastric antrum. **(E)** Autoradiograph showing total binding. **(F)** Autoradiograph of adjacent section showing nonspecific binding. Specific binding was localized to the circular muscle layer in both segments. CM, circular muscle; LM logitudinal muscle; M mucosa. From Moran et al. (1988a), with permission of the publisher.

Zarbin et al., 1985). BN-binding sites were also found in the pancreas and gastrointestinal tract. Early work from Jensen's laboratory (Jensen et al., 1978) demonstrated the presence and biological relevance of BN-binding sites on rat pancreatic acinar cells. Although homogenate-binding experiments reported similar K_ds for BN in brain and pancreas, differences were evident in the rank order potencies of various BN-like peptides for inhibiting BN binding in brain and pancreas, providing the original support for the idea that BN-binding sites were not homogeneous. Furthermore, BN-binding sites in a rat pituitary cell line demonstrated additional differences in affinity for BN-like peptides (Westendorf and Schonbrunn, 1983).

To identify potential sites in the gastrointestinal tract at which BN's feeding inhibitory actions could be mediated, we mapped the distribution of BN-binding sites in the rat gastrointestinal tract using in vitro auto-

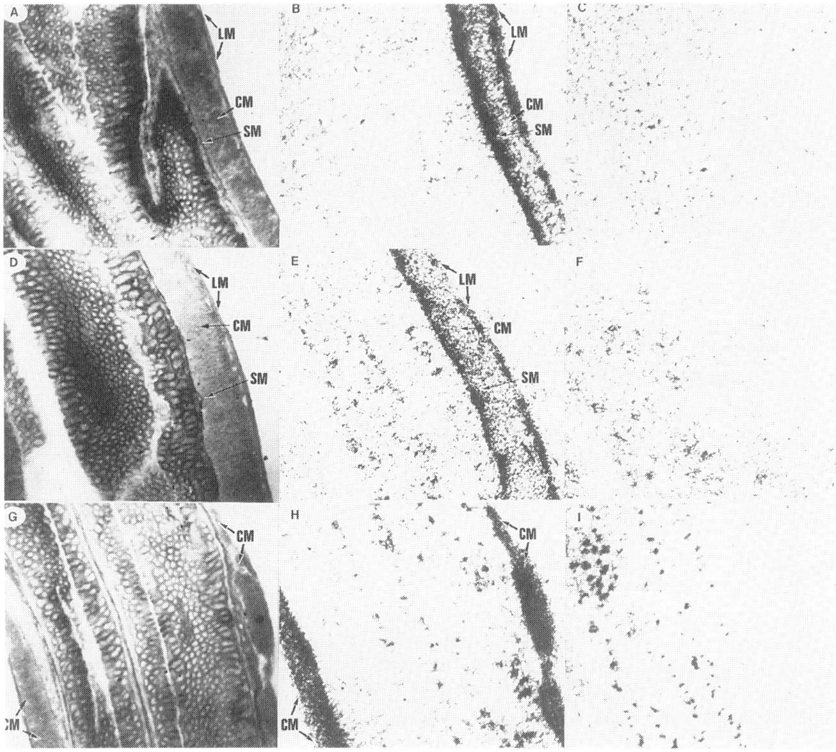

FIGURE 6.9. Photomicrographs of stained intestinal sections and corresponding autoradiographs showing total BN binding. **(A)** Cross section from duodenum. **(B)** Autoradiograph of section showing total ^{125}I-Tyr$_4$ BN binding. **(C)** Autoradiograph of adjacent section showing nonspecific binding. Specific binding is localized to the submucosal and circular muscle layers. **(D)** Cross section of jejunum. **(E)** Autoradiograph of section showing total ^{125}I-Tyr$_4$ BN binding. **(F)** Autoradiograph of adjacent section showing nonspecific binding. Specific binding is localized to the submucosal layer. **(G)** Cross section from ileum. **(H)** Autoradiograph of section showing total ^{125}I-Tyr$_4$ BN binding. **(I)** Autoradiograph of adjacent section showing nonspecific binding. Specific binding is localized to the submucosal layer. CM, circular muscle; LM, longitudinal muscle, SM, submucosa. From Moran et al. (1988a), with permission of the publisher.

radiography with ^{125}I-Tyr$_4$ BN. These studies demonstrated that specific BN-binding sites were widely distributed throughout the gastrointestinal tract. As shown in Figures 6.8 to 6.10, BN-binding sites were localized in the circular muscle layer of the gastric fundus and antrum and submucosal layer of the duodenum, jejunum, and ileum and were present in the longitudinal and circular muscle layers of the ascending, transverse, and descend-

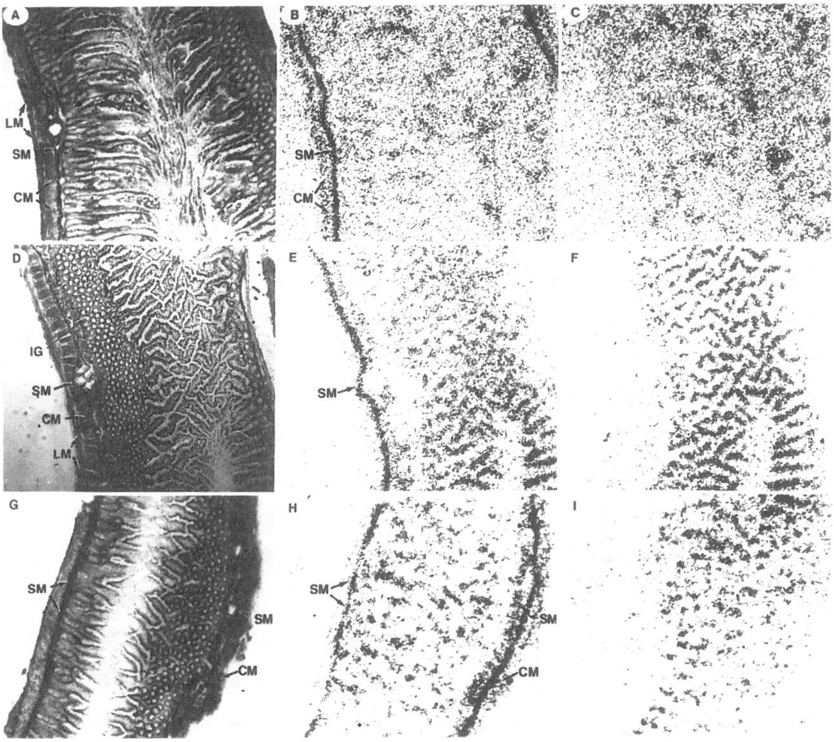

FIGURE 6.10. Photomicrographs of stained large intestinal sections and corresponding autoradiographs showing total BN binding. **(A)** Section of ascending colon. **(B)** Autoradiograph of section showing total ^{125}I-Tyr$_4$ BN binding. **(C)** Autoradiograph of adjacent section showing nonspecific binding. Specific binding is localized to the circular and longitudinal muscles and the submucosal layers. **(D)** Section of transverse colon. **(E)** Autoradiograph of section showing total ^{125}I-Tyr$_4$ BN binding. **(F)** Autoradiograph of adjacent section showing nonspecific binding. Specific binding is localized to the circular and longitudinal muscles and the submucosal layers. **(G)** Section of descending colon. **(H)** Autoradiograph of section showing total ^{125}I-Tyr$_4$ BN binding. **(I)** Autoradiograph of adjacent section showing nonspecific binding. Specific binding is localized to the circular muscle layer. CM, circular muscle; IG, intestinal glands; LM, longitudinal muscle; SM, submucosa. From Moran et al. (1988a), with permission of the publisher.

ing colon (Moran et al., 1988a). Furthermore, contraction studies demonstrated that the fundic and small intestinal sites represented functional receptors as BN produced changes in isometric tension in isolated tissue preparations (Margolis et al., 1989). Pharmacological studies of a representative of these gastrointestinal binding sites taken from the rat fundus demonstrated again that the rank order potency for inhibiting binding was

TABLE 6.2.
IC_{50}s (nM) for Inhibiting the Binding of ^{125}I-Tyr_4 BN

Analogue	IC50			
	Pancreas[a]	Brain[b]	Pituitary[c]	Fundus[d]
Bombesin	4	7	1	20
Litorin	40	60	1	1
Ranatensin	12	40	1	8
GRP	10	16	3	2

[a]Jensen et al. (1978).
[b]Moody et al. (1978).
[c]Westendorf and Schonbrunn (1983).
[d]Moran et al. (1988a).

different from those that had been shown previously for rat pancreas, forebrain, or pituitary (Table 6.2). From these results we suggested that these fundic-binding sites may represent an additional subtype of BN receptors that was different from those that have previously been characterized in brain, pancreas, and pituitary.

BN receptor heterogeneity was clearly demonstrated and more fully characterized in work examining the difference between esophageal and pancreatic BN-binding sites. von Schrenk et al. (1989) compared the binding of ^{125}I-labeled Bolton Hunter NMB and ^{125}I-Tyr_4 BN to tissue sections from the rat esophagus and pancreas. In each tissue, binding was saturable and dependent on pH, time, and temperature. The esophagus bound both tracers well, while the pancreas bound only the ^{125}I-Tyr_4 BN. In the esophagus, the relative potency for inhibiting the binding of both tracers was NMB > BN > GRP = $GRP_{18\text{-}27}$. Similar potencies were found for contracting muscle strips from the whole esophagus and from isolated esophageal muscularis mucosa. In contrast, in pancreatic sections and on dispersed pancreatic acini, the relative potencies of inhibitions of ^{125}I-Tyr_4 BN were BN > GRP = $GRP_{18\text{-}27}$ » NMB. That is, while the esophageal site had high affinity for BN, GRP, and NMB, the pancreatic sites had much less affinity for NMB (Fig. 6.11). We interpreted the results of this study as demonstrating that the esophageal binding sites represented a previously unidentified class of BN-binding sites in that they had a higher affinity for NMB than for BN and that these binding sites were likely to represent NMB receptors. In contrast, the pancreatic binding site had a much higher affinity for BN and GRP than for NMB and probably represented a GRP receptor subtype. These ideas were consistent with the view that BN was essentially an am-

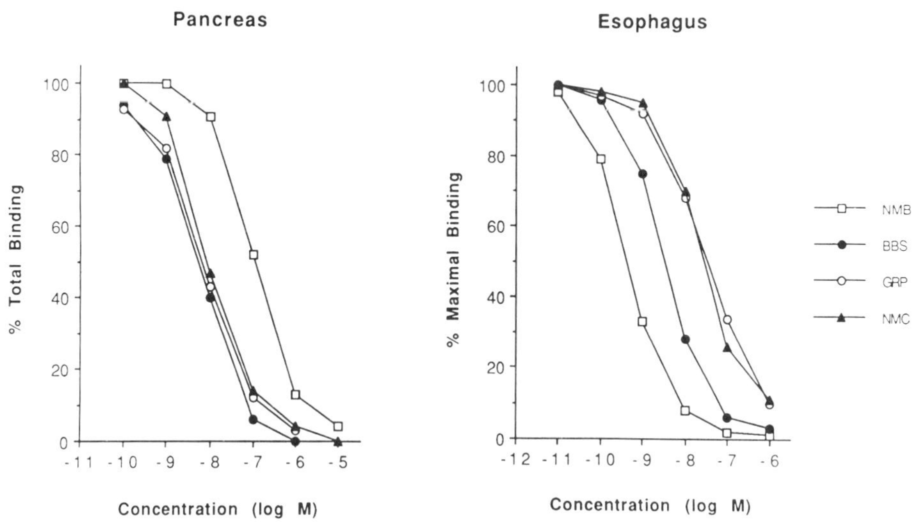

FIGURE 6.11. Pharmacological characterization of ^{125}I-Tyr$_4$ BN binding to sections from rat pancreas **(left)** and rat esophagus **(right).** NMB has a significantly higher affinity for binding sites in the esophagus than in the pancreas. Esophageal sites are NMB receptors and pancreatic sites are GRP receptors. Data from von Schrenck et al. (1989).

phibian peptide that exerted its biological activity in mammals by interacting with a variety of mammalian peptide systems. Receptor heterogeneity for BN-like peptides has been confirmed with the identification and cloning of separate genes for the two receptor subtypes (Battey et al., 1991; Wada et al., 1991).

As with CCK, combining autoradiographic methods with a pharmacological characterization identified brain sites that contain each of these two distinct BN receptor subtypes (Ladenheim et al., 1990). These experiments depended on the comparisons of the abilities of ^{125}I-Tyr$_4$ BN and ^{125}I-NMB to bind to rat brain sections and the abilities of various BN analogues to inhibit the binding of the these two ligands. The results demonstrated two distinct binding patterns. One mimicked that found in the pancreas, and the second was similar to that found in the esophagus. Thus, the overall pharmacological characterization from rat brain as determined in original homogenate studies was the outcome of binding to a mixture of GRP and NMB receptors.

Subsequent experiments used these differentiations to map the distributions of these two distinct receptor subtypes throughout the rat brain (Figs. 6.12 to 6.14) (Ladenheim et al., 1992, 1993b). This work demonstrated that both GRP and NMB receptor subtypes were widely and distinctly distrib-

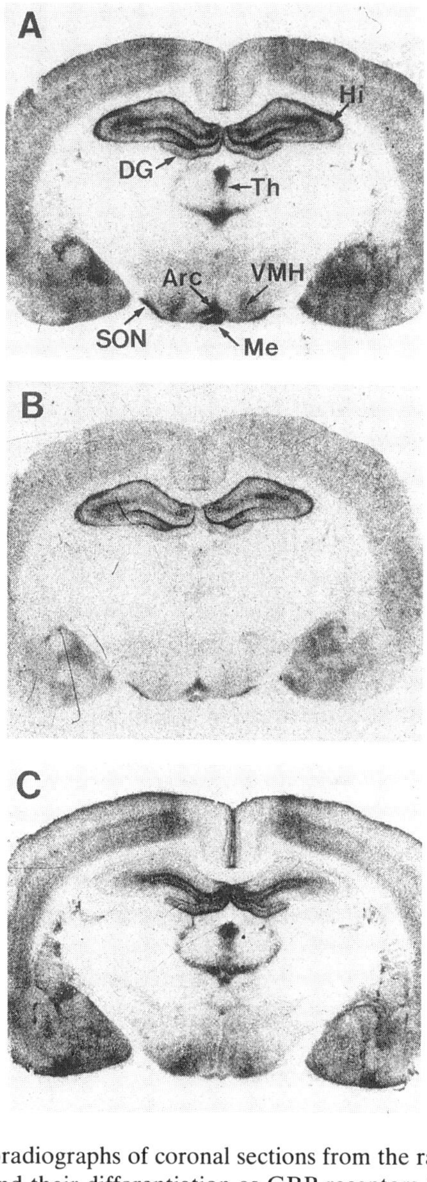

FIGURE 6.12. Autoradiographs of coronal sections from the rat brain showing BN-binding sites (A) and their differentiation as GRP receptors (B) and NMB receptors (C). **(A)** Total binding using ^{125}I-Tyr$_4$ BN. High densities of binding were present in the SON, Th, Arc,Hi, DG, Me, and VMH. **(B)** ^{125}I-Tyr$_4$ BN binding in the presence of 100 nM NMB to demonstrate GRP receptor sites. These are localized to the SON, Arc, and Hi. **(C)** Total binding using ^{125}I-D-Tyr$_0$ NMB. High densities of binding in Th, DG, VMH, and the cortex. SON, supraoptic nucleus; DG, dentate gyrus, Hi, hippocampus; VMH, ventromedial hypothalamus; Arc, arcuate nucleus, Th, thalamus; Me, median eminence. From Ladenheim et al. (1992), with permission of the publisher.

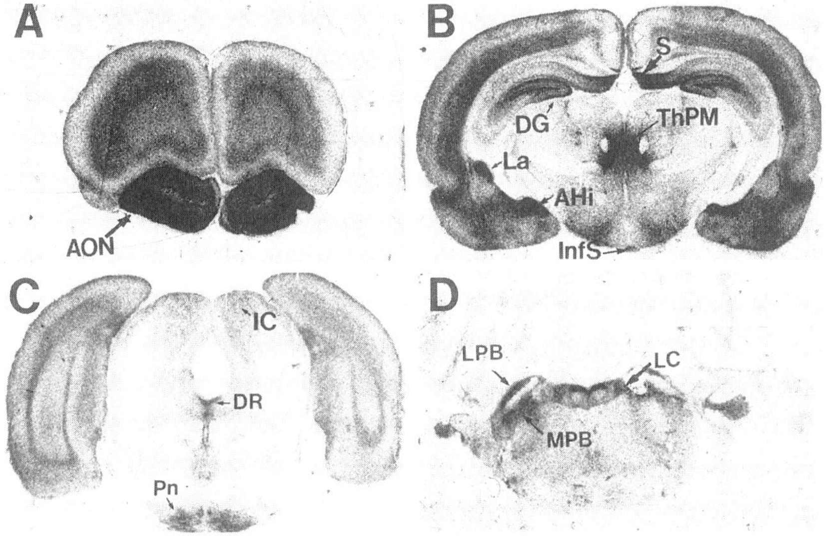

FIGURE 6.13. Autoradiographs of ^{125}I-D-Tyr$_0$ NMB binding to rat coronal brain sections demonstrating NMB-binding sites. **(A–D)** Sections running from forebrain to hindbrain. AHi, amygdalohippocampal area; AON, anterior olfactory nucleus; DG, dentate gyrus; DR, dorsal raphe; IC, inferior colliculus; InfS, infundibular stem; La, lateral amygdaloid nucleus; LC, locus coeruleus; LPB, lateral parabrachial nucleus; MPB, medial parabrachial nucleus; Pn, pontine nuclei; S, subiculum; ThPM, posteromedian thalamic nucleus. From Ladenheim et al. (1992), with permission of the publisher.

uted in the rat brain. The dominant receptor type in rat brain is the NMB receptor. Particularly high densities of this receptor were found in the olfactory bulb, olfactory nucleus, and various cortical layers and regions, especially in the pyriform cortex. High densities of NMB receptors were also found in various thalamic nuclei, in the dorsal tegmental area, in dorsal medial aspects of the interpeduncular nuclei, and in hindbrain regions, such as the parabrachial nuclei, and the locus coeruleus. Furthermore, as is the case with CCK_A receptors, they are found in the nucleus of the solitary tract. In contrast, GRP receptors were found with high density in the nucleus of the lateral olfactory tract, the bed nucleus of the stria terminalis, and in various hypothalamic nuclei. GRP-binding sites were also found in the dentate gyrus and in the substantia nigra. In the hindbrain, they were localized to the inferior olive and the spinal trigeminal nucleus. There were a few brain sites that appeared to contain heterogeneous mixtures of both receptor subtypes. These structures included the medial aspects of the bed nucleus of the stria terminalis, cortical

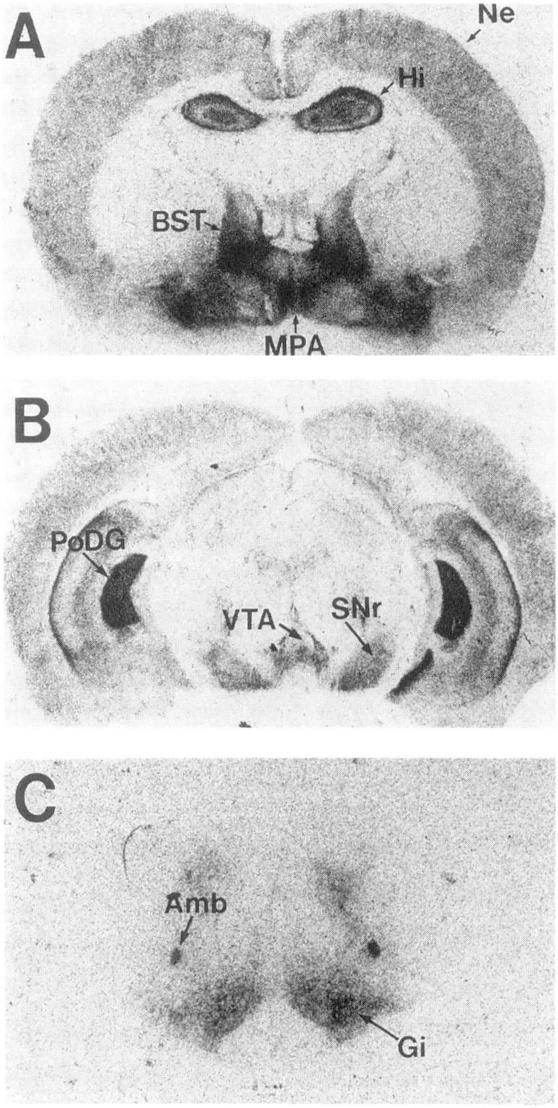

FIGURE 6.14. Autoradiographs of ^{125}I-Tyr$_4$ BN in the presence of 100 nM NMB demonstrating GRP-binding sites. **(A–C)** Sections running from forebrain to hindbrain. Amb, nucleus ambiguus; BST, bed nucleus of the stria terminalis; Gi, gigantocellular reticular nucleus; Hi, hippocampus; MPA, medial preoptic area; Ne, neocortex; PoDG, polymorphic layer of the dentate gyrus; SNr, substantia nigra, reticular; VTA, ventral tegmental area. From Ladenheim et al. (1992), with permission of the publisher.

layers I–III, the nucleus accumbens, the medial amygdala, and the medial preoptic area. BN-binding sites have also been found in the spinal cord (Zarbin et al., 1985). These have not yet been pharmacologically characterized. Distinct distributions for these two receptor subtypes is also confirmed by the distinct distributions of mRNA for their expression (Wada et al., 1990).

We have taken the same approach in attempting to provide a more complete characterization of BN-binding sites in the gastrointestinal tract. As recently demonstrated by Ladenheim et al. (1992), BN-binding sites in the circular muscle layer of the gastric corpus and fundus appear to be GRP receptors. Binding sites localized to the submucosal layer of the duodenum, jejunum, and ileum demonstrate a high affinity for NMB, while the colon appears to contain both receptor subtypes, GRP receptors localized to the circular muscle layer and NMB receptors localized to the submucosal layer. Although binding can be associated with these cellular layers, we do not have sufficient resolution with autoradiography to identify the specific cells that contain the binding sites. For example, the binding sites localized to the submucosal layer may be directly on secretory cells, enteric neurons, or sympathetic fibers innervating these areas.

We have also investigated whether BN binding sites are present and transported in the rat vagus nerve (Ladenheim et al., 1993a). Using an identical ligation procedure to that used for CCK receptors, there was no evidence of BN binding around a ligature site. Similarly, there was no change in BN binding in the NTS following unilateral nodosectomy. Thus the vagus does not appear to contain BN receptors. This result also implies that the high density of BN binding sites in the NTS are not of vagal origin and are located postsynaptic to vagal afferent inputs.

Pharmacological Characterization of the Satiating Action of Bombesin-Like Peptides

As detailed in Chapter 5, BN inhibits food intake following both peripheral and central administration. Since BN has the ability to interact with two mammalian receptor populations with high affinity, the satiating actions of peripheral or central BN could depend on interactions with one or both receptor subtypes.

Peptides that are structurally related to BN have been shown to inhibit food intake following both central and peripheral administration. Peripheral administration of litorin, ranatensin, GRP, acetylated GRP, $GRP_{18\text{-}27}$, and NMB have been shown to reduce food intake in a manner similar to BN with differing relative potencies (Kulkosky and Gibbs, 1982; Kulkosky et al.,

1982; Stein and Woods, 1983; DeCaro et al., 1984; DiPaola and Gibbs, 1985; Hostetler et al., 1989). A direct comparison of the satiating potency of peripheral BN, GRP, $GRP_{18\text{-}27}$, and NMB has demonstrated that BN is more potent than GRP and $GRP_{18\text{-}27}$, which were both more potent than NMB, suggesting a satiating action mediated by GRP-preferring receptors (Ladenheim et al., 1996). However, a close examination of the dose–response curves for these agonists suggested an alternative interpretation. The–dose response curve to GRP and $GRP_{18\text{-}27}$ has three components, a partial effect at low doses, a plateau at middle doses, and a further suppression at high doses. The molar difference between the two inhibitory components of the dose–response curve conforms with the differing affinities of GRP and $GRP_{18\text{-}27}$ for the two receptor subtypes. Thus, GRP and $GRP_{18\text{-}27}$ could be affecting intake at low doses through an interaction with the GRP receptor, and the further inhibition at higher doses could be the result of interaction with the NMB site (Ladenheim et al., 1996). This conclusion is also supported by data demonstrating that doses of GRP and NMB combine to produce inhibitions of food intake comparable to those produced by BN (Ladenheim et al., 1996). While more work is necessary, it appears that peripherally administered BN is affecting intake through an interaction with both GRP and NMB receptors (see Chapter 5 for further discussion).

Some comparisons have also been made of the ability of BN-like peptides to inhibit food intake and other behaviors following central administration. The relative abilities of the 32 amino acid form of NMB (NMB-32), the 10 amino acid form of NMB (NMB-10), and BN to inhibit ingestion and induce grooming when administered in the third ventricle were compared (Piggins et al., 1989). The results demonstrated that the NMB compounds are roughly one half to one log unit less potent than BN in affecting food intake. However, third ventricular administration of all these compounds also resulted in grooming and did so at lower dosages than those that inhibited food intake. Thus, the specificity of the feeding inhibitory action of these peptides following third ventricular administration is open to question. As discussed in Chapter 5, fourth ventricular administration of BN is significantly more potent than either third or lateral ventricular administration (Ladenheim and Ritter, 1988; Flynn, 1989). The feeding inhibitory actions of fourth ventricular BN administration appears to be behaviorally specific in that feeding is inhibited at doses that do not produce behavioral activation or grooming. As yet, however, no complete comparisons of the potencies of BN, GRP, and NMB to inhibit intake following fourth ventricular administration have been reported. Thus, the agonist pharmacology of the feeding inhibitory action of central BN-like peptides remains to be determined.

Recently, various BN antagonists have become available, and their relative affinity for the GRP and NMB receptor subtypes in the periphery and brain have been determined (Coy et al., 1988; von Schrenck et al., 1990; Ladenheim et al., 1993a). Two types of antagonists have been produced. One type demonstrates nanomolar affinity for the GRP receptor, but only high micromolar affinity for the NMB receptor (von Schrenck et al., 1990). The other has relative specificity for the NMB receptor (Orbuch et al., 1993). Use of these compounds in feeding studies has been limited and again no within-study comparisons have been made. Peripheral administration of GRP antagonists has been demonstrated to block the satiating actions of peripheral BN and GRP (Kirkham et al., 1994). Furthermore, ventricular administration of specific GRP antagonists blocks or attenuates the feeding inhibitory effects of both centrally (Merali et al., 1988) and peripherally (Ladenheim et al., 1993c; Motamedi et al., 1993) administered BN. In contrast, the peripheral administration of NMB antagonists blocked the feeding inhibitory action of peripheral NMB but was ineffective against $GRP_{18\text{-}27}$, again suggesting an independent role for NMB receptors in the suppression of intake by BN-like peptides (Ladenheim et al., 1994).

Central or peripheral administration of GRP-specific antagonists in the absence of exogenous agonist have been reported to increase food intake (Merali et al., 1990; Reidelberger et al., 1991; Flynn, 1992; Stratford et al., 1995). No reports on the ability of a specific NMB antagonist to increase food intake have appeared.

Summary

The sites and the pharmacological characteristics of the receptor subpopulation through which the feeding inhibitory actions of BN-like peptides are mediated have not yet been identified. However, considerable progress in characterizing the potential systems underlying the behavioral actions of these peptides has been made. It has been demonstrated that BN interacts with at least two receptor populations, and these receptors have been characterized as GRP and NMB sites. The pharmacological characteristics of these sites have been identified using both agonist and antagonist compounds. The genes for these receptors have been differentiated and cloned. The distinct distributions of these receptors in both the brain and the gastrointestinal tract have been displayed. A pharmacological characterization of the feeding inhibitory actions of BN-like peptides has begun, and it appears likely that the feeding inhibitory actions of BN-like peptides are mediated through the activation of both the GRP and NMB receptor subtypes. The specific location of the critical receptor populations for the feeding inhibitory actions of either central or peripherally administered BN-like peptides are yet to be determined.

Conclusions

The strategy of identifying receptor subpopulations through which the feeding inhibitory actions of CCK and BN-like peptides are mediated has been fruitful. Receptor subpopulations have been recognized based on anatomical location and pharmacological characteristics. Identification of these subpopulations has significantly advanced the understanding of the physiological role of these peptides.

Progress in identifying primary candidate receptor populations for the mediation of peptidergic satiating actions is significantly more advanced for CCK than for BN-like peptides. However, the mammalian systems underlying BN's feeding effects have now been identified, and the pharmacological tools for identifying the specific mediation of the actions of these peptides in food intake are now available.

ACKNOWLEDGMENTS

This work was supported by NIH grants DK19302 and DK46448.

REFERENCES

Asin, K. E., and L. Bednarz. Differential effects of CCK–JMV-180 on food intake in rats and mice. *Pharmacol. Biochem. Behav.* 42:291–295, 1992.

Battey, J. F., J. M. Way, M. H. Corjay, H. Shapira, K. Kusano, R. Harkins, J. M. Wu, T. Slattery, E. Mann, and R. I. Feldman. Molecular cloning of the bombesin/gastrin releasing peptide receptor in Swiss 3T3 cells. *Proc. Natl. Acad. Sci. U.S.A.* 88:395–399, 1991.

Blackshaw, L. A., and D. Grundy. Effects of cholecystokinin (CCK-8) on two classes of gastroduodenal vagal afferent fibre. *J. Auton. Nerv. Syst.* 31:191–202, 1990.

Brenner, L. A., and R. C. Ritter. CCK-A but not CCK-B antagonists attenuates suppression of food intake by exogenous CCK or intestinal oleate. *Soc. Neurosci.* 16:978, 1990.

Bylund, D. B. Analysis of receptor binding data. In *Receptor Binding Techniques*. Bethesda, MD: Society for Neuroscience, 1980, pp. 70–99.

Corp, E. S., J. McQuade, T. H. Moran, and G. P. Smith. Characterization of type A and type B CCK receptor binding sites in rat vagus nerve. *Brain Res.* 623:161–166, 1993.

Coy, D. H., P. Heinz-Erian, N.-Y. Jiang, J. E. Taylor, J.-P. Moreau, J. D. Gardner, and R. T. Jensen. Progress in the development of competitive bombesin antagonists. *Ann. N.Y. Acad. Sci.* 547:150–157, 1988.

Crawley, J. N. Clarification of the behavioral functions of peripheral and central cholecystokinin. *Peptides* 6(Suppl 2):129–136, 1985.

Crawley, J. N., D. W. Hommer, and L. R. Skirboll. Topographical analysis of nucleus accumbens sites at which cholecystokinin potentiates dopamine induced hyperlocomotion. *Brain Res.* 355:337–346, 1985.

Davison, J. S., and G. D. Clarke. Mechanical properties and sensitivity to CCK of vagal gastric slowly adapting mechanoreceptors. *Am. J. Physiol.* 255:G55–G61, 1988.

DeCaro, G., M. Massi, L. G. Micossi, and M. Perfumi. Drinking and feeding inhibition by ICV pulse injection or infusion of bombesin, ranatensin and litorin to rats. *Peptides* 5:607–613, 1984.

Della-Fera, M. A., and C. A. Baile. Cholecystokinin octapeptide: Continuous picomole injections into the cerebral ventricles of sheep suppress feeding. *Science* 212:687–688, 1979.

DiPaola, J. A., and J. Gibbs. Neuromedin C inhibits food intake in rats. *Soc. Neurosci.* 11:38, 1985.

Dockray, G. J. Immunochemical evidence of cholecystokinin-like peptides in brain. *Nature* 264:568–570, 1976.

Dourish, C. T., A. C. Ruckert, F. D. Tattersall, and S. D. Iversen. Evidence that decreased feeding induced by systemic injection of cholecystokinin is mediated by CCK-A receptors. *Eur. J. Pharmacol.* 173:233–234, 1989a.

Dourish, C. T., W. Rycroft, and S. D. Iversen. Postponement of satiety by blockade of brain cholecystokinin receptors. *Science* 245:1509–1511, 1989b.

Edwards, G. L., E. E. Ladenheim, and R. C. Ritter. Dorsomedial hindbrain participation in cholecystokinin-induced satiety. *Am. J. Physiol.* 251:R971–R977, 1986.

Enna, S. J. Basic receptor methods. In *Receptor Binding Techniques.* Bethesda, MD: Society for Neuroscience, 1980, pp. 33–52.

Flynn, F. W. Fourth ventricle bombesin injection suppresses ingestive behaviors in rats. *Am. J. Physiol.* 256:R590–R596, 1989.

Flynn, F. W. Fourth ventricular injection of selective bombesin receptor antagonists facilitates feeding in rats. *Am. J. Physiol.,* 264:R218–R221, 1992.

Galas, M. C., M. F. Lignon, M. Rodriguez, C. Mendre, P. Fulcrand, J. Laur, and J. Martinez. Structure activity relationship studies on cholecystokinin: Analogues with partial agonist activity. *Am. J. Physiol.* 254:G176–G182, 1988.

Gibbs, J., R. C. Young, and G. P. Smith. Cholecystokinin elicits satiety in rats with open gastric fistulas. *Nature* 245:323–325, 1973.

Gross, P. M., K. M. Wall, J. J. Pang, S. W. Shaver, and D. S. Wainman. Microvascular specializations promoting rapid interstital solute dispersion in the nucleus tractus solitarius. *Am. J. Physiol.* 259:R1131–R1138, 1990.

Hewson, G., G. E. Leighton, R. G. Hill, and J. Hughes. The cholecystokinin receptor antagonist L364,718 increases food intake in the rat by attenuation of the action of endogneous cholecystokinin. *Br. J. Pharmacol.* 93:79–84, 1988.

Hill, D. R., N. J. Campbell, T. M. Shaw, and G. N. Woodruff. Autoradiographic localization and biochemical characterization of peripheral type CCK receptors in rat CNS using highly selective nonpeptide CCK antagonists. *J. Neurosci.* 7:2967–2976, 1987.

Hill, D. R., T. M. Shaw, and G. N. Woodruff. Species differences in the localization of "peripheral" type cholecystokinin receptors in rodent brain. *Neurosci. Lett.* 79:286–289, 1988a.

Hill, D. R., T. M. Shaw, and G. N. Woodruff. Binding sites for ^{125}I-cholecystokinin in primate spinal cord are CCK A subclass. *Neurosci. Lett.* 89:133–139, 1988b.

Hill, R. D. and G. N. Woodruff. Differentiation of central cholecystokinin receptor binding sites using the non-peptide antagonists MK-329 and L-365,260. *Brain Res.* 526:276–283, 1990.

Hostetler, A. M., P. R. McHugh, and T. H. Moran. Bombesin affects feeding independent of a gastric mechanism or site of action. *Am. J. Physiol.* 257:R1219–1224, 1989.

Innis, R. B., and S. H. Snyder. Distinct cholecystokinin receptors in brain and pancreas. *Proc. Natl. Acad. Sci. U.S.A.* 77:6917–6921, 1980.

Jensen, T. T., G. F. Lemp, and J. D. Gardner. Interaction of cholecystokinin with specific membrane receptors on pancreatic acinar cells. *Proc. Natl. Acad. Sci. U.S.A.* 77:2079–2083, 1980.

Jensen, R. T., T. Moody, J. E. Pert, J. E. Rivier, and J. D. Gardner. Interaction of bombesin and litorin with specific membrane receptors on pancreatic acinar cells. *Proc. Natl. Acad. Sci. U.S.A.* 75:6139–6143, 1978.

Kirkham, T. C., C. A. Walsh, J. Gibbs, G. P. Smith, J. Leban, and J. McDermed. A novel bombesin receptor antagonist selectively blocks the satiety action of peripherally administered bombesin. *Pharmacol. Biochem. Behav.* 48:809–811, 1994.

Kulkosky, P. J., and J. Gibbs. Litorin suppresses food intake in rats. *Life Sci.* 31:685–692, 1982.

Kulkosky, P. J., J. Gibbs, and G. P. Smith. Behavioral effects of bombesin administration in rats. *Physiol. Behav.* 28:505–512, 1982.

Ladenheim, E. E., R. T. Jensen, S. A. Mantey, P. R. McHugh, and T. H. Moran. Receptor heterogeneity for bombesin-like peptides in the rat central nervous system. *Brain Res.* 537:233–240, 1990.

Ladenheim, E. E., R. T. Jensen, S. A. Mantey, and T. H. Moran. Distinct distributions of two bombesin receptor subtypes in the rat central nervous system. *Brain Res.* 593:168–178, 1992.

Ladenheim, E. E., R. T. Jensen, S. A. Mantey, and T. H. Moran. Bombesin binding sites in the nucleus of the solitary tract are not of vagal origin. *Soc. Neurosci.,* 19:1270, 1993a.

Ladenheim, E. E., R. T. Jensen, S. A. Mantey, J. E. Taylor, D. H. Coy, and T. H. Moran. Bombesin receptor antagonists differentiate receptor subtypes in rat brain. *Eur. J. Pharmacol.* 235:121–125, 1993b.

Ladenheim, E. E., and R. C. Ritter. Low-dose fourth ventricular bombesin selectively suppresses food intake. *Am. J. Physiol.* 255:R988–R992, 1988.

Ladenheim, E. E., R. C. Speth, and R. C. Ritter. Reduction of CCK-8 binding in the nucleus of the solitary tract in unilaterally nodosectomized rats. *Brain Res.* 474:125–129, 1988.

Ladenheim, E. E., J. E. Taylor, and T. H. Moran. Peripheral bombesin satiety is partially blocked by a central GRP antagonist. *ICPFFI* 11:57, 1993c.

Ladenheim, E. E., J. E. Taylor, D. H. Coy, and T. H. Moran. Blockade of feeding inhibition by neuromedin B using a selective receptor antagonist. *Eur. J. Pharmacol.*, 271:R7–R9, 1994.

Ladenheim, E. E., K. Wirth, and T. H. Moran. Receptor subtype mediation of feeding suppression by bombesin-like peptides. *Pharmacol. Biochem. Behav.* 54:705–711, 1996.

Larsson, L. I., and J. F. Rehfeld. Localization and molecular heterogeneity of cholecystokinin in the central and peripheral nervous system. *Brain Res.* 165:201–218, 1978.

Lorenz, D. N., G. Kreielsheimer, and G. P. Smith. Effect of cholecystokinin, gastrin, secretin and GIP on sham feeding in rats. *Physiol. Behav.* 23:1065–1072, 1979.

Margolis, R. L., T. H. Moran, and P. R. McHugh. In vitro response of rat gastrointestinal segments to cholecystokinin and bombesin. *Peptides* 10:157–161, 1989.

Matozaki, T., J. Martinez, and J. A. Williams. A new CCK analogue differentiates two functionally distinct CCK receptors in rat and mouse pancreatic acini. *Am. J. Physiol.* 257:G594–G600, 1989.

Merali, Z, T. W. Moody, P. Kateb, and H. Piggins. Antagonism of satiety and grooming effects of bombesin by antiserum to bombesin and by [Tyr^4, D-Phe^{12}] bombesin: Central versus peripheral effects. *Ann. N.Y. Acad. Sci.* 547:489–492, 1988.

Merali, Z., T. W. Moody, and D. H. Coy. Blockade of the central bombesin receptors enhances food intake in prefed rats. *Soc. Neurosci.* 16:978, 1990.

Mercer, J. G., and C. B. Lawrence. Selectivity of cholecystokinin receptor antagonists, MK-329 and L-365,260, for axonally transported CCK binding sites in the rat vagus nerve. *Neurosci. Lett.* 137:229–231, 1992

Moody, T. W., C. B. Pert, J. Rivier, and M. R. Brown. Bombesin: Specific binding to rat brain membranes. *Proc. Natl. Acad. Sci. U.S.A.* 75:5372–5376, 1978.

Moran, T. H., P. J. Ameglio, G. J. Schwartz, and P. R. McHugh. Blockade of type A, not type B, CCK receptors attenuated satiety actions of exogenous and endogenous CCK. *Am. J. Physiol.* 262:R46–R50, 1992.

Moran, T. H., R. J. Crosby, and P. R. McHugh. Effects of pylorectomy on cholecystokinin (CCK) induced inhibition of liquid gastric emptying. *Am. J. Physiol.* 261:R531–R535, 1991a.

Moran, T. H., R. Kornbluh, K. Moore, and G. J. Schwartz. Cholecystokinin inhibits gastric emptying and contracts the pyloric sphincter in rats by interacting with low affinity CCK receptor sites. *Regul. Pept.* 52:165–172, 1994.

Moran, T. H., and P. R. McHugh. Anatomical and pharmacological differentiation of pyloric vagal and brain stem cholecystokinin receptors. In Wang R. Y., and R. Schoenfeld (eds.): *Cholecystokinin Antagonists.* New York: Alan R. Liss, 1987, pp. 117–132.

Moran, T. H., and P. R. McHugh. Vagal receptor transport. In Ritter, R. C., S. Ritter, and C. D. Barnes (eds.): *Neuroanatomy and Physiology of Abdominal Vagal Afferents.* Boca Raton: CRC Press, 1992, pp. 157–178.

Moran, T. H., T. W. Moody, A. M. Hostetler, P. H. Robinson, M. Goldrich, and P. R. McHugh. Distribution of bombesin binding sites in the rat gastrointestinal tract. *Peptides* 9:643–649, 1988a.

Moran, T. H., R. Norgren, R. J. Crosby, and P. R. McHugh. Central and peripheral vagal transport of cholecystokinin binding sites occurs in afferent fibers. *Brain Res.* 526:95–102, 1990.

Moran, T. H., P. H. Robinson, M. S. Goldrich, and P. R. McHugh. Two brain cholecystokinin receptors: Implications for behavioral actions. *Brain Res.* 362:175–179, 1986.

Moran, T. H., L. Shnayder, G. J. Schwartz, and P. R. McHugh. Pyloric cholecystokinin receptors. In Adler G., and C. Beglinger (eds.): *Cholecystokinin Antagonists in Gastroenterology: Basic and Clinical Status.* Berlin: Springer Verlag, 1991b, pp. 159–164.

Moran, T. H., L. Shnayder, A. M. Hostetler, and P. R. McHugh. Pylorectomy reduces the satiety action of cholecystokinin. *Am. J. Physiol.* 255:R1059–R1063, 1988b.

Moran, T. H., G. P. Smith, A. M. Hostetler, and P. R. McHugh. Transport of cholecystokinin (CCK) binding sites in subdiaphragmatic vagal branches. *Brain Res.* 415:149–152, 1987.

Motamedi, F., A. Rashidy-Pour, M. R. Zarrindast, and M. Badavi. Bombesin-induced anorexia requires central bombesin receptor activation: Independence from interaction with central catecholamine systems. *Psychopharmacology* 110:193–197, 1993.

Murphy, R. B., L. H. Schneider, and G. P. Smith. Peripheral loci for the mediation of cholecystokinin-induced satiety. In Wang, R., and R. Schoenfeld (eds.): *CCK Antagonists.* New York: Alan R. Liss, 1988, 173–191.

Murphy, R. B., G. P. Smith, and J. Gibbs. Pharmacological examination of cholecystokinin (CCK-8)-induced contractile activity in the rat isolated pylorus. *Peptides* 8:127–134, 1987.

Orbuch, M., J. E. Taylor, D. H. Coy, J. E. Mrozinski, S. A. Mantey, J. F. Battey, J. P. Moreau, and R. T. Jensen. Discovery of a novel class of neuromedin B antagonists, substituted somatostatin analogs. *Mol. Pharmacol.* 44:841–850, 1993.

Passaro, E., H. Debas, W. Oldendorf, and T. Yamada. Rapid appearance of intraventricularly administered neuropeptides in the peripheral circulation. *Brain Res.* 241:338–340, 1982.

Piggins, H., D. Lafreniere, and Z. Merali. Effects of neuromedin B-32, neuromedin B-10 and bombesin on ingestive and grooming behaviors. *Soc. Neurosci.* 15:712, 1989.

Reidelberger, R. D., and T. E. Solomon. Comparative effects of CCK-8 on feeding, sham feeding, and exocrine pancreatic secretion in rats. *Am. J. Physiol.* 251:R97–R105, 1986.

Reidelberger, R. D., and M. F. O'Rourke. Potent cholecystokinin antagonist L-364,718 stimulates food intake in rats. *Am. J. Physiol.* 257:R1512–R1518, 1989.

Reidelberger, R. D., G. Varga, and T. E. Solomon. Comparative effects of CCK-A receptor antagonist L-364,718 and CCK-B receptor antagonist L-365,260 on solid food intake in rats. *Soc. Neurosci.* 16:978, 1990.

Reidelberger, R. D., G. Varga, T. Castonguay, and D. H. Coy. Potent bombesin receptor antagonist D-Phe6 bombesin (6-13)-methyl ester stimulates food intake in rats. *Soc. Neurosci.* 17:543, 1991.

Robinson, P. H., P. R. McHugh, T. H. Moran, and J. D. Stephenson. Gastric control of food intake. *J. Psychosom. Res.* 32:593–606, 1988.

Saito, A., I. D. Goldfine, and J. A. Williams. Characterization of receptors for cholecystokinin and related peptides in mouse cerebral cortex. *J. Neurochem.* 37:483–490, 1981.

Saito, A., H. Sankaran, I. D. Goldfine, and J. A. Williams. Cholecystokinin receptors in the brain: Characterization and distribution. *Science* 208:1155–1156, 1980.

Sankaran, H., I. D. Goldfine, C. W. Deveney, K. Y. Wong, and J. A. Williams. Binding of cholecystokinin to high affinity receptors on isolated rat pancreatic acinin. *J. Biol. Chem.* 255:1849–1853, 1980.

Sankaran, H., I. D. Goldfine, A. Bailey, V. Licko, and J. A. Williams. Relationship of cholecystokinin receptor binding to regulation of biological functions in pancreatic acini. *Am. J. Physiol.* 242:G250–G257, 1982.

Schick, R. R., T. L. Yaksh, and V. L. W. Go. Intracerebroventricular injections of cholecystokinin suppress feeding in rats—Pharmacological characterization of this action. *Regul. Pept.* 14:277–291, 1986.

Schwartz, G. J., P. R. McHugh, and T. H. Moran. Integration of vagal afferent responses to gastric loads and cholecystokinin in rats. *Am. J. Physiol.* 261: R64–R69, 1991.

Schwartz, G. J., P. R. McHugh, and T. H. Moran. Gastric loads and cholecystokinin synergistically stimulate rat gastric vagal afferents. *Am. J. Physiol.* 265:R872–R876, 1993.

Schwartz, G. J., P. R. McHugh, and T. H. Moran. Pharmacological dissociation of responses to CCK and gastric loads in rat mechanosensitive vagal afferents. *Am. J. Physiol.* 267:R303–R308, 1994.

Schwartz, G. J., G. Tougas, and T. H. Moran. Integration of vagal afferent responses to duodenal loads and exogenous CCK in rats. *Peptides* 16:707–711, 1995.

Silver, A. J., J. F. Flood, A. M. Song, and J. E. Morley. Evidence for a physiological role for CCK in the regulation of food intake in mice. *Am. J. Physiol.* 256:R646–R652, 1989.

Smith, G. P., J. Falasco, T. H. Moran, K. M. S. Joyner, and J. Gibbs. CCK-8 decreases food intake and gastric emptying after pylorectomy or pyloroplasty. *Am. J. Physiol.* 255:R113–116, 1988.

Smith, G. P., C. Jerome, B. J. Cushin, R. Eterno, and K. J. Simansky. Abdominal vagotomy blocks the satiety effect of cholecystokinin in the rat. *Science* 213:1036–1037, 1981a.

Smith, G. P., C. Jerome, and G. P. Gibbs. Abdominal vagotomy does not block the satiety effect of bombesin in the rat. *Peptides* 2:409–411, 1981b.

Smith, G. P., C. Jerome, and R. Norgren. Afferent axons in abdominal vagus mediate satiety effect of cholecystokinin in rats. *Am. J. Physiol.* 249:R638–R641, 1985.

Smith, G. T., T. H. Moran, J. T. Coyle, M. J. Kuhar, T. L. O'Donahue, and P. R. McHugh. Anatomical localization of cholecystokinin receptors in the pyloric sphincter. *Am. J. Physiol.* 246:R127–R130, 1984.

Stark, H. A., C. M. Sharp, V. E. Sutliff, J. Martinez, R. T. Jensen, and J. D. Gardner. CCK-JMV-180: A protein that distinguishes high affinity cholecystokinin receptors from low affinity cholecystokinin receptors. *Biochem. Biophys. Acta* 1010:145–150, 1989.

Steigerwalt, R. W., and J. A. Williams. Characterization of cholecystokinin receptors on rat pancreatic membranes. *Endocrinology* 109:1746–1753, 1981.

Stein, L. J. and S. C. Woods. GRP reduces meal size in rats. *Peptides* 3:833–835, 1983.

Stratford, T. B., J. Gibbs, D. H. Coy, and G. P. Smith. Fourth ventricular injection of the bombesin receptor antagonist [D-Phe^6]Bombesin(6-13)methyl ester, but not BW2258U89, increases food intake in rats. *Pharmacol. Biochem. Behav.* 50:463–471, 1995.

Strohmayer, A. J., D. Greenberg, T. H. Moran, and P. R. McHugh. CCK receptor binding in the pylorus of genetically obese mice (ob/ob). *Soc. Neurosci.* 13:587, 1987.

Vanderhaegen, J. J., J. C. Signeau, and W. Gepts. New peptide in the vertebrate CNS reacting with antigastrin antibodies. *Nature* 257:603–605, 1975.

Van der Kooy, D. Area postrema: Site where cholecystokinin acts to decrease food intake. *Brain Res.* 295:345–347, 1984.

von Schrenck, T., P. Heinz-Erian, T. H. Moran, S. A. Mantey, J. D. Gardner, and R. T. Jensen. Neuromedin B receptor in esophagus: Evidence for subtypes of bombesin receptors. *Am. J. Physiol.* 256:G747–G758, 1989.

von Schrenck, T., L. H. Wang, D. H. Coy, M. L. Villanueva, S. Mantey, and R. T. Jensen. Potent bombesin receptor antagonists distinguish receptor subtypes. *Am. J. Physiol.* 259:G468–G473, 1990.

Wada, E., J. Way, A. M. Lebacq-Verheyden, and J. F. Battey. NMB and gastrin releasing peptide mRNAs are differentially distributed in the rat nervous system. *J. Neurosci.* 10:2917–2930, 1990.

Wada, E., J. Way, H. Shapira, K. Kusano, A. M. Lebacq-Verheyden, D. Coy, R. Jensen, and J. Battey. cDNA cloning, characterization and brain region-specific expression of a neuromedin-B preferring bombesin receptor. *Neuron* 6:421–430, 1991.

Wank, S. A., R. Harkins, R. T. Jensen, H. Shapira, A. de Weerth, and T. Slattery. Purification, molecular cloning and functional expression of the cholecystokinin receptor from rat pancreas. *Proc. Natl. Acad. Sci. U.S.A.* 89:3125–3129, 1992a.

Wank, S. A., J. R. Pisegna, and A. de Weerth. Brain and gastrointestinal cholecystokinin receptor family: Structure and functional expression. *Proc. Natl. Acad. Sci. U.S.A.* 89:8691–8695, 1992b.

Weatherford, S. C., W. B. Laughton, J. Salabarria, W. Danho, J. W. Tilley, L. A. Netterville, G. J. Schwartz, and T. H. Moran. CCK satiety is differentially mediated by high and low affinity CCK receptors in the mouse and rat. *Am. J. Physiol.* 264:R244–R249, 1993.

Westendorf, J. M., and A. Schonbrunn. Characterization of bombesin receptors in rat pituitary cell line. *J. Biol. Chem.* 258:7527–7535, 1983.

White, F. J., and R. Y. Wang. Interactions of cholecystokinin and dopamine on nucleus accumbens. *Brain Res.* 300:161–166, 1984.

Wolf, S. S., and T. W. Moody. Receptors for GRP/bombesin-like peptides in the rat forebrain. *Peptides* 6:111–114, 1985.

Young, W. S., and M. J. Kuhar. A new method for receptor autoradiography: [^{3}H]opioid receptors in the rat brain. *Brain Res.* 179:255–270, 1979.

Young, W. S., and M. J. Kuhar. The light microscopic radiohistochemistry of drug and neurotransmitter receptors using diffusable ligands. In Johnson, J. E. Jr. (ed.): *Current Trends in Morphological Techniques.* Boca Raton, FL: CRC Press, 1980, pp. 119–136.

Zarbin, M. A., R. B. Innis, J. K. Wamsley, S. H. Snyder, and M. J. Kuhar. Autoradiographic localization of CCK receptors in the guinea pig brain. European *J. Pharmacol.* 71:877–906, 1981a.

Zarbin, M. A., R. B. Innis, J. K. Wamsley, S. H. Snyder, and M. J. Kuhar. Autoradiographic localization of cholecystokinin receptors in the rodent brain. *J. Neurosci.* 3:877–906, 1983.

Zarbin, M. A., M. J. Kuhar, T. L. O'Donohue, S. S. Wolf, and R. W. Moody. Autoradiographic localization of (^{125}I-Tyr4) bombesin binding sites in the rat brain. *J. Neurosci.* 5:429–437, 1985.

Zarbin, M. A., J. K. Wamsley, R. B. Innis, and M. J. Kuhar. CCK receptors: Presence and axonal flow in the rat vagus nerve. *Life Sci.* 29:697–705, 1981b.

7

Glucagon and the Control of Meal Size

NORI GEARY

By 1960 it was known that intramuscular or subcutaneous administration of glucagon decreased food intake in rats and humans (Salter, 1960; Schulman et al., 1957). Glucagon's inhibitory action on feeding, however, appeared to require large doses and to develop slowly in comparison to its metabolic effects (for reviews, see Geary, 1990; Penick and Smith, 1964). Furthermore, although dramatic inhibitions of feeding were produced by repeated glucagon administration, they were often accompanied by visceral and metabolic side effects. For these reasons, glucagon was not considered likely to participate in the normal physiology of appetite, and basic research on the problem languished. This situation changed with the publication of two watershed studies in 1977.

The first of these studies was Martin and Novin's report (1977) that hepatic-portal injections of 35–65 μg/kg glucagon just before refeeding 4-hour food-deprived rats produced a dose-related reduction of food intake. The inhibition was clearest 30–120 minutes postinjection and subsequently waned. The effect appeared behaviorally specific because water intake during the test was unaffected and because similar glucagon injections did not produce conditioned taste aversions for flavored water in water-deprived rats. These results suggested that *exogenous glucagon* can pharmacologically increase satiety in rats.

The second study, by de Jong et al. (1977), showed that plasma glucagon increases immediately after meal onset in rats. This occurred during consumption of foods that contained varying proportions of fat and carbohydrate and relatively little protein, which is usually considered the most potent glucagon secretogogue. Several other studies have confirmed the presence of prandial glucagon secretion during mixed nutrient meals in humans (Day et al., 1978; Denker et al., 1975; Holst, 1983b; Muller et al., 1971) and rats (Berthoud and Jeanrenaud, 1982; Langhans et al., 1984).

This is probably a cephalic phase reflex because it appears during the first minute of meals in rats (de Jong et al., 1977), occurs during sham feeding in dogs (Nilsson and Uvnas-Wallenstein, 1977), and is absent in diabetic-uremic patients after kidney–pancreas transplantation (Secchi et al., 1995). Furthermore, glucagon secretion can be produced by stimulation of the hypothalamus or the dorsal motor nucleus of the vagus (de Jong et al., 1977; Laughton and Powley, 1987; Palmer and Porte, 1983). Hepatic extraction of glucagon (Denker et al., 1975; Langhans et al., 1984) as well as hepatic glycogenolysis and glucose production (Langhans et al., 1982a) also increase during rats' meals. The existence of prandial glucagon secretion suggests the hypothesis that *endogenous glucagon* is a signal for terminating a meal.

This chapter reviews the current status of these two hypotheses concerned with the satiating effects of exogenous and endogenous glucagon, beginning with research involving exogenous glucagon.

Exogenous Glucagon and Meal Size in Rats

Test Meals

Martin and Novin's work (1977) did not reveal how exogenous glucagon affects meal patterns. Behavioral observations indicated that glucagon administration just before meals selectively reduced meal size. In rats refed at dark onset after food deprivation during the light phase, intraperitoneal injection of 120–360 μg glucagon just before refeeding reduced meal size by about 40%–50% (Geary et al., 1981). Similarly, in chow-fed rats offered a test meal of sweetened condensed milk during the light phase, intraperitoneal injection of 25–800 μg/kg glucagon reduced meal size by about 20%–40% and did not affect the subsequent intermeal interval (Geary and Smith, 1982a). Reductions in test meal size after glucagon administration were also observed in several similar tests in rats (Langhans et al., 1987; Strubbe et al., 1989; VanderWeele et al., 1986; Weick and Ritter, 1986a) (but see VanderWeele et al., [1980] for a different result) and in a variety of other animals (Geary, 1990).

Pretest food deprivation, circadian timing, and other experimental variables affect the potency of exogenous glucagon's effect on meal size. For example, although glucagon has consistently inhibited feeding in diurnal tests of mildly deprived or nondeprived rats, it failed to inhibit feeding in similar tests done at dark onset (Geary and Smith, 1982a; Martin and Novin, 1977; Stein et al., 1985; VanderWeele et al., 1986; Weatherford and

Ritter, 1986). Conversely, glucagon inhibited feeding in dark-onset tests of rats deprived of food for 6 hours or more, but not in diurnal tests after similar periods of food deprivation (Geary et al., 1987; Stein et al., 1985; VanderWeele et al., 1986; Weatherford and Ritter, 1986). This marked degree of situational variability probably accounts for glucagon's rejection as a possible satiety signal in influential early experiments that tested the feeding effects of gut peptides on diurnal test meals after overnight food deprivation (Schally et al., 1967), including work from the Bourne Laboratory (Smith and Gibbs, 1975). The mechanisms that produce this contextual specificity may be related to peripheral metabolic or endocrine variables or to differences in the central processing of the peripheral signal produced by glucagon.

Behavioral Specificity

The inhibitory effect of exogenous glucagon on test meal size appears to mimic natural postprandial satiety. Rats began to feed at the normal rate after intraperitoneal glucagon injections, but slowed and stopped sooner and displayed the normal sequence of postprandial satiety behaviors after meals despite eating less than usual (Geary and Smith, 1982a; Hinton et al., 1987a; Ritter et al., 1986). When glucagon reduced food intake, it did not produce anomalous behaviors, signs of toxicity, or conditioned taste aversions (Geary and Smith, 1982a,b; Hinton et al., 1987a; Martin and Novin, 1977). Glucagon also did not affect other appetitive behaviors: Glucagon failed to inhibit drinking in thirsty rats (Geary and Smith, 1982a; Hinton et al., 1987a) or running in rats given scheduled access to activity wheels (Geary et al., 1992a). Thus, there is no evidence that glucagon's inhibitory effect on feeding results from a nonspecific or toxic action. This is consistent with the hypothesis that exogenous glucagon accelerates the normal physiological mechanisms of satiety during the meal.

Spontaneous Meals

The situational variability of glucagon's potency to reduce test meal size, especially the lack of effect of glucagon on nocturnal test meals after brief food deprivation, raised doubts about glucagon's physiological relevance. Le Sauter and Geary (1991) therefore investigated glucagon's effect on spontaneous meals during the dark. Remotely controlled hepatic-portal infusions of 3.4–14 μg glucagon per meal dose-dependently reduced the size of nocturnal spontaneous meals by 20%–60% (Fig. 7.1). Note that spontaneous meals were reduced by smaller glucagon doses than typically required to

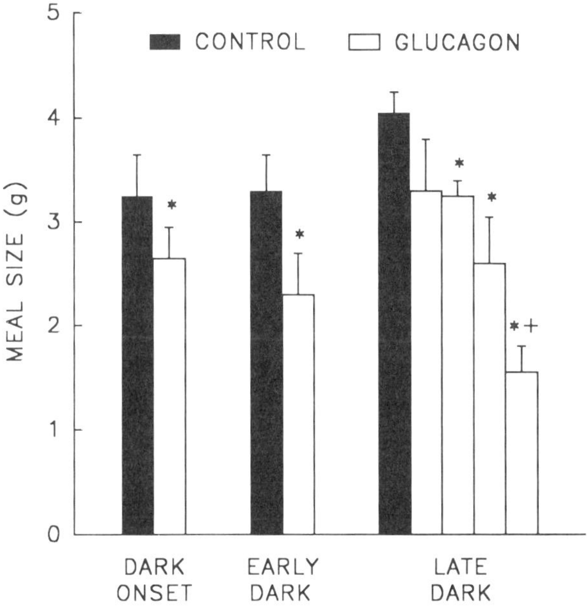

FIGURE 7.1. Hepatic-portal glucagon infusion reduces spontaneous meal size in ad libitum fed rats. Glucagon was infused for the first 2 minutes of the first spontaneous meal after dark onset **(left)**, the first spontaneous meal beginning at least 3 hours after dark onset **(middle)**, or the first spontaneous meal beginning at least 9.5 hours after dark onset **(right)**. Data are meal sizes, mean ± SEM. Dark onset and early dark glucagon doses were 13.6 μg glucagon/meal; late dark doses were, from left to right, 1.7, 3.4, 6.8, and 13.6 μg glucagon/meal. *Different from control, $p < .05$; +different from effect of other glucagon doses, $p < .05$. Data are from Le Sauter and Geary (1991).

inhibit feeding under other conditions (see above). Furthermore, hepatic-portal infusion of glucagon reduced spontaneous meal size both early and late in the dark phase, the times when rats initiate meals most frequently (Rosenwasser et al., 1981). These results not only extended the analysis of exogenous glucagon's effect on meal size to the more naturalistic behavioral and physiological context of spontaneous feeding, but also demonstrated that the effect is more robust under these conditions.

Physiological Dose

The relative magnitude of the increase in endogenous glucagon secretion during meals and the amount of exogenous glucagon required to reduce meal size in rats is not clear. As reviewed below, glucagon appears to act in the liver to inhibit feeding. Hepatic-portal glucagon levels produced by glucagon treatments that inhibit feeding have been directly compared to prandial endogenous glucagon levels only once. Langhans et al. (1987) administered glucagon to rats intraperitoneally after 12 hours of food deprivation. Plasma glucagon concentration in the hepatic-portal vein 15 minutes after meal onset was 447 ± 112 pg/ml after control injections and 12,710 ± 3,078 pg/ml after injection of the minimum effective glucagon dose for a significant inhibition of feeding, which in this experiment was 480 μg/kg. Thus, the satiating action of intraperitoneally injected glucagon is pharmacological.

When glucagon is infused into the hepatic-portal vein, glucagon doses that are more than 100 times less than Langhans et al.'s minimum effective dose (1987) are sufficient to inhibit feeding. In the spontaneous feeding situation that we have tested (Geary et al., 1992c; Le Sauter and Geary, 1991), hepatic-portal infusion of only 1.7–3.4 μg glucagon during the first 2 minutes of meals significantly reduced spontaneous meal size, and in a study by Weick and Ritter (1986a), continuous, 30-minute infusion of 0.33 μg/kg/min glucagon inhibited deprivation-induced feeding. The effects of these treatments on plasma glucagon levels have not been determined, but estimates of plasma glucagon concentration based on glucagon's metabolic clearance rate (or on the plasma half-life values derived from it) suggest that they also result in plasma glucagon concentrations over 1,000 pg/ml. These calculations result in gross overestimates, however, because they fail to account for the dynamics of hepatic extraction of glucagon. When the amount of glucagon reaching the liver increases, as occurs during meals, there is a large, transient increase in hepatic extraction of glucagon from basal levels of about 5%–15% to 30%–70% (Hildebrandt et al., 1991a; Ishida et al., 1982, Ishida and Field, 1983).

The dynamics of hepatic glucagon extraction has two consequences relevant to interpretation of plasma glucagon levels. First, brief hepatic-portal glucagon infusions will result in smaller increases in hepatic-portal glucagon concentration than suggested by glucagon's metabolic clearance rate. This is because the increased hepatic extraction of glucagon decreases the amount of glucagon that survives the first pass through the liver to recirculate into the portal vein. Second, acute increases in hepatic glucagon extraction, together with the dilution of the portal circulation into the systemic circulation, will increase the typically small basal difference between portal vein and systemic glucagon concentrations. This can be clearly seen during meals when both hepatic-portal vein and posthepatic plasma glucagon levels are measured (Denker et al., 1975; Langhans et al., 1984). In one study in rats, meals increased hepatic-portal concentration of glucagon by 300 pg/ml with *no* change in the concentration of glucagon in the hepatic vein (Langhans et al., 1984). Clearly, measurements of systemic glucagon concentration cannot be used to estimate prandial hepatic-portal glucagon concentration.

Three additional issues should also be considered in experiments directed at this question. First, the infusion parameters that minimize the threshold dose of glucagon for inhibition of feeding have not been established. The reduction in spontaneous meal size by 2-minute, meal-onset glucagon infusions was a graded function of the concentration of glucagon infused (Le Sauter and Geary, 1991). The total amount infused appeared irrelevant because a fixed glucagon concentration reduced meal size similarly whether it was infused during the first 2 minutes of meals, during the first 4 minutes of meals, or throughout entire meals, which lasted about 6–12 minutes. Whether this would also be true of even briefer infusions has not been tested.

Second, glucagon infusions that match the pattern of endogenous glucagon release, which may oscillate with periods of only 0.5–4 minutes (Hildebrandt et al., 1991b; Opara and Go, 1991), may inhibit feeding more effectively than continuous infusions. Pulsatile glucagon infusion has been shown to increase hepatic glucagon extraction (Hildebrandt et al., 1991a,b, 1992) and to increase hepatic glucose production under some conditions (Lefèbvre et al., 1987; Paolisso et al., 1989; Weigle, 1987). A preliminary study (Geary, 1996) failed to support the hypothesis that pulsatile delivery of glucagon would increase glucagon's satiating potency, but this possibility warrants more investigation.

Third, exogenous glucagon adds to the hypothesized satiating action of endogenous glucagon released during meals. Thus, any satiating action of endogenous glucagon would have to be taken into account in order to evaluate the physiological status of infused glucagon. These considerations indicate that additional research is required to determine whether hepatic-portal infusion of physiological doses of glucagon are sufficient to inhibit feeding.

Interactions

Exogenous glucagon appears to accelerate satiation, but it does not appear sufficient to end a meal in the absence of other postingestive mechanisms of satiety. This is because glucagon fails to inhibit sham feeding in rats with open gastric cannulas (Geary and Smith, 1982b). This phenomenon suggests that glucagon signals satiety only in the presence of gastric distension or some postgastric food stimulus. Among the postingestive mechanisms, cholecystokinin may be adequate. When glucagon was injected simultaneously with doses of cholecystokinin that had little or no effect when injected alone, sham feeding was inhibited markedly (Le Sauter and Geary, 1987, 1990). This functional interaction of cholecystokinin and glucagon may be synergistic because combined peptide injections inhibited feeding more than the sum of the individual effects.

Whether a similar synergy occurs during real feeding is unclear. It occurred in one test in rats (Hinton et al., 1986), but not in another (Hinton et al., 1987b), and it did not occur in dogs (Kalogeris et al., 1991) or humans (Geary et al., 1992b).

Glucagon may also interact positively with other controls of meal size. Simultaneous intraperitoneal injections of glucagon and bombesin produced a supra-additive inhibition of real feeding (Hinton et al., 1986), although not of sham feeding (Le Sauter and Geary, 1987). In one test, intraperitoneal injection of secretin also increased glucagon's effect (Hinton et al., 1987b), but this has been difficult to replicate (Geary, unpublished data).

Satiating Potency of Exogenous Glucagon in Humans

The effect of intrameal infusion of glucagon on human food intake has been recently demonstrated (Geary et al., 1992b). A 10-minute infusion of 3 ng/kg/min glucagon or saline into an arm vein began after young, lean (mean BMI, 21.6 kg/m^2) males finished a cup of soup and 5 minutes before they were offered a main course of macaroni and beef in tomato sauce. Twelve volunteers were tested using a double-blind procedure. Glucagon reduced intake of the main course significantly, from 803 to 672 g (standard error of the difference, 29 g). The subjects were not aware when glucagon was infused or that they ate less after glucagon. They reported equal degrees of fullness and satiation after meals, and extensive psychophysical ratings gave no indication of side effects (Fig. 7.2). Thus, under these conditions glucagon appears to reduce food intake by accelerating satiation. The men reacted to the glucagon behaviorally and subjectively as if it represented ingested food. This is what an endogenous satiety signal should do.

The smallest dose of glucagon that reduced meal size in this study was 3 ng/kg/min for 10 minutes. Because plasma volume is 47 ml/kg in men of this age and weight (Diem and Lentner, 1970), glucagon's plasma half-life is 5.1 minutes (Polonsky et al., 1983), and peripheral venous infusion avoids first-pass hepatic extraction, plasma glucagon levels in both the hepatic-

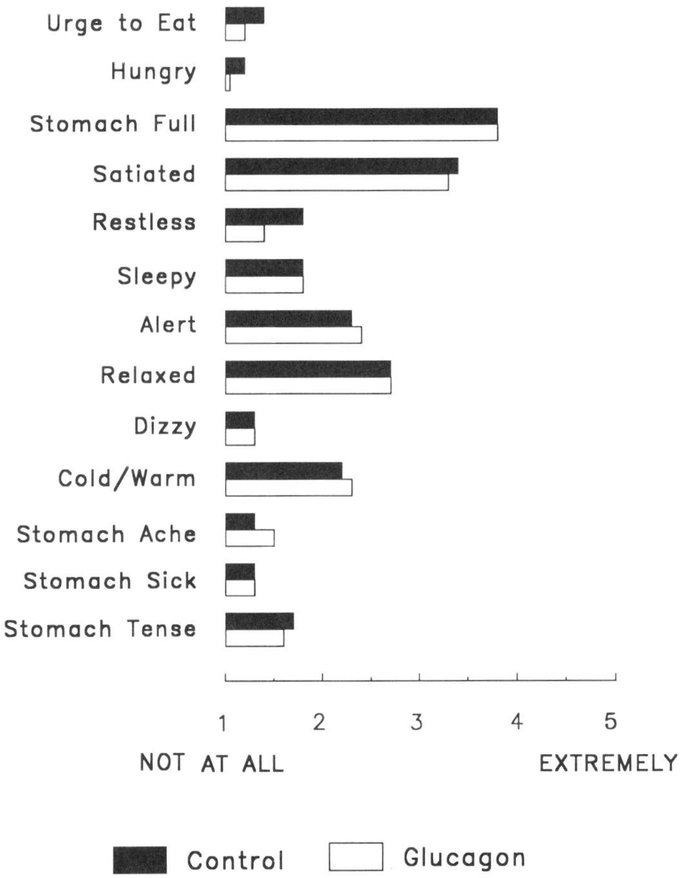

FIGURE 7.2. Glucagon infusions produce a subjective-specific reduction in meal size in humans. Data are mean category scale (1, not at all; to 5, extremely) estimates of the 13 internal perceptions indicated done immediately after meals in 12 men. The standard errors of the differences were 0.1–0.2. There were no significant differences in postmeal perceptions after meals during which infusion of 3 ng/kg/min glucagon significantly decreased meal size, as described in the text, vs. after control infusions. Visual analogue scale ratings of hunger and satiety also failed to detect any nonspecific effect of glucagon. Furthermore, during interviews after the last test, the men were unaware that they had eaten different amounts and could not distinguish test and control days. Data from Geary et al. (1992).

portal and the systemic circulation were probably elevated about 250 pg/ml at the beginning of the main course, 380 pg/ml 5 minutes later when the infusion was ended, and 190 pg/ml 5 minutes after that, which was about when meals ended. These increases appear to be in or very near the normal prandial range. Prandial glucagon level in the hepatic-portal vein has been measured in only three patients recovering from colon cancer (Denker et al., 1975). Plasma glucagon increased 20–100 pg/ml in the hepatic-portal vein, but not at all in the periphery, within 5–10 minutes of the onset of a small (670 kcal) meal. But prandial increases in hepatic-portal glucagon levels are probably often much larger. This is because systemic increases of up to about 100 pg/ml glucagon have been reported during meals in healthy humans (Denker et al., 1975; Holst, 1983b; Muller et al., 1971), and, as discussed above, hepatic-portal glucagon levels are several times higher than systemic glucagon levels during periods of increased glucagon secretion.

To summarize, a dose of glucagon that appears to be in the physiological range decreased meal size in healthy males without side effects.

Endogenous Glucagon and Meal Size

The research reviewed in the previous sections strongly supports the hypothesis that *exogenous* glucagon can reduce meal size by accelerating postprandial satiety. This raises the question of the degree to which *endogenous* glucagon secreted during meals normally contributes to postprandial satiety. This section addresses this question.

The crucial test of the physiological role of a hypothesized endogenous satiety signal is the demonstration that its antagonism increases meal size. Four tests of this type have been done using glucagon antibodies. Each has produced results supporting a role for endogenous glucagon in satiety.

Langhans et al. (1982b) tested the effect of intraperitoneal injection of a highly specific, polyclonal glucagon antibody in rats refed at dark onset after food deprivation during the light phase. An antibody dose sufficient to neutralize 5 ng glucagon in vitro increased meal size 63%. This was the first report that antagonism of any putative peptide satiety signal could stimulate feeding.

Le Sauter et al. (1991) extended this by demonstrating that, in spontaneous feeding rats, remote hepatic-portal infusion of the same glucagon antibody in a dose sufficient to neutralize 1.5 ng glucagon in vitro increased meal size 73% at dark onset and 57% 9.5 hours after dark onset (Fig. 7.3). In a second study under the same conditions (Geary et al., 1992c), a dose of glucagon antibody capable of neutralizing 1 ng glucagon also increased late dark spontaneous meal size (Fig. 7.4, right). Note that the increased meal size produced by glucagon antibody in these experiments was blocked by hepatic vagotomy (Fig. 7.4). This is consistent with other evidence reviewed below for the liver as the site of the satiating action of glucagon.

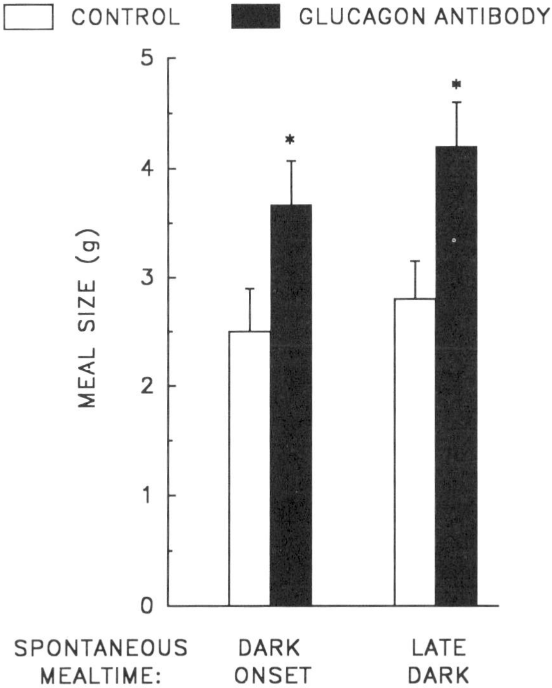

FIGURE 7.3. Hepatic-portal infusion of glucagon antibodies increases spontaneous meal size in ad libitum fed rats. Glucagon antibodies were infused for the first 2 minutes of either the first spontaneous meal after dark onset **(left)** or the first spontaneous meal beginning at least 9.5 hours after dark onset **(right)**. Data are meal sizes, mean ± SEM. The antibody dose was sufficient to neutralize 1.5 ng glucagon *in vitro*. *Different from control, $p < 0.05$. Data from Le Sauter et al. (1991).

These two tests provided the first evidence for the participation of a putative peptide satiety signal in the control of *spontaneous* feeding. Interestingly, larger antibody doses did not affect meal size (Geary et al., 1992c). The reason for this dose reversal is unclear.

The fourth test with an antibody of glucagon produced only small increases in feeding (McLaughlin et al., 1986), possibly due to procedural differences, including use of a less specific antibody and of Zucker rather than Sprague-Dawley rats.

The increase in meal size after acute antagonism of glucagon by a specific antibody in the experiments by Langhans, Le Sauter, Geary and their colleagues (Geary et al., 1992c; Langhans et al., 1982b; Le Sauter et al., 1991) provides compelling evidence that endogenous glucagon participates in the physiological control of meal size in rats.

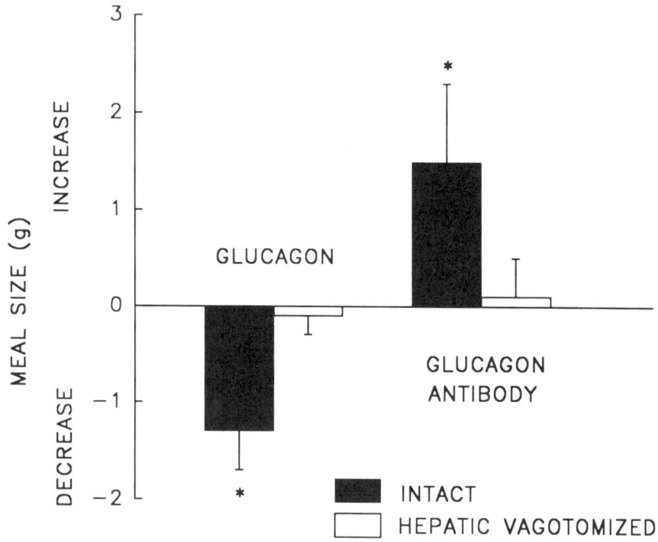

FIGURE 7.4. Hepatic vagotomy blocks the inhibitory effect of glucagon infusion **(left)** and the stimulatory effect of glucagon antibody infusion **(right)** during rats' spontaneous meals. Data are differences between control meal sizes and meal sizes during glucagon or glucagon antibody infusions, mean ± SEM, for infusions done during the first spontaneous meal beginning 9.5 hours after dark onset. Similar results were obtained in tests of spontaneous meals early in the dark phase. Glucagon dose was 13.6 μg/meal, and glucagon antibody dose was 1 ng neutralizing capacity. *Significant change in meal size, $p < 0.05$. Data from Geary et al. (1992).

Site of Action

Glucagon's satiating effect originates in the liver. The most direct evidence is that glucagon infusions that reduced rats' spontaneous meal size when infused into the hepatic-portal vein did not affect feeding significantly when infused into the inferior vena cava near the junction of the hepatic vein (Geary et al., 1992c). Because the liver efficiently extracts glucagon from the plasma (Ishida and Field, 1983; Langhans et al., 1984), vena-caval glucagon infusions increase extrahepatic glucagon levels more than hepatic-portal glucagon infusions. Thus, the greater satiating potency of hepatic-portal glucagon infusions shows that glucagon acts in the liver to decrease spontaneous meal size. No such differential effect was observed in a comparison of hepatic-portal and systemic glucagon infusions on test meal size in 14-hour food-deprived rats (Strubbe et al., 1989). This difference may be related to the smaller glucagon doses, the larger rates of glucagon infusion, or the spontaneous feeding situations tested by Geary et al. (1992c).

Further evidence that the satiating effect of glucagon originates in the liver results from tests with the cytotoxin alloxan (Ritter et al., 1986). Glucagon failed to reduce meal size in rats that had received hepatic-portal injections of a dose of alloxan that appeared to produce toxic effects only in the liver.

Finally, as reviewed below, the satiating effects of both exogenous (Geary and Smith, 1983; MacIsaac and Geary, 1985; Weatherford and Ritter, 1986) and endogenous (Geary et al., 1992c) glucagon have been eliminated by resection of the hepatic branch of the vagus nerve (Fig. 7.4). This is consistent with a hepatic site of action for glucagon, but does not prove it because the hepatic branch fibers also reach the stomach, pylorus, small intestine, and cecum (Berthoud et al., 1991, 1992; Phillips et al., 1995; Prechtl and Powley, 1987).

Glucagon therefore appears to act peripherally, in the liver, to limit meal size. This suggests that glucagon's satiating effect is hormonal, as proposed by the classic gut hormone satiety hypothesis of Gibbs et al. (1973).

Transduction of the Satiating Effect of Glucagon

Glucagon Receptors

Glucagon's metabolic actions are initiated by specific binding to cell surface receptors (Holst, 1983a; Rodbell, 1983). Hepatocyte glucagon receptors have recently been cloned in rats (Jelinek et al., 1993) and humans (Lok et al., 1994; MacNeil et al., 1994). These receptors activate both the cyclic AMP-dependent protein kinase and the calcium-phosphatidylinositol intrahepatocyte signaling systems (Jelinek et al., 1993). These alternate signaling mechanisms probably do not represent distinct subtypes of glucagon receptors (Wakelam et al., 1986), but may be differentially activated by lesser or greater amounts of receptor occupation. Glucagon's interaction with the hepatocyte receptor is complex.

The carboxyl-terminal serine in position 16, basic groups in positions 17 and 18, aspartic acid in position 9 of a backbone conformation in the 9–12 region, and histidine in position 1 of a message sequence in the 1–5 region appear necessary for glucagon's full biological activity. Extensive structure-function studies of cAMP activation in vitro, however, have begun to parse out differential roles in receptor binding and signal transduction (Hruby et al., 1993; Post et al., 1993; Unson and Merrifield, 1994; Unson et al., 1993). This work has resulted in the synthesis of a highly potent glucagon receptor antagonist, des-His[1]-[Nle[9]-Ala[11]-Ala[16]]glucagon amide (Unson et al., 1994). Applications of this antagonist to the problem of glucagon's satiating action should be productive.

Glycogenolysis and glucose metabolism

Increases in plasma glucagon rapidly stimulate hepatic glucose production (Lefèbvre, 1983, 1995). Two results from S. Ritter's laboratory provided the strongest support for the idea that glucagon's satiating action results from this action of glucagon or from the stimulation of glucose metabolism that may result from it. First, as mentioned above, administration of alloxan, which is toxic to some glucoreceptive cells, blocked exogenous glucagon's satiating action (Ritter et al., 1986). Second, the magnitude of exogenous glucagon's inhibitory effect on feeding was related to the magnitude of increases in plasma insulin level (Weick and Ritter, 1986a,b). These results suggested that insulin-dependent increases in glucose utilization may initiate glucagon's satiating action. Other results were consistent with this idea. For example, in 12-hour fasted rats tested at dark onset, the reduction in meal size after glucagon injection and the increase in meal size after glucagon antibody injection were associated, respectively, with accelerated and decelerated increases in hepatic glycogenolysis and hepatic vein glucose level (Geary et al., 1981; Langhans et al., 1982b).

Several recent studies, however, indicate that these relations are only correlational: (1) If glucagon's satiating effect were mediated by insulin-stimulated glucose utilization, then co-infusion of insulin with glucagon should increase glucagon's satiating effect, co-infusion of insulin antibodies with glucagon should decrease glucagon's satiating effect, or both. But neither insulin nor insulin antibody co-infused with glucagon during spontaneous meals influenced glucagon's satiating potency (Geary et al., 1995). (2) Chronic administration of glucagon decreased meal size as much in streptozotocin-diabetic rats as in healthy rats (deCastro et al., 1978). (3) In normal rats, reducing the duration of pretest food deprivation increased the stimulation of hepatic glucose production produced by glucagon, but decreased glucagon's satiating potency (Geary et al., 1987). (4) Glucagon failed to inhibit sham feeding despite increasing insulin secretion and hepatic glucose production (Geary and Smith, 1982b; Berthoud and Jeanrenaud, 1982). These and other dissociations of glucagon's satiating and hepatic glucose production effects (Geary, 1990) are persuasive evidence that glucagon's satiating effect does not depend on increased glucose availability or metabolism.

Hepatic Fuel Oxidation

Accumulating evidence links hepatic substrate oxidation to the control of feeding (Langhans and Scharrer, 1992; Scharrer and Langhans, 1986; Tordoff et al., 1991). An elegant series of studies by Langhans and Scharrer (1992) demonstrated that administration of substrates that are preferentially oxidized in hepatic mitochondria inhibits feeding, whereas pharmacological antagonism of mitochondrial fuel oxidation, especially β-fatty acid oxidation, stimulates feeding. Although the temporal pattern of these metabolic feeding effects was typically more suggestive of signals maintaining postprandial satiety through the intermeal interval, antagonism of fatty acid oxidation affected feeding rapidly enough to be considered a control of meal size under some conditions (Stockinger and Geary, 1990). Mitochondrial oxidation seems a plausible mechanism for glucagon satiety because glucagon potently stimulates oxidation of long chain fatty acids in liver mitochondria (Foster, 1984; McGarry and Foster, 1983; Seifter and Englard, 1993). This idea has not yet received a direct test.

Hepatocyte Membrane Potential

Russek (1981) originally proposed that hepatic afferents may sense changes in hepatocyte membrane potential produced by metabolic events. Langhans and Scharrer's observation (1987) that feeding was stimulated by ouabain, which depolarizes the membrane by inactivating Na^+/K^+-ATPase, supports this idea. In one test, ouabain pretreatment also antagonized the inhibitory effect of intraperitoneal injection of glucagon in 12-hour food-deprived rats refed at dark onset (Duss, 1986). This seems a promising avenue for exploration. The complexity of the relationship between hepatic metabolism and membrane potential, however, will complicate this work. For example, hepatocytes were similarly hyperpolarized by superfusion with glucose or fructose, which can be metabolized, and with 2-deoxy-D-glucose, which cannot be metabolized (Meyer and Scharrer, 1991). Alternatively, because glucagon may have a direct effect on hepatocyte membrane potential, hepatocyte polarization could mediate glucagon's satiating effect independent of metabolic changes (Petersen, 1974).

Endocrine Effects

Exogenous glucagon's numerous endocrine effects (Lefèbvre, 1983) provide few attractive candidates for mediation of glucagon's satiating action (for review, see Geary, 1990). Amylin, a recently discovered peptide that is produced by pancreatic β-cells (Balasubramaniam et al., 1991; Cooper,

1994; Wang et al., 1993), is an interesting exception. Amylin is co-released with insulin under all conditions tested so far (Cooper, 1994; Wang et al., 1993), including during meals (Butler et al., 1990). Amylin administration potently inhibits feeding (Balasubramaniam et al., 1991; Lutz et al., 1994; Morley and Flood, 1991), apparently mainly by decreasing meal size (Lutz et al., 1995). Lutz et. al. (1996) recently tested whether amylin mediates glucagon's satiating action using the amylin receptor blocker calcitonin-gene-related peptide 8–37 ($CGRP_{8-37}$). Intraperitoneal injection of 10 μg/kg $CGRP_{8-37}$ in 24-hour food deprived rats abolished the satiating action of intraperitoneal injection of 400 μg/kg glucagon. Under other feeding conditions, however, $CGRP_{8-37}$ did not antagonize glucagon's satiating action. $CGRP_{8-37}$ alone did not affect meal size. These results indicate that further research is required to determine the extent to which amylin participates in the satiating actions of exogenous and endogenous glucagon.

Smooth Muscle Effects

Exogenous glucagon relaxes visceral and vascular smooth muscle in the liver and elsewhere in the gut (Blackmore et al., 1991; D'Almeida and Lautt, 1991), and the majority of vagal afferents in the liver innervate the biliary tree and the hepatic vasculature (Berthoud et al., 1992; Sawchenko and Friedman, 1979). At present, however, no evidence implicates any of glucagon's smooth muscle effects in satiety, and two results suggest that they are not involved. The first is the absence of any effect on test meal size of intraperitoneal injection of glucagon 1–21 (Geary, 1987), a peptide that is apparently equipotent with glucagon for most smooth muscle effects, but lacks metabolic effects (Jørgenson et al., 1983). The second is that intraperitoneal injection of 400 μg/kg glucagon decreased the size of evaporated milk test meals, but did not affect the amount of milk emptied from the stomach in 20 minutes (Stockinger and Geary, 1989). It is also relevant that recent work indicates that prandial increases in glucagon appear not to mediate the increase in splanchnic blood flow associated with meals (Braatvedt et al., 1993).

Peripheral Neural Mechanisms

Afferent Mechanism

Hepatic Afferents and the Satiating Effect of Exogenous Glucagon

The satiating effect of intraperitoneally injected glucagon on diurnal, scheduled, test meals was blocked by selective lesion of the hepatic branch of the abdominal vagus under several (Geary and Smith, 1983; MacIsaac and Geary, 1985; Weatherford and Ritter, 1986), although not all (Bellinger and Williams, 1986; Weatherford and Ritter, 1986), conditions. This suggests that information about glucagon's hepatic satiating effect may reach the central nervous system via the vagus nerve. The hepatic branch appeared sufficient to mediate glucagon's satiating action because abdominal vagotomies that spared only the hepatic branch did not affect it (Geary and Smith, 1983).

Surgical resection of the hepatic artery combined with phenol denervation of the surfaces of the portal vein and bile duct also failed to affect the satiating action of glucagon (MacIsaac and Geary, 1985). This result has two implications. First, it indicates that this action of glucagon does not depend on hepatic vagal fibers that project through the liver to innervate more distal sites (Berthoud et al., 1991, 1992; Phillips et al., 1995; Prechtl and Powley, 1987). This is important because recent anatomical work indicates that as many as two-thirds of hepatic branch fibers traverse the liver via the hepatic artery (Berthoud et al., 1992). Second, this hepatic denervation indicates that spinal visceral afferents, which reach the liver via these routes, are not necessary for glucagon's satiating action.

The hepatic vagal contribution to glucagon's satiating effect appears to be sensory. First, antagonism of peripheral postganglionic muscarinic receptors with atropine methyl nitrate did not block it (Geary and Smith, 1983). Second, according to a preliminary report (Ritter and Weatherford, 1986), lesion of unmyelinated peripheral afferents by systemic administration of capsaicin did. More evidence would be welcome to confirm these suggestions that loss of hepatic afferents blocks glucagon's satiating action after hepatic branch vagotomy.

The hepatic vagal contribution to exogenous glucagon's satiating effect on test meals exhibits some situational variability. Weatherford and Ritter (1986) found that hepatic vagotomy blocked the satiating effect of intraperitoneal glucagon injections in 8- or 9-hour food-deprived rats tested 3 hours after dark onset, but failed to block glucagon's effect in 6-hour food-deprived rats refed 0–90 minutes after dark onset. Hepatic vagotomy also failed to block the synergistic satiating effect of glucagon plus cholecystokinin on sham feeding (Le Sauter and Geary, 1990). Total abdominal vagotomy, in contrast, blocked both these satiating effects. The identity of the nonhepatic vagal signals contributing to glucagon's satiating effect, and the stimuli that recruit them, have not been identified.

We recently investigated involvement of the hepatic vagus in the satiating actions of glucagon on spontaneous feeding in rats (Geary et al., 1992c). Selective hepatic vagotomy blocked the satiating effect of hepatic-portal glucagon infusions during spontaneous meals both early (first and second hours) and late (after the ninth hour) in the dark phase (Fig. 7.4, left). Note that these data taken together with those of Weatherford and Ritter (1986) suggest that different vagal mechanisms are involved in glucagon's satiating actions on spontaneous and deprivation-induced meals early in the dark.

Hepatic Vagotomy and the Satiating Effect of Endogenous Glucagon

As noted above, hepatic vagotomy blocked the stimulatory effect of hepatic-portal infusions of glucagon antibodies on spontaneous feeding (Geary et al., 1992c). Hepatic-portal infusion of a dose of glucagon antibody sufficient to neutralize 1 ng glucagon in vitro increased late-dark spontaneous meals approximately 40% in neurally intact rats, but not at all in hepatic vagotomized rats (Fig. 7.4, right). This is the only evidence suggesting that hepatic vagal afferents mediate the satiating effect of endogenous as well as exogenous glucagon.

If endogenous glucagon normally controls meal size via a hepatic vagal afferent mechanism, then hepatic vagotomy should increase meal size. This prediction has not been confirmed convincingly. Hepatic vagotomy occasionally increased food intake, but more often did not (for review, see Langhans and Scharrer, 1992). The failure of hepatic denervation to increase food intake may indicate that hepatic signals are not necessary for the control of meal size, due perhaps to redundant mechanisms. This argument, however, is difficult to reconcile with the glucagon antagonism data that reveal a hepatic-vagal signal necessary for the normal control of meal size (Geary et al., 1992c).

Perhaps chronic denervation recruits dynamic responses that gradually replace the missing effect of glucagon. If so, then food intake may increase only transiently following hepatic vagotomy. Friedman and Sawchenko's observation (1984) that food intake increased within hours of hepatic vagotomy accomplished by retracting a previously implanted suture is consistent with this.

Conditioned satiety responses also may mask hepatic vagal effects. This possibility is suggested by Langhans et al.'s observation (1989) that hepatic vagotomy increased the intake of a novel diet introduced after surgery, but did not affect intake of the maintenance diet.

Finally, introduction of intravenous total parenteral nutrition (TPN) also unmasked an effect of hepatic vagotomy (Yang et al., 1992). The inhibitory effect of TPN on food intake was reduced in vagotomized rats due to significantly smaller reductions in meal size and in meal frequency in comparison to the effects of TPN in sham-operated control rats.

HEPATIC VAGOTOMY AND THE MECHANISM OF GLUCAGON SATIETY

Studies of hepatic vagotomies may help to evaluate the mechanism transducing the satiating effect of glucagon. For example, reports that selective hepatic vagotomy blocked the feeding effects of manipulations of mitochondrial fatty acid oxidation and of ouabain (Langhans and Scharrer, 1987, 1992) are consistent with the hypothesis that mitochondrial metabolic events initiate glucagon satiety.

Whether hepatocytes receive direct vagal innervation remains controversial. Horseradish peroxidase and DiI fiber tracing revealed that the predominant targets of vagal afferents in the rat liver are the bile ducts and the paraganglia of the nerve plexus, with less dense innervation of the portal vein and apparently no innervation of liver parenchyma (Berthoud et al., 1992). Neurofilament protein staining, however, has produced evidence for innervation of rat hepatocytes (Friedman et al., 1994; Lee et al., 1992). Intralobular nerve fibers are also associated with hepatocytes in human liver (Bioulac-Sage et al., 1990). Whether any of these neural elements are vagal afferent fibers remains unclear. Even if there is no direct vagal afferent innervation of the hepatocytes, however, it remains possible that changes in hepatocyte membrane potential elicited by glucagon reach remotely located vagal afferent terminals by electrotonic conduction through gap junctions, which can span as much as 500 μm in the liver (Berthoud et al., 1992; Graf and Petersen, 1978; Langhans and Scharrer, 1992).

Vagal Efferent Mechanism

The cephalic phase of prandial glucagon secretion may be vagally mediated. The evidence that suggests this is that electrical stimulation of the dorsal motor nucleus of the vagus or of the gastric or hepatic branches of the vagus elicits glucagon secretion (Berthoud et al., 1990; Laughton and Powley, 1987). If this is so, selective hepatic branch or selective gastric branch vagotomy may block the satiating action of endogenous glucagon by disrupting prandial glucagon secretion. This leads to the interesting prediction that gastric vagotomy should block the stimulatory effect of glucagon antagonism on feeding but not affect the inhibitory effect of glucagon on feeding.

Central Processing of Peripheral Glucagon Signals

The dorsal hindbrain appears to be the initial site involved in the central processing and elaboration of glucagon's satiating effect, as predicted by its mediation by vagal afferents. Lesions of the medial division of the nucleus of the solitary tract and the area postrema blocked the satiating effect of intraperitoneal injections of 200 μg/kg glucagon on a scheduled test meal (Weatherford and Ritter, 1988). Unfortunately, the effect of lesions of only the area postrema was not evaluated. The lesions did not attenuate the satiating effect of cholecystokinin, although more extensive lesions of this region did (Edwards et al., 1986). This suggests that if the synergistic effect of glucagon and CCK on sham feeding (Le Sauter and Geary, 1987) depends on neural convergence of the two signals, the convergence may not occur at glucagon's first central projection site. On the other hand, some integration of the glucagon signal with other signals affecting feeding may occur in the nucleus of the solitary tract because hepatic-portal infusion of glucagon decreased the multiunit electrophysiological activity of neurons in the nucleus of the solitary tract elicited by oral glucose (Giza et al., 1993). This effect was specific for sweet stimuli because hepatic-portal infusion of glucagon did not change the electrophysiological responses to NaCl, HCl, or quinine.

Little is known about the involvement of more rostral neural structures in glucagon's satiating action. The ventromedial hypothalamus is not necessary for the processing or expression of the glucagon signal because intraperitoneal injections of glucagon inhibited feeding in rats in the hyperphagic phase of the obesity syndrome produced by bilateral lesion of the ventromedial hypothalamus (Geary et al., 1986).

Glucagon Family Peptides in Periphery and Brain

Periphery

The discoveries of glucagon's amino acid sequence in 1956 (Bromer et al., 1957) and of a glucagon radioimmunoassay in 1959 (Unger et al., 1959) led to the development of the concept of a glucagon family of peptides with similar amino acid sequences or immunological reactivities (Lefèbvre, 1995). This family includes secretin, vasoactive intestinal polypeptide, gastric inhibitory polypeptide, and growth hormone-releasing factor (GHRF). Although careful structure–activity studies have not been done, there is little indication that any of these peptides shares glucagon's satiating effect (Conover et al., 1989; Gibbs et al., 1973; Grovum, 1981; Hinton et al., 1987b; Lorenz et al., 1979; Meyer and Scharrer, 1991; Vaccarino and Hayward, 1988), presumably as a result of the specificity of the receptors for the different peptides. For example, secretin, which with 15 of 29 amino acid identities is most similar to glucagon in this family, is almost completely ineffective as a competitor to glucagon for binding to the hepatic glucagon receptor (Jelinck et al., 1993).

The recent identification of the biosynthetic relationships among glucagon-like peptides has permitted the replacement of the traditional kinship identification (Conlon, 1988; Holst and Orskov, 1994; Lefèbvre, 1995). Preproglucagon (179 amino acid polypeptide) and its cleavage product proglucagon (158 amino acids in humans and 160 in rats) are synthesized in pancreas, intestines, brain, and other organs. Post-translational processing of proglucagon is organ specific. In humans, rats, and most other mammals, pancreatic α-cells produce, sequester, and release glucagon (proglucagon 33–61) but not the "gut glucagons" glicentin (proglucagon 1–69), oxyntomodulin (proglucagon 33–69), glucagon-like peptide-1 (GLP-1; proglucagon 78–107 amide), and glucagon-like peptide-2 (GLP-2; proglucagon 126–158), which are produced by gut L-cells. The physiology of glucagon-like peptides appears mechanistically distinct from that of glucagon (Bataille et al., 1986; Göke et al., 1991; Holst and Orskov, 1994; Lefèbvre, 1983). The hepatocyte glucagon receptor has little affinity for any glucagon-like peptides, including those that contain the entire glucagon sequence (Hjorth et al., 1994), and no peripheral glucagon-family peptide besides glucagon has been implicated in the control of appetite.

Brain

Glucagon and glucagon receptors are present in the brain as well as the periphery (Conlon, 1988; Göke et al., 1991, 1995; Jin, 1988; Uttenthal et al., 1992). Proglucagon mRNA and glucagon, GLP-1, and glicentin immunoreactivities occur only in the caudal one third of the nucleus of the solitary tract and in the medullary reticular nucleus of the rat, but fibers and terminals immunoreactive to these peptides and specific binding sites for them appear in many sites from cortex to caudal hindbrain, notably including the hypothalamus and sensory circumventricular organs. GLP-1 seems to fulfill many of the criteria for a neurotransmitter (Calvo et al., 1995; Kreymann et al., 1989; Uttenthal et al., 1992). With the exception of extensive work with GHRF, little is known about the biology of the glucagon-family peptides in the brain.

Because peripheral glucagon's satiating effect appears to originate in the liver and because the blood–brain barrier is impermeable to peptides of glucagon's size, peripheral glucagon probably does not act directly in the brain to inhibit feeding. Hoebel's idea (1988) of integrative peptides suggests another possibility, that central glucagon participates in the central interneuronal networks that process the afferent signals initiated by peripheral glucagon. Central administration of glucagon has been reported to decrease feeding as well as to affect hypothalamic unit activity, peripheral plasma glucose and glucagon levels, sympathetic activity, and respiration quotient (Atrens and Menendez, 1993; Inokuchi et al., 1984, 1986; Marubashi et al., 1985; Nagai et al., 1991; Shimizu et al., 1993). The mechanisms, physiological significance, and possible relationships of these effects to glucagon's satiating action are not yet clear.

It has recently been reported that intracerebroventricular, but not intraperitoneal, administration of GLP-1 inhibited feeding and that intracerebroventricular administration of GLP-1 antagonists increased feeding in both food-deprived and free-feeding rats (Lambert et al., 1993; Schick et al., 1992; Tang-Christensen et al., 1994; Turton et al., 1996). The effects of GLP-1 on feeding were dose related and potent. Threshold intracerebroventricular doses for inhibition of deprivation-induced feeding were 1–3 μg, and large doses reduced food intake by 75% or more (Tang-Christensen et al., 1994; Turton et al., 1996). Central administration of GLP-1 also inhibited water intake in water-deprived rats (Tang-Christensen et al., 1994). Intraperitoneal administration of GLP-1, however, also inhibited drinking. The receptor sites mediating these responses have not been identified. They are not likely to be directly related to central glucagon because, as noted, GLP-1 does not interact with glucagon receptors. GLP-1 receptors are densely expressed in the circumventricular organs (Göke et al., 1995). Because they may be accessible to GLP-1 both in the cerebrospinal fluid and in the blood, these receptors are interesting candidates for the mediation of GLP-1's inhibitory effect on drinking.

Unlike glucagon and GLP-1, central administration of GHRF does not decrease food intake. Instead, GHRF has been reported to increase food intake after administration into the lateral ventricle or the medial preoptic/suprachiasmatic nucleus region of the hypothalamus (Vaccarino et al., 1985; Vaccarino and Hayward, 1988).

Further work is required to determine if these effects are pharmacological or physiological actions of central glucagon-family peptides. Investigation of the central actions of GLP-1 on food and water intake appears especially important because the stimulatory effect of central GLP-1 antagonism suggests that its actions may be physiological and because it is one of the few candidate inhibitory controls of water intake.

Pathophysiology and Therapeutic Potential

Obesity

No dramatic alteration in glucagon physiology has been established in any human eating disorder or animal model of obesity (Holst, 1983b; Johnson et al., 1991). The research in this area is not extensive, however, and the lack of prominent effects may in part reflect the heterogeneity of the disorders. For example, in one study, only those obese patients who were hyperinsulinemic and glucose intolerant displayed increased glucagon responses to arginine (Hamaguchi et al., 1991). In genetically obese *fa/fa* Zucker rats, glucagon metabolism appeared relatively normal (Holst, 1983b), but exogenous glucagon's inhibitory effect on feeding was reduced in one test (McLaughlin et al., 1986). Active immunization against glucagon decreased feeding slightly in both lean and obese Zucker rats, perhaps due to the increase in glucagon secretion stimulated by the procedure (McLaughlin et al., 1984). This problem requires more investigation.

A recent report by Raben et al. (1994) raised the intriguing possibility that glucagon may be involved in the persistent difficulties of dieters to maintain weight loss. The effects of a 50% fat, 3500 kJ breakfast was compared between formerly obese women who had maintained target body weights and fat mass for at least 3 months after a mean weight loss of 24 kg and a closely matched group of women with no history of obesity. Postprandial glucagon secretion, postprandial fatty acid oxidation, and subjective ratings of satiety were all less in the postobese women than in the controls.

If the satiating action of glucagon is demonstrated in obese humans, its therapeutic use should be evaluated. Early tests of chronic glucagon administration on food intake in animals produced mixed results. In rats, a 3-day regimen of twice-daily subcutaneous injections of protamine zinc glucagon reduced meal size and total daily intake (deCastro et al., 1978). In three more recent studies, however, chronic subcutaneous infusion of glucagon via osmotic minipumps failed to decrease daily food intake (Billington et al., 1987, 1991; Nagai et al., 1991).

In normal weight men, intramuscular injection of 1 mg glucagon produced a delayed decrease in food intake that was accompanied by glycosuria and negative nitrogen balance, but not by malaise (Penick and Smith, 1964). A larger glucagon dose, 2 mg/meal, produced nausea, anorexia, and rapid weight loss in an obese adolescent girl (Clayton and Librik, 1963). These results question the safety and the efficacy of chronically administered glucagon. Nevertheless, in view of the dependency of glucagon's satiating potency on route of administration and of the acute reduction of human meal size by microgram doses of glucagon (Geary et al., 1992b), studies aimed at establishing effective doses that are devoid of side effects are warranted. Because current formulations of glucagon are not effective orally, novel methods of drug delivery may be needed. Progress has already been made in this area (Langer, 1990). For example, glucagon eye drops have successfully countered hypoglycemia (Chiou and Chuang, 1988; Stenninger and Aman, 1993). The development of orally active agents to stimulate endogenous glucagon release is another possibility that should be pursued.

Other Disorders

The glucagonoma syndrome includes extremely elevated plasma glucagon levels and severe anorexia (Madsen et al., 1995; Wynick et al., 1993). Increased plasma glucagon concentrations and increased glucagon/insulin ratios also occur in association with anorexia in patients with other types of cancer (Chance et al., 1983). Bartlett et al. (1995) have recently shown that, in rats inoculated with metastasizing mammary adenocarcinoma, an insulin-somatostatin treatment that normalized glucagon and increased the insulin/glucagon ratio reversed cachexia. Anorexia did not appear to contribute importantly to cachexia in this model, however, so it remains unclear whether glucagon pathophysiology associated with cancer directly affects appetite.

Postprandial glucagon levels are increased in noninsulin-dependent diabetes mellitus (NIDDM) and may contribute to hyperglycemia in both NIDDM and insulin-dependent diabetes mellitus (Brand et al., 1994; Dinneen et al., 1995; Gutniak et al., 1986). A missense mutation in the glucagon receptor gene resulting in the expression of a glucagon receptor with greatly reduced affinity has been reported in late-onset NIDDM (Hager et al., 1995). If this defect reduces the satiating potency of endogenous glucagon, it may account for some of the obesity associated with NIDDM.

Conclusions

Work over the last 15 years has confirmed the hypotheses that exogenous and endogenous glucagon signal satiety. In rats, exogenous glucagon produces a dose-related, behaviorally specific reduction in meal size in a variety of conditions, including spontaneous nocturnal meals in free-feeding rats. Antagonist studies under similar conditions indicated that endogenous glucagon is part of the normal control of meal size. These data constitute compelling evidence that glucagon is a physiological satiety signal in rats. It may also contribute to satiety in humans because intravenous infusion of a dose of glucagon that is apparently in or near the physiological range produced a subjectively specific decrease in meal size in male volunteers.

There has also been progress in understanding the mechanism of the satiating action of glucagon. Recent work indicates that this action originates in the liver and that the hepatic branch of the abdominal vagus provides the afferent route of the signal to the brain. The transduction mechanisms in the liver that produce this neural signal and its processing in the brain are unknown.

Glucagon's prandial release, peripheral action, vagal afferent mechanism, and potent effect on meal size parallel the satiating effects of other peptides, especially cholecystokinin. But there are also sharp contrasts. For example, whereas glucagon's satiating action appears to be an endocrine effect in the rat, that of cholecystokinin now appears to be a paracrine or neurocrine effect (Smith and Gibbs, 1992).

Challenges for future glucagon research lie on four fronts. First, what physiological transduction mechanism initiates an afferent signal? Second, what brain mechanisms elaborate this signal into the perceptions and behaviors of postprandial satiety? Third, how does the glucagon signal interact with other signals controlling meals, with the physiological context (nutritional adaptation, reproductive status, etc.), and with the environmental and ecological context? Fourth, does glucagon have a role, or therapeutic potential, in normal or disordered human appetite? These are formidable problems that require application of increasingly sophisticated techniques available in behavioral neuroscience, including cellular and molecular tools.

ACKNOWLEDGEMENT

I thank Dr. Gerard P. Smith for his scientific guidance and his friendship, two of the most wonderful gifts that I have ever received. I also thank the National Institute of Diabetes and Digestive and Kidney Diseases for research support.

REFERENCES

Atrens, D. M., and J. A. Menendez. Glucagon and the paraventricular hypothalamus: Modulation of energy balance. *Brain Res.* 630:245–251, 1993.

Balasubramaniam, A., V. Renugopalakrishan, M. Stein, J. F. Fisher, W. T. Chance.

Synthesis, structures and anorectic effects of human and rat amylin. *Peptides* 12:919–924, 1991.

Bartlett, D. L., S. L. Charland, and M. H. Totosian. Reversal of tumor-associated hyperglucagonemia as treatment for cancer cachexia. *Surgery* 118:87–97, 1995.

Bataille, D., C. Jarrousse, A. Kervran, C. Depigny, and M. Dubrasquet. The biological significance of "enteroglucagon." Present status. *Peptides* 7:37–42, 1986.

Bellinger, L. L., and F. E. Williams. Glucagon and epinephrine suppression of food intake in liver denervated rats. *Am. J. Physiol.* 251:R349-R358, 1986.

Berthoud, H.-R., N. R. Carlson, and T. L. Powley. Topography of efferent vagal innervation of the rat gastrointestinal tract. *Am. J. Physiol.* 260:R200-R207, 1991.

Berthoud, H.-R., E. A. Fox, and T. L. Powley. Localization of vagal preganglionics that stimulate insulin and glucagon secretion. *Am. J. Physiol.* 258:R160-R168, 1990.

Berthoud, H.-R., and B. Jeanrenaud. Sham feeding-induced cephalic phase insulin release in the rat. *Am. J. Physiol.* 242:E280-E285, 1982.

Berthoud, H.-R., M. Kressel, and W. L. Neuhuber. An anterograde tracing study of the vagal innervation of rat liver, portal vein and biliary system. *Anat. Embryol.* 186:431–442, 1992.

Billington, C. J., T. J. Bartness, J. E. Briggs, A. S. Levine, and J. E. Morley. Glucagon stimulation of brown adipose tissue growth and thermogenesis. *Am. J. Physiol.* 252:R160-R165, 1987.

Billington, C. J., J. E. Briggs, J. G. Link, and A. S. Levine. Glucagon in physiological concentrations stimulates brown fat thermogenesis *in vivo*. *Am. J. Physiol.* 261:R501-R507, 1991.

Bioulac-Sage, P., M. E. Lafon, J. Saric, and C. Balabaud. Nerves and perisinusoidal cells in human liver. *J. Hepatol.* 10:105–112, 1990.

Blackmore, P. F., S. Mojsov, J. H. Exton, and J. F. Habener. Absence of insulinotropic glucagon-like peptide-1 (7–37) receptors on isolated rat liver hepatocytes. *FEBS Lett.* 283:7–10, 1991.

Braatvedt, G. D., A. Stanners, P. G. Newrick, M. Halliwell, and R. J. M. Corrall. Evidence against a putative role for glucagon as a physiological splanchnic vasodilator in man. *Clin. Sci.* 84:193–199, 1993.

Brand, C. L., B. Rolin, P. N. Jørgenson, I. Svendsen, J. S. Kristensen, and J. J. Holst. Immunoneutralization of endogenous glucagon with monoclonal glucagon antibody normalizes hyperglycaemia in moderately streptozotocin-diabetic rats. *Diabetologia* 37:985–993, 1994.

Bromer, W. W., A. Staub, E. R. Diller, H. L. Bird, L. G. Sinn, and O. K. Behrens. The amino acid sequence of glucagon. I. The amino acid composition and terminal amino acid analyses. *J. Am. Chem. Soc.* 79:2794–2798, 1957.

Butler, P. C., J. Chou, W. B. Carter, Y.-N. Wang, B.-H. Bu, D. Chang, J.-K. Chang, and R. A. Rizza. Effects of meal ingestion on plasma amylin concentration in NIDDM and nondiabetic humans. *Diabetes* 39:752–756, 1990.

Calvo, J. C., B. Yusta, F. Mora, and E. Blazquez. Structural characterization by affinity cross-linking of glucagon-like peptide-1 (7–36) amide receptor in rat brain. *J. Neurochem.* 64:299–306, 1995.

Chance, W. T., F. M. Van Lameren, M. H. Chen, R. F. Murphy, S. N. Joffe, and

J. E. Fischer. Alteration in plasma levels of insulin and glucagon associated with cancer anorexia. *Surg. Forum* 34:441–443, 1983.

Chiou, G. C. Y., and C. Y. Chuang. Treatment of hypoglycemia with glucagon eye drops. *J. Ocular. Pharm.* 4:179–186, 1988.

Clayton, G. W., and L. Librik. Therapy of exogenous obesity in childhood and adolescence. *Pediatr. Clin. North Am.* 10:99–107, 1963.

Conlon, J. M. Proglucagon-derived peptides: Nomenclature, biosynthetic relationships and physiological roles. *Diabetologia* 31:562–566, 1988.

Conover, K. L., S. M. Collins, and H. P. Weingarten. Gastric emptying changes are neither necessary nor sufficient for CCK-induced satiety. *Am. J. Physiol.* 256:R56-R62, 1989.

Cooper, G. J. Amylin compared with calcitonin gene-related peptide: Structure, biology and relevance to metabolic disease. *Endocrine Rev.* 15:163–201, 1994.

D'Almeida, M. S., and W. W. Lautt. Glucagon pharmacodynamics and modulation of sympathetic nerve and norepinephrine-induced constrictor responses in the superior mesenteric artery of the cat. *J. Pharmacol. Exp. Ther.* 259:118–123, 1991.

Day, J., L. K. Johansen, O. P. Ganda, J. S. Soeldner, R. E. Gleason, and W. Medgley. Factors governing insulin and glucagon response during normal meals. *Clin. Endocrinol.* 9:443–454, 1978.

deCastro, J. M., S. K. Paullin, and G. M. DeLugas. Insulin and glucagon as determinants of body weight set point and microregulation in rats. *J. Comp. Physiol. Psychol.* 92:571–579, 1978.

de Jong, A., J. H. Strubbe, and A. B. Steffens. Hypothalamic influence on insulin and glucagon release in the rat. *Am. J. Physiol.* 233:E380-E388, 1977.

Denker, H., P. Hedner, J. Holst, and K. G. Tranberg. Pancreatic glucagon response to an ordinary meal. *Scand. J. Gastroenterol.* 10:471–474, 1975.

Diem, K., and C. Lentner. *Scientific Tables. 7th Ed.* Ardsley, NY: Geigy Pharmaceuticals, 1970.

Dinneen, S., A. Alzaid, D. Turk, and R. Rizza. Failure of glucagon suppression contributes to postprandial hyperglycaemia in IDDM. *Diabetologia* 38:337–343, 1995.

Duss, M. Untersuchungen zur Saettingungswirkung von Glucagon. Inaugural Dissertation, Universitaet Zürich. Konstanz, Switzerland: Maus, 1986.

Edwards, G. L., E. E. Ladenheim, and R. C. Ritter. Dorsal hindbrain participation in cholecystokinin-induced satiety. *Am. J. Physiol.* 251:R971-R977, 1986.

Foster, D. From glycogen to ketones—and back. *Diabetes* 33:1188–1199, 1984.

Friedman, M. I., M. Ketchum, P. Ulrich, P. A. S. Breslin, and C. DellaCorte. Evidence for innervation of the hepatic parenchyma in the rat. *Soc. Neurosci. Abstr.* 20:1226, 1994.

Friedman, M. I., and P. E. Sawchenko. Evidence for hepatic involvement in control of ad libitum food intake in rats. *Am. J. Physiol.* 247:R106-R113, 1984.

Geary, N. Glucagon 1–21 fails to inhibit feeding. *Peptides* 8:943–945, 1987.

Geary, N. Pancreatic glucagon signals postprandial satiety. *Neurosci. Biobehav. Rev.* 14:323–338, 1990.

Geary, N. A failure of pulsatile infusion to increase glucagon's satiating potency. *Physiol. Behav.* 59:613–616, 1996.

Geary, N., L. Asarian, and W. Langhans. Insulin does not increase glucagon's potency to reduce meal size in spontaneously feeding rats. *Soc. Neurosci. Abstr.* 21:8, 1995.

Geary, N., N. Farhoody, and A. Gersony. Food deprivation dissociates pancreatic glucagon's effects on satiety and hepatic glucose production at dark onset. *Physiol. Behav.* 39:507–511, 1987.

Geary, N., J. Fudge, and J. Le Sauter. Scheduled running wheel activity indexes the specificity of pharmacological anorexia. *Behav. Neur. Biol.* 68:1–7, 1992a.

Geary, N., H. R. Kissileff, F. X. Pi-Sunyer, and V. Hinton. Individual, but not simultaneous, glucagon and cholecystokinin infusions inhibit feeding in men. *Am. J. Physiol.* 262:R975-R980, 1992b.

Geary, N., W. Langhans, and E. Scharrer. Metabolic concomitants of glucagon-induced suppression of feeding in the rat. *Am. J. Physiol.* 241:R330-R335, 1981.

Geary, N., J. Le Sauter, and N. Noh. Glucagon acts in the liver to control spontaneous meal size in rats. *Am. J. Physiol.* 264:R116-R122, 1992c.

Geary, N., and G. P. Smith. Pancreatic glucagon and postprandial satiety in the rat. *Physiol. Behav.* 28:313–322, 1982a.

Geary, N., and G. P. Smith. Pancreatic glucagon fails to inhibit sham feeding in the rat. *Peptides* 3:163–166, 1982b.

Geary, N., and G. P. Smith. Selective hepatic vagotomy blocks pancreatic glucagon's satiety effect. *Physiol. Behav.* 31:391–394, 1983.

Geary, N., G. P. Smith, and J. Gibbs. Pancreatic glucagon and bombesin inhibit meal size in ventromedial hypothalamus-lesioned rats. *Regul. Pept.* 15:261–268, 1986.

Gibbs, J., R. C. Young, and G. P. Smith. Cholecystokinin decreases food intake in rats. *J. Comp. Physiol. Psychol.* 84:488–495, 1973.

Giza, B. K., R. O. Deems, D. A. VanderWeele, and T. R. Scott. Pancreatic glucagon suppresses gustatory responsiveness to glucose. *Am. J. Physiol.* 265:R123-R127, 1993.

Göke, R., H.-C. Fehmann, and B. Göke. Glucagon-like peptide-1 (7–36) amide is a new incretin/enterogastrone candidate. *Eur. J. Clin. Invest.* 21:135–144, 1991.

Göke, R., P. J. Larsen, J. D. Mikkelsen, and S. P. Sheikh. Distribution of GLP-1(7–36) binding sites in the rat brain. *Eur. J. Neurosci.* 7:2294–2300, 1995.

Graf, J., and O. H. Petersen. Cell membrane potential and resistance in the liver. *J. Physiol. Lond.* 284:105–126, 1978.

Grovum, W. L. Factors affecting the voluntary intake of food by sheep. 3. The effect of intravenous infusions of gastrin, cholecystokinin and secretin on motility of the reticulo-rumen and intake. *Br. J. Nutr.* 45:183–201, 1981.

Gutniak, M., V. Grill, and S. Efendic. Effect of composition of mixed meals—low- versus high-carbohydrate content—on insulin, glucagon, and somatostatin release in healthy humans and in patients with NIDDM. *Diabetes Care* 9:244–249, 1986.

Hager, J., L. Hansen, C. Vaisse, N. Vionnet, A. Philippi, W. Poller, G. Velho, C. Carcassi, L. Contu, C. Julier, F. Cambien, P. Passa, M. Lathrop, W. Kindsvogel, F. Demenais, E. Nishimura, and P. Froguel. A missense mutation in the glucagon receptor gene is associated with non-insulin-dependent diabetes mellitus. *Nature Genet.* 9:299–304, 1995.

Hamaguchi, T., H. Fukushima, M. Uehara, S. Wada, T. Shirotani, H. Kishikawa,

K. Ichinose, K. Yamaguchi, and M. Schichiri. Abnormal glucagon response to arginine and its normalization in obese hyperinsulinaemic patients with glucose intolerance: importance of insulin action on pancreatic alpha cells. *Diabetologia* 34:801–806, 1991.

Hildebrandt, W., W. Blech, and K.-D. Kohnert. Kinetic studies on hepatic handling of glucagon using the model of non-recirculating perfused rat livers. *Horm. Metab. Res.* 23:410–413, 1991a.

Hildebrandt, W., W. Blech, and K.-D. Kohnert. Hepatische Reaktion auf pulsatile Anderungen Pankreasglukagonkonzentration. *Wiss. Zeit. der Univ. Halle* 40:137–149, 1991b.

Hildebrandt, W., W. Blech, and K.-D. Kohnert. Amplitude smoothing and phase shifting effect of the isolated perfused rat liver on oscillating insulin and glucagon. *Horm. Metab. Res.* 24:345–346, 1992.

Hinton, V., M. Esguerra, N. Farhoody, J. Granger, and N. Geary. Epinephrine inhibits feeding nonspecifically in the rat. *Physiol. Behav.* 40:109–115, 1987a.

Hinton, V., B. Goldberg, M. K. Albertson, K. Jean-Baptiste, K. Couret, and N. Geary. Interactive effects of gut peptides on postprandial satiety in the rat. *EPA Abstr.* 58:32, 1987b.

Hinton, V., M. Rosofsky, J. Granger, and N. Geary. Combined injection potentiates the satiety effects of pancreatic glucagon, cholecystokinin, and bombesin. *Brain Res. Bull.* 17:615–619, 1986.

Hjorth, S. A., K. Adelhorst, B. B. Pedersen, O. Kirk, and T. W. Schwartz. Glucagon and glucagon-like petpide 1: selective receptor recognition via distinct peptide epitopes. *J. Biol. Chem.* 269:30121–30124, 1994.

Hoebel, B. G. Neuroscience and motivation: pathways and peptides that define motivational systems. In R. C. Atkinson, R. J. Herrnstein, G. Lindzey, and R. D. Luce (eds.): *Stevens' Handbook of Experimental Psychology, 2nd Ed., Vol. 1: Perception and Motivation, 2nd Ed.* New York: John Wiley & Sons, Inc., 1988, pp. 547–625.

Holst, J. J. Radioreceptorassays for glucagon. In Lefèbvre, P. J. (ed.): *Handbook of Experimental Pharmacology, Vol. 66/I, Glucagon.* Berlin: Springer-Verlag, 1983a, pp. 245–261.

Holst, J. J. Glucagon in obesity. In Lefèbvre, P. J. (ed.): *Handbook of Experimental Pharmacology, Vol. 66/II, Glucagon.* Berlin: Springer-Verlag, 1983b, pp. 507–521.

Holst, J. J., and C. Orskov. Glucagon and other proglucagon-derived peptides. In Walsh, J. H., and G. J. Dockray *Gut Peptides: Biochemistry and Physiology.* New York: Raven Press, Ltd., 1994, pp. 305–340.

Hruby, V. J., B. Gysin, D. Trivedi, and D. G. Johnson. New glucagon analogues with conformational restrictions and altered amphiphilicity: Effects on binding, adenylate cyclase and glycogenolytic activities. *Life Sci.* 52:845–855, 1993.

Inokuchi, A., Y. Oomura, and H. Nishimura. Effect of intracerebroventricularly infused glucagon on feeding behavior. *Physiol. Behav.* 33:397–400, 1984.

Inokuchi, A., Y. Oomura, N. Shimazu, and T. Yamamoto. Central action of glucagon in rat hypothalamus. *Am. J. Physiol.* 250:R120-R126, 1986.

Ishida, T., J. Chou, R. M. Lewis, C. J. Hartley, M. Entman, and J. B. Field. The effect of ingestion of meat on hepatic extraction of insulin and glucagon and hepatic glucose output in conscious dogs. *Metabolism* 32:558–567, 1982.

Ishida, R., and J. B. Field. Hepatic handling of glucagon. In Lefèbvre, P. J. (ed.): *Handbook of Experimental Pharmacology, Vol. 66/II, Glucagon.* Berlin: Springer-Verlag, 1983, pp. 361–388.

Jelinek, L. J., S. Lok, G. B. Rosenberg, R. A. Smith, F. J. Grant, S. Biggs, P. A. Bensch, J. L. Kuijper, P. O. Sheppard, C. A. Sprecher, P. J. O'Hara, D. Foster, K. M. Walker, L. H. J. Chen, P. A. Mckernan, and W. Kindsvogel. Expression cloning and signaling properties of the rat glucagon receptor. *Science* 259:1614–1616, 1993.

Jin, S. L. Distribution of glucagon-like peptide-1 (GLP-1), glucagon, and glicentin in the rat brain. *J. Comp. Neurol.* 271:519–532, 1988.

Johnson, P. R., M. R. C. Greenwood, B. A. Horwitz, and J. S. Stern. Animal models of obesity: Genetic aspects. *Annu. Rev. Nutr.* 11:325–353, 1991.

Jørgenson, K. D., J. U. Weis, and B. Diamant. Dissociation of the spasmolytic and metabolic effects of glucagon. *Eur. J. Pharmacol.* 90:315–323, 1983.

Kalogeris, T. J., R. D. Reidelberger, V. E. Mendel, and T. E. Solomon. Interaction of cholecystokinin-8 and pancreatic glucagon in control of food intake in dogs. *Am. J. Physiol.* 260:R688-R692, 1991.

Kreymann, B., M. A. Ghatei, P. Burnet, G. Williams, S. Kanse, A. R. Diani, and S. R. Bloom. Characterization of glucagon-like peptide-1-(7–36)amide in the hypothalamus. *Brain Res.* 502:325–331, 1989.

Lambert, P. D., J. P. H. Wilding, M. A. Ghatei, and S. R. Bloom. A role for GLP-1(7–36)NH_2 in the central control of feeding behavior. *Digestion* 54:360–361, 1993.

Langer, R. New methods of drug delivery. *Science* 249:1527–1533, 1990.

Langhans, W., M. Duss, and E. Scharrer. Decreased feeding and supraphysiological plasma levels of glucagon after glucagon injection in rats. *Physiol. Behav.* 41:31–35, 1987.

Langhans, W., N. Geary, and N. Scharrer. Liver glycogen content decreases during meals in rats. *Am. J. Physiol.* 243:R450-R453, 1982a.

Langhans, W., U. Kunz, and E. Scharrer. Hepatic vagotomy increases consumption of a novel-tasting diet presented immediately after surgery. *Physiol. Behav.* 46:671–678, 1989.

Langhans, W., K. Pantel, W. Müller-Schell, E. Eggengerger, and E. Scharrer. Hepatic handling of pancreatic glucagon and glucose during meals in rats. *Am. J. Physiol.* 247:R827-R832, 1984.

Langhans, W., and E. Scharrer. Evidence for a role of the sodium pump of hepatocytes in the control of food intake. *J. Auton. Nerv. Syst.* 20:199–205, 1987.

Langhans, W., and E. Scharrer. Metabolic control of eating, energy expenditure and the bioenergetics of obesity. In Simopoulos, A. P. (ed.): *World Review of Nutrition and Dietetics.* Basel, Switzerland: S. Karger, 1992, pp. 1–67.

Langhans, W., U. Zieger, E. Scharrer, and N. Geary. Stimulation of feeding in rats by intraperitoneal injection of antibodies to glucagon. *Science* 218:894–896, 1982b.

Laughton, W. B., and T. L. Powley. Localization of efferent function in the dorsal motor nucleus of the vagus. *Am. J. Physiol.* 252:R13-R25, 1987.

Lee, J. A., Q. Ahmed, and A. D. Burt. Disappearance of hepatic parenchymal nerves in human liver cirrhosis. *Gut* 33:87–91, 1992.

Lefèbvre, P. J. (ed.). *Handbook of Experimental Pharmacology, Vol. 66/I-II, Glucagon*. Berlin: Springer-Verlag, 1983.

Lefèbvre, P. J. Glucagon and its family revisited. *Diabetes Care* 18:715–730, 1995.

Lefèbvre, P. J., G. Paolisso, A. J. Scheen, and J. C. Henquin. Pulsatility of insulin and glucagon release: Physiological significance and pharmacological implications. *Diabetologia* 30:443–452, 1987.

Le Sauter, J., and N. Geary. Pancreatic glucagon and cholecystokinin synergistically inhibit sham feeding in rats. *Am. J. Physiol.* 253:R719-R725, 1987.

Le Sauter, J., and N. Geary. Redundant vagal mediation of the synergistic satiety effect of pancreatic glucagon and cholecystokinin in sham feeding rats. *J. Auton. Nerv. Syst.* 30:13–22, 1990.

Le Sauter, J., and N. Geary. Hepatic portal glucagon infusion decreases spontaneous meal size in rats. *Am. J. Physiol.* 261:R154-R161, 1991.

Le Sauter, J., U. Noh, and N. Geary. Hepatic portal infusion of glucagon antibodies increases spontaneous meal size in rats. *Am. J. Physiol.* 261:R162-R165, 1991.

Lok, S., J. L. Kuijper, L. J. Jelinek, J. M. Kramer, T. E. Whitmore, C. A. Sprecher, S. Mathewes, F. J. Grant, S. H. Biggs, G. B. Rosenberg, P. O. Sheppard, P. J. O'Hara, D. C. Foster, and W. Kindsvogel. The human glucagon receptor encoding gene: Structure, cDNA sequence and chromosomal localization. *Gene* 140:203–209, 1994.

Lorenz, D. N., G. Kreielsheimer, and G. P. Smith. Effect of cholecystokinin, gastrin, secretin and GIP on sham feeding in the rat. *Physiol. Behav.* 23:1065–1072, 1979.

Lutz, T. A., E. Del Prete, and E. Scharrer. Reduction of food intake in rats by intraperitoneal injection of low doses of amylin. *Physiol. Behav.* 55:891–895, 1994.

Lutz, T. A., E. Del Prete, M. M. Szabady, and E. Scharrer. Attenuation of the anorectic effects of glucagon, cholecystokinin and bombesin by the amylin receptor antagonist CGRP-8–37. *Peptides* 17:119–124, 1996.

Lutz, T. A., N. Geary, M. M. Szabady, E. Del Prete, and E. Scharrer. Amylin decreases meal size in rats. *Physiol. Behav.* 58:1197–1202, 1995.

MacIsaac, L., and N. Geary. Partial liver denervations dissociate the inhibitory effects of pancreatic glucagon and epinephrine on feeding. *Physiol. Behav.* 35:233–237, 1985.

MacNeil, D. J., J. L. Occi, P. J. Hey, C. D. Strader, and M. P. Graziano. Cloning and expression of a human glucagon receptor. *Biochem. Biophys. Res. Commun.* 198:328–334, 1994.

Madsen, O. D., C. Karlsen, N. Blume, H. I. Jensen, L.-I. Larsson, and J. J. Holst. Transplantable glucagonomas derived from pluripotent rat islet tumor tissue cause severe anorexia and adipsia. *Scand. J. Clin. Lab. Invest.* 55(Suppl 220):27–36, 1995.

Martin, J. R., and D. Novin. Decreased feeding in rats following hepatic-portal infusion of glucagon. *Physiol. Behav.* 19:461–466, 1977.

Marubashi, S., M. Tominaga, T. Katagiri, K. Yamatani, Y. Yawata, M. Hara, and H. Sasaki. Hyperglycaemic effect of glucagon administered intracerebroventricularly in the rat. *Acta Endocrinol.* 108:6–10, 1985.

McGarry, J. D., and D. W. Foster. Glucagon and ketogenesis. In Lefèbvre, P. J. (ed.): *Handbook of Experimental Pharmacology, Vol.66/I, Glucagon*. Berlin: Springer-Verlag, 1983, pp. 383–398.

McLaughlin, C. L., R. L. Gingerich, and C. A. Baile. Decreased food intakes and body weights in rats immunized against pancreatic glucagon. *Physiol. Behav.* 33:723–727, 1984.

McLaughlin, C. L., R. L. Gingerich, and C. A. Baile. Role of glucagon in the control of food intake in Zucker obese and lean rats. *Brain Res. Bull.* 17:419–426, 1986.

Meyer, A. H., and E. Scharrer. Hyperpolarization of the cell membrane of mouse hepatocytes by metabolizable and nonmetabolizable monosaccharides. *Physiol. Behav.* 50:351–355, 1991.

Morley, J. E., and J. F. Flood. Amylin decreases food intake in mice. *Peptides* 12:865–869, 1991.

Muller, W. A., G. R. Faloona, and R. H. Unger. The influence of the antecedent diet upon glucagon and insulin secretion. *N. Engl. J. Med.* 285:1450–1454, 1971.

Nagai, K., L. Thibault, K. Nishikawa, A. Hashida, K. Ootani, H. Nakagawa. Effects of glucagon in macronutrient self-selection: Glucagon-enhanced protein intake. *Brain Res. Bull.* 27:409–415, 1991.

Nilsson, G., and K. Uvnas-Wallenstein. Effect of teasing and sham feeding on plasma glucagon concentration in dogs. *Acta Physiol. Scand.* 100:298–302, 1977.

Opara, E. C., and V. L. W. Go. Effect of nerve blockade on pulsatile insulin and glucagon secretion in vitro. *Pancreas* 6:653–658, 1991.

Palmer, J. P., and D. Porte, Jr. Neural control of glucagon secretion. In Lefèbvre, P. J. (ed.): *Handbook of Experimental Pharmacology, Vol. 66/II, Glucagon.* Berlin: Springer-Verlag, 1983, pp. 115–132.

Paolisso, G., A. J. Scheen, A. Albert, and P. J. Lefèbvre. Effects of pulsatile delivery of insulin and glucagon in humans. *Am. J. Physiol.* 257:E686-E696, 1989.

Penick, S. B., and G. P. Smith. The effect of glucagon on food intake and body weight in man. *J. Obesity* 1:1–5, 1964.

Petersen, O. H. The effect of glucagon on the liver cell membrane potential. *J. Physiol. Lond.* 239:647–656, 1974.

Phillips, R. J., E. Baronowsky, and T. L. Powley. Hepatic vagal afferent innervation of the GI tract. *Soc. Neurosci. Abstr.* 21:1674, 1995.

Polonsky, K. S., J. B. Jaspan, and A. H. Rubenstein. The metabolic clearance rate of glucagon. In Lefèbvre, P. J. (ed.): *Handbook of Experimental Pharmacology, Vol. 66/II, Glucagon.* Berlin: Springer-Verlag, 1983, pp. 353–359.

Post, S. R., P. G. Rubinstein, and H. S. Tager. Mechanism of action of des-His1-[Glu9]glucagon amide, a peptide antagonist of the glucagon receptor system. *Proc. Natl. Acad. Sci. U.S.A.* 90:1662–1666, 1993.

Prechtl, J. C., and T. L. Powley. A light and electron microscopic examination of the vagal hepatic branch of the rat. *Anat. Embryol.* 176:115–126, 1987.

Raben, A., H. B. Andersen, N. J. Christensen, J. Madsen, J. J. Holst, and A. Astrup. Evidence for an abnormal postprandial response to a high-fat meal in women predisposed to obesity. *Am. J. Physiol.* 267:E549-E559, 1994.

Ritter, S., and S. C. Weatherford. Capsaicin pretreatment blocks glucagon-induced suppression of food intake. *Appetite* 7:291, 1986.

Ritter, S., S. C. Weatherford, and S. L. Stone. Glucagon-induced inhibition of feeding is impaired by hepatic portal alloxan injection. *Am. J. Physiol.* 250:R682-R690, 1986.

Rodbell, M. The actions of glucagon at its receptor: Regulation of adenylate cyclase. In Lefèbvre, P. J. (ed.): *Handbook of Experimental Pharmacology, Vol. 66/I, Glucagon.* Berlin: Springer-Verlag, 1983, pp. 263–290.

Rosenwasser, A. M., Z. Boulos, and M. Terman. Circadian organization of food intake and meal patterns in the rat. *Physiol. Behav.* 27:33–39, 1981.

Russek, M. Current status of the hepatostatic theory of food intake control. *Appetite* 2:137–143, 1981.

Salter, J. M. Metabolic effects of glucagon in the Wistar rat. *Am. J. Clin. Nutr.* 8:535–539, 1960.

Sawchenko, P. E., and M. I. Friedman. Sensory functions of the liver—a review. *Am. J. Physiol.* 236:R5-R20, 1979.

Schally, A. V., T. W. Redding, H. W. Lucien, and J. Meyer. Enterogastrone inhibits eating by fasted mice. *Science* 157:210–211, 1967.

Scharrer, E., and W. Langhans. Control of food intake by fatty acid oxidation. *Am. J. Physiol.* 250:R1003-R1006, 1986.

Schick, R. R., J. P. Zimmerman, T. vorm Walde, V. Schusdziarra, and M. Classen. Glucagon-like peptide (GLP)-1-(7–36)-amide: A central suppressor of food intake in fasted rats. *Gastroenterology* 102:A756, 1992.

Schulman, J. L., J. L. Carleton, G. Whitney, and J. C. Whitehorn. Effect of glucagon on food intake and body weight in man. *J. Appl. Physiol.* 11:419–421, 1957.

Secchi, A., R. Caldara, A. Caumo, L. D. Monti, D. Bonfatti, V. DiCarlo, and G. Possa. Cephalic-phase insulin and glucagon release in normal subjects and in patients receiving pancreas transplantation. *Metabolism* 44:1153–1158, 1995.

Seifter, S., and S. Englard. Energy metabolism. In I. M. Arias, W. B. Jakoby, H. Popper, D. Schachter, and D. A. Shafritz (eds.): *The Liver: Biology and Pathobiology.* New York: Raven Press, 1993, pp. 279–315.

Shimizu, H., M. Egawa, H. Yoshimatsu, and G. A. Bray. Glucagon injected in the lateral hypothalamus stimulates sympathetic activity and suppresses monoamine metabolism. *Brain Res.* 630:95–100, 1993.

Smith, G. P., and J. Gibbs. Cholecystokinin: a putative satiety signal. *Pharmacol. Biochem. Behav.* 3:135–138, 1975.

Smith, G. P., and J. Gibbs. The development and proof of the CCK hypothesis of satiety. In C. T. Dourish, S. J. Cooper, S. D. Iversen, and L. L. Iversen (eds.): *Multiple Cholecystokinin Receptors in the CNS.* Oxford: Oxford University Press, 1992, pp. 166–182.

Stein, L. J., L. O'Farrell, and D. Novin. Circadian variation of hepatic glycogen: Effect of glucagon on food intake and glycogen concentration. *Soc. Neurosci. Abstr.* 11:58, 1985.

Stenninger, E., and J. Aman. Intranasal glucagon treatment relieves hypoglycaemia in children with type 1 (insulin-dependent) diabetes mellitus. *Diabetologia* 36:931–935, 1993.

Stockinger, Z., and N. Geary. Pancreatic glucagon does not alter intrameal gastric emptying of milk in the rat. *Physiol. Behav.* 45:1259–1261, 1989.

Stockinger, Z., and N. Geary. 2-Mercaptoacetate stimulates sham feeding in rats. *Physiol. Behav.* 47:1283–1285, 1990.

Strubbe, J. H., J. G. Wolsink, A. M. Schutte, and A. J. A. Prins. Hepatic-portal and cardiac infusion of CCK-8 and glucagon induce different effects on feeding. *Physiol. Behav.* 46:643–646, 1989.

Tang-Christensen, M., R. Göke, M. Moller, and S. P. Sheikh. GLP-1 and its receptors in the central nervous system: Powerful regulators of autonomic functions in the rat brain. *Soc. Neurosci. Abstr.* 20:1375, 1994.

Tordoff, M. G., N. Rawson, and M. I. Friedman. 2,5-Anhydro-D-mannitol acts in the liver to initiate feeding. *Am. J. Physiol.* 261:R283-R288, 1991.

Turton, M. D., D. O'Shea, I. Gunn, S. A. Beak, C. M. B. Edwards, K. Meeran, S. J. Choi, G. M. Taylor, M. M. Heath, P. D. Lambert, J. P. H. Wilding, D. M. Smith, M. A. Ghatei, J. Herbert, and S. R. Bloom. A role for glucagon-like peptide-1 in the central regulation of feeding. *Nature* 379:69–72, 1996.

Unger, R. H., A. Eisentraut, M. S. McCall, S. Keller, H. C. Lanz, and L. L. Madison. Glucagon antibodies and their use for immunoassay for glucagon. *Proc. Soc. Exp. Biol.* 102:621–623, 1959.

Unson, C. G., D. Macdonald, and R. B. Merrifield. The role of histidine-1 in glucagon action 1. *Arch. Biochem. Biophys.* 300:747–750, 1993.

Unson, C. G., and R. B. Merrifield. Identification of an essential serine residue in glucagon: Implication for an active site triad. *Proc. Natl. Acad. Sci.* U.S.A. 91:454–458, 1994.

Unson, C. G., C.-R. Wu, K. J. Fitzpatrick, and R. B. Merrifield. Multiple-site replacement analogs of glucagon. *J. Biol. Chem.* 269:12548–12551, 1994.

Uttenthal, L. O., A. Toledano, and E. Blazquez. Autoradiographic localization of receptors for glucagon-like peptide-1(7–36) amide in rat brain. *Neuropeptides* 21:143–146, 1992.

Vaccarino, F. J., F. E. Bloom, J. Rivier, W. Vale, and G. F. Koob. Stimulation of food intake in rats by centrally administered hypothalamic growth hormone-releasing factor. *Nature* 314:167–168, 1985.

Vaccarino, F. J., and M. Hayward. Microinjections of growth hormone-releasing factor into the medial preoptic area/suprachiasmatic nucleus region of the hypothalamus stimulate food intake in rats. *Regul. Pept.* 21:21–28, 1988.

VanderWeele, D. A., E. Haraczkiewicz, and M. A. DiConti. Pancreatic glucagon administration, feeding, glycemia, and liver glycogen in rats. *Brain Res. Bull.* 5 (Suppl 4):17–21, 1980.

VanderWeele, D. A., B. L. Macrum, and R. L. Oetting. Glucagon, satiety from feeding and liver/pancreatic interactions. *Brain Res. Bull.* 72:539–543, 1986.

Wakelam, M. J. O., G. J. Murphy, V. J. Hruby, and M. D. Houslay. Activation of two signal-transduction systems in hepatocytes by glucagon. *Nature Lond.* 323:68–71, 1986.

Wang, Z.-L., W. M. Bennet, M. A. Ghatei, P. G. H. Byfield, D. M. Smith, and S. R. Bloom. Influence of islet amyloid polypeptide and the 8–37 fragment of islet amyloid polypeptide on insulin release from perifused rat islets. *Diabetes* 42:330–335, 1993.

Weatherford, S. C., and S. Ritter. Glucagon satiety: Diurnal variation after hepatic branch vagotomy or intraportal alloxan. *Brain Res. Bull.* 17:545–549, 1986.

Weatherford, S. C., and S. Ritter. Lesion of vagal afferent terminals impairs glucagon-induced suppression of food intake. *Physiol. Behav.* 43:645–650, 1988.

Weick, B. G., and S. Ritter. Dose-related suppression of feeding by intraportal glucagon infusion in the rat. *Am. J. Physiol.* 250:R676-R681, 1986a.

Weick, B. G., and S. Ritter. Stimulation of insulin release and suppression of feeding by hepatic portal glucagon. *Physiol. Behav.* 38:531–536, 1986b.

Weigle, D. S. Pulsatile secretion of fuel-regulatory hormones. *Diabetes* 36:764–775, 1987.

Wynick, D., P. Hammond, and S. Bloom. The glucagonoma syndrome. *Clin. Dermatol.* 11:93–97, 1993.

Yang, Z.-J., C. Ratto, J. R. Gleason, R. Bellantone, F. Crucitti, and M. M. Meguid. Influence of anterior subdiaphragmatic vagotomy and TPN on rat feeding behavior. *Physiol. Behav.* 51:919–926, 1992.

8

Insulin as a Satiating Signal

DENNIS A. VANDERWEELE

My work on the satiating effect of pancreatic insulin was initiated in response to an apparent paradox. The paradox was that although endogenous insulin was secreted in response to ingested food (McIntyre et al., 1965), and its plasma concentration was elevated when satiation occurred and the meal ended, exogenous insulin produced overeating and obesity (MacKay et al., 1940; Lovett and Booth, 1970; Panksepp et al., 1975). Because I was impressed more by the temporal correlation between the endogenous secretion of intake and the satiation of eating, I decided to pursue the hypothesis that insulin had a satiating effect under physiological conditions.

This hypothesis was consistent with a number of observations reported around that time. For example, Lovett and Booth (1970) demonstrated that one of the four effects of a single injection of insulin was to decrease short-term food intake, presumably by increasing glucose uptake and utilization in insulin-dependent tissues. This idea was supported by Rowland and Nicolaidis (1976), who reported that intravenous infusion of insulin and glucose produced more inhibition of food intake than the infusions of glucose alone.

The inhibitory effect of exogenous insulin on food intake further predicted that when endogenous insulin was eliminated or markedly reduced, as in experimental diabetes produced by treatment with streptozotocin, meal size would increase. This prediction was confirmed (de Castro and Balagura, 1975; Panksepp and Ritter, 1975; Panksepp et al., 1975).

Finally, Steffens, Strubbe, and their coworkers had made four relevant observations: First, insulin release began as soon as orosensory stimulation occurred (Strubbe and Steffens, 1975). This initial (cephalic) phase of insulin secretion was subsequently shown to be particularly sensitive to oral glucose (Grill et al., 1984) and was mediated by vagal efferent fibers to the pancreas (Kaneto et al., 1967; Berthoud and Jeanrenaud, 1982). Second, insulin was also released during the ingestion of carbohydrate-free meals (Strubbe et al., 1977). Third, at the termination of a period of lateral hypothalamic stimulation that produced obesity, rats did not resume eating normal amounts of food until the plasma concentration of insulin had fallen to near pre-obese levels (Steffens, 1975). Fourth, the initiation of spontaneous meals occurred when the concentration of insulin was low (Strubbe et al., 1977).

With this information in mind, my colleagues and I began to test the hypothesis that the release of pancreatic insulin during a meal provided a negative-feedback signal for the satiation of that meal. The initial study assessed spontaneous food intake and body weight changes in male Holtzman rats fed ad libitum that received continuous intraperitoneal infusions of insulin from implanted osmotic minipumps (VanderWeele et al., 1980). Animals had been observed for baseline measurements of food and water intake and body weight and were allowed Purina Chow only. When rats received approximately 1 or 2 U of insulin over a 24-hour period, food intake decreased over the 7-day infusion period. At the 2 U per 24-hour dose, insulin also reduced body weight gain significantly. When insulin administration ceased, food intake began to return to the baseline level and body weight increased. Analysis of the meal pattern in a separate group of similarly infused rats on a similar diet revealed that the hypophagia was specifically due to a reduction in meal size (average = −27%) without an increase in meal number.

It is important to note that at least three factors seemed crucial for this effect to be observed: First, animals had to remain euglycemic. Rats that received higher doses (6 U/24-hour) showed a marked hyperphagia, characteristic of many earlier cited studies, that was probably due to the induction of hypoglycemia by such large doses. Second, rats had to be free feeding or only minimally deprived and fed a high-carbohydrate diet. Long periods of deprivation or a decrease in utilizable carbohydrates appeared to attenuate the effects. Third, the insulin infusions had to be prolonged. Bolus injections of identical amounts of insulin increased food intake slightly or had no effect.

As a result of the above studies, we began a series of experiments aimed at assessing insulin as a putative prandial satiety factor and identifying its mechanism of action. I review the results of these investigations in the next two sections. In the final two sections, I discuss the status of insulin as a physiological satiating hormone and its potential for the treatment of human hyperphagia in obesity and eating disorders.

The Satiating Effect of Peripheral Insulin

Real-Feeding Studies

We replicated the initial result in male rats and extended it to females (VanderWeele, et al., 1982b; VanderWeele, 1993). Most recently, we demonstrated that meal-contingent infusions of insulin reduced meal size (VanderWeele, 1994). This study is discussed below (see Decreased Release of Insulin).

The effect was not limited to rats. Acute intrajugular infusions of insulin in miniswine (Anika et al., 1980) and intraportal infusions in sheep (Deetz et al., 1980) also reduced short-term food intake.

Not all attempts to replicate our results in rats have been successful. Larue-Achagiotis and LeMagnen (1985) administered similar and slightly higher doses of insulin continuously into the peritoneal cavity or intravenously in rats and did not observe a suppression of intake. A host of differences might contribute to their failure to replicate our results, including type of food offered, placement of infusing pumps and catheters, genetic strain differences in insulin sensitivity, conditioning variables involved in free-feeding animals that often affect meal intake and, finally, type of insulin and stabilizing fluids used, as the vehicle appears crucial in maintaining insulin's activity.

In contrast to Larue-Achagiotis and LeMagnen (1985), DeStefano et al. (1991) attempted to replicate our studies as closely as possible with a wider range of dosages and a variation of the stabilizing fluid as vehicle. They also reported no significant effects on ingestion or body weight. Close examination of the food intake at the lowest dose (0.5 U/day), however, revealed a modest trend toward reduced intake. No increase in ingestion or body weight was reported by these authors until the dosage of insulin reached 5.0 U per day; this was quite different from the results reported by Larue-Achagiotis and LeMagnen (1985; see above).

Thus, the results of peripheral infusions of insulin on food intake during real feeding are mixed. Sustained peripheral infusions of modest doses of insulin (0.5–2.0 U per day) produced either no effect or a suppression of food intake in rats freely feeding and well adapted to the testing procedures, provided carbohydrates comprised a substantial portion of the diet offered.

Sham-Feeding Studies

To determine if other postingestive mechanisms are necessary for the inhibitory effect of insulin on meal size, we measured the effect of bolus injec-

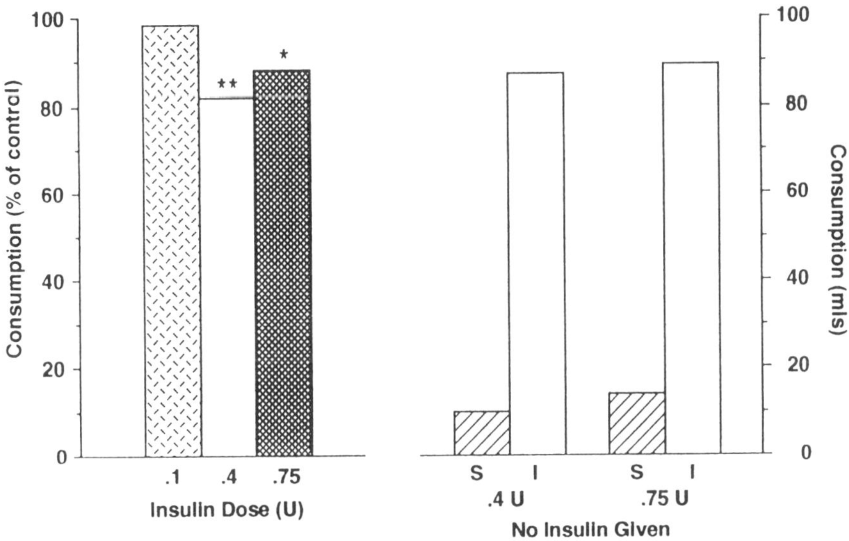

FIGURE 8.1. Sweetened condensed milk sham fed by rats following various doses of insulin **(left)**, over 60 minutes, expressed as a percentage of control consumption (following saline injection). **(Right)** Preference for milk flavors paired with insulin over those paired with saline, over 60 minutes, expressed as absolute amount consumed. No injections were made in this test of these animals; however, all had received a minimum of 6 prior paired training days with flavors associated separately with injections of insulin at dose specified or saline. Asterisks indicate significant suppression from saline injection, $*p < 0.05$ and $**p < 0.02$. From VanderWeele et al. (1990) with permission of the publisher.

tions of insulin (0.4 or 0.75 U) on intake of sweetened, condensed milk (0.4 and 0.75 U/rat) during sham feeding. (See Chapter 2 for a discussion of the rationale of using sham feeding for this purpose.) Insulin produced a small, but reliable, decrease in intake beginning in the second 15 minutes of sham feeding. The inhibition lasted throughout the remainder of the hour (Oetting and VanderWeele, 1985; VanderWeele et al., 1990) (Fig. 8.1). This inhibitory effect of bolus injection on sham feeding was not observed in real feeding, confirming our earlier experiences (see above).

Specificity of the Inhibitory Effect of Insulin

It is not clear from these experiments whether the effect of insulin to reduce sham-fed intake was a satiating effect or was due to a nonspecific or toxic effect. To address this issue, we combined the conditioned taste aversion paradigm with sham feeding in both of the studies discussed above. Rats were sham fed a flavored diet paired with insulin or saline administration. When offered both flavored diets simultaneously, rats preferred the flavor paired with insulin (Fig. 8.1). Because we did not demonstrate a conditioned taste aversion when insulin was administered exogenously in these doses, we concluded that the inhibitory effect of insulin on sham feeding was a satiating effect.

Other evidence supports that conclusion. When animals were administered insulin via osmotic minipump, water intakes were continuously observed and no reliable changes were noted. Rats also did not appear sick or lose marked amounts of weight (only the first study found a statistically reliable reduction in body weight gain; VanderWeele et al., 1980). Furthermore, insulin only decreased meal size in our experiments; it had no effect on the number of meals. In contrast, when rats are made sick with lithium chloride, meal size and meal number are reduced (VanderWeele et al., 1982a).

Endogenous Release of Insulin and Satiation

All of the above studies employed injections of insulin into animals in various feeding paradigms. A more important consideration is whether the release of endogenous hormone will influence food intake. We used drugs to modify the release of endogenous insulin. Our hypothesis predicts that increasing release of endogenous insulin by a drug should increase the satiating effect of ingested food, and decreasing the release of insulin should decrease the satiating effect of ingested food.

Increased Release of Insulin

Our initial studies sought to increase insulin release when animals were initiating a meal. Separate groups of animals were deprived of food for 2 or 3 hours. Deprivation terminated at light offset to ensure a spontaneous meal. Just prior to the meal, tolbutamide (5–50 mg/kg) was injected. Tolbutamide (5 and 15 mg/kg) significantly elevated insulin at 10 minutes postinjection and produced a reliable decrease in meal size (VanderWeele et al., 1988). At larger doses, particularly 50 mg/kg, a larger insulin release occurred, and this produced hypoglycemia and an increase in meal size.

The second study used the α-adrenergic blocker phentolamine (Barnett et al., 1988). That study was more compelling because a potential role for aversiveness of the treatment was also investigated. Groups of rats received phentolamine (300 mg/kg) just prior to a meal after 4 hours of food deprivation. Phentolamine reduced food intake significantly in the first hour postinjection. Intake was reliably decreased during the second and third 20-minute intervals after the injection, but decreased only slightly in the first 20 minutes. Total reduction in food intake over the 1-hour period in separate experiments and groups of rats ranged from 30% to 59%.

A second experiment utilized the same deprivation, injection, and feeding measurement schedules, but used a high-carbohydrate milk diet differentially flavored to investigate whether the inhibition of food intake was due to sickness or satiation. Flavored diets were consistently paired with lithium chloride, phentolamine, or saline injection. As expected, an aversion was formed to the milk flavors paired with lithium chloride, but no significant reductions of intake of milk flavors that had been previously paired with saline or phentolamine were observed. These experiments demonstrated that phentolamine (300 mg/kg) reduced food intake without producing a demonstrable aversion. Thus, the decreased food intake was probably the result of a satiating effect of insulin.

In a follow-up experiment to the phentolamine effects on eating, separate groups of rats were allowed to eat (without measurement), and blood was withdrawn for the measurement of plasma insulin and glucose at the times coinciding with observed decreases in food ingestion. During the initial 20 minutes postinjection of phentolamine, plasma insulin increased reliably compared with saline-injected control subjects. Plasma insulin level returned to that of control subjects in the 20–40- and 40–60-minute intervals. Plasma glucose levels, however, were decreased compared with controls only during the 20–40-minute interval. This effect is most probably attributable to increased glucose clearance from the blood and might account for the maximal suppression (83%) of food ingestion during this interval.

Taken together, these studies of tolbutamide and phentolamine suggest that pharmacologically induced, increased release of endogenous insulin during a meal can decrease meal size. In the case of phentolamine, the effect is associated with increased glucose clearance from the blood that is not sufficient to produce hypoglycemia, and the effect is probably not aversive.

Decreased Release of Insulin

Decreased release of endogenous insulin is a consequence of chemically induced diabetes mellitus, and it has been shown that diabetic animals increase meal size (deCastro and Balagura, 1975). One must realize, however, that severe diabetes produces a variety of physiological alterations in the animal in addition to decreased release of insulin. To test the possibility that more limited impairment of insulin, by reduced injections of the diabetogen streptozotocin, would permit a clearer demonstration of the effects of decreased insulin availability upon eating, Lindberg et al. (1984)

analyzed eating in severely and mildly diabetic rats. Their studies demonstrated an increased food intake in mildly diabetic rats after 21 hours of food deprivation.

Recently, I replicated and extended these findings (VanderWeele, 1993). Working with male, Sprague-Dawley rats, I induced mild insulin deficiency and glycosuria by injecting small doses of streptozotocin into the tail vein (20–22 mg/rat or approximately 45 mg/kg body weight). The subjects had acquired a bar-press response for food reinforcement and earned all of their meals in this manner. Their spontaneous meal patterns were displayed on an event recorder. When the meal patterns of animals with mild impairments of insulin release were compared with the patterns prior to streptozotocin, it was clear that the hyperphagia of insulin-impaired animals was produced by an increase in meal size. No reliable changes occurred in rate of eating within a meal or in number of meals initiated per day. It is probably important to note that these animals did not have reduced body weights, and, thus, a loss of body weight was not responsible for the increased meal size.

Summary

The complementary results predicted by our hypothesis were observed. Rats with increased insulin secretion decreased meal size and food intake. Rats with decreased insulin release increased meal size and food intake.

The Effect of Intraportal Insulin

A critical but difficult distinction to draw for satiety factors is whether the factors act within a physiological framework or, alternatively, only reduce food intake when pharmacological manipulations are produced. This has been an important and difficult issue to satisfy for other peptide hormones, such as Cholecystokinin (CCK) and pancreatic glucagon (see Chapters 5 and 7). In our most recent experiments, we addressed this issue for insulin.

The rat is notoriously insensitive to the porcine form of insulin used in most studies; hence, it is difficult to compare amounts of exogenous insulin required for an effect to levels of endogenous insulin that occur during the ingestion of a spontaneous meal. Estimates of insulin release at feeding in the rat range from 30 to 80 uU/ml of plasma in Strubbe's work, depending on diet (Strubbe et al., 1977), and up to 120 uU/ml in my own hands (unpublished observations) when rats are drinking sweet solutions. To approximate this more physiological range, I have recently adopted the techniques that West and his colleagues (1984) used in their work with CCK.

In 12 animals that had been trained to bar-press for food and assessed

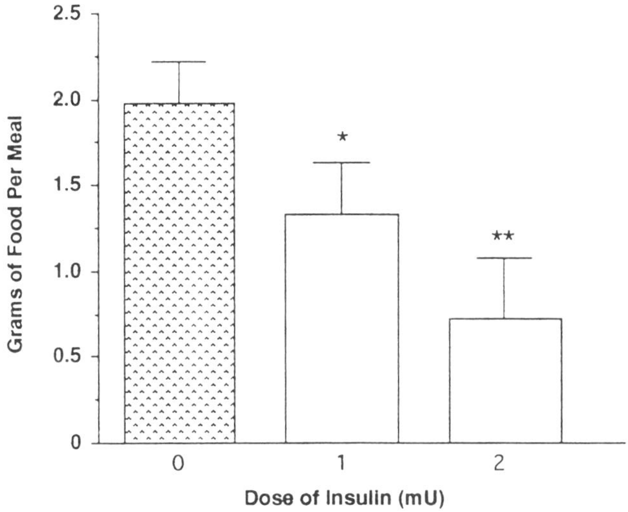

FIGURE 8.2. Average meal size when insulin or physiological saline was co-infused with spontaneously taken meals in free-feeding rats. Amounts given were dissolved in 0.15-ml aliquots and began 1–2 minutes into the meal. N = 12 rats for 0 mU, 7 for 1 mU, and 5 for 2 mU. $^{*}p < 0.01$; $^{**}p < 0.001$. From VanderWeele (1994) with permission of thepublisher.

for baseline meal patterns, I implanted hepatic portal catheters for insulin infusion (VanderWeele, 1985, 1994). The animals were allowed to recover from surgery and then reintroduced to the operant chambers for re-establishment of baseline meal patterns. The food offered was the standard Noyes pellet (45-mg pellet with 52.9% carbohydrate, 23.9% protein, and 5.6% fat), and animals were free feeding at all times during infusions. When once again stable, animals received microinfusions of 1 or 2 mU of regular porcine insulin once a meal had been initiated, in a volume of 0.15 ml of sterile, isosmotic diluting fluid; a similar volume of vehicle was used as a control. The rats were often tested with insulin at these dosages over several days in succession. Regardless of dosages, insulin infused at the beginning of a spontaneous meal reduced the average size of those meals between 5% and 89% compared with vehicle infusions (Fig. 8.2). While there was a trend to increase meal frequency over days, that effect did not reach statistical significance. As only diurnal infusions were given and animals were free to eat at night, the reductions in meal size observed accounted for decreases in total food intake of only 1% to 20% in various animals. Also, there were no significant changes in rate of body weight gain compared with noninfused animals in this paradigm.

I believe that this dosage range for insulin, particularly given the insensitivity of the rat to porcine insulin, is close to or within the prandial physiological range of the rat. I should point out that the size of all meals was not clearly reduced, and the inhibitory effect tended to appear during the first day of infusion, but not always at the first meal. First meals were reduced, but the effect was of borderline significance ($p > 0.05$ and < 0.10).

This claim that the satiating effect of insulin is a physiological effect of the hormone is supported by the results of Surina-Baumgartner et al. (1995). They reported that meal-contingent infusion of one dose of insulin (2 mU) into the hepatic-portal vein decreased the size of the first meal in the dark ($p = 0.05$) but not in the light. Other doses (1, 4, 8, or 16 mU) had no effect. This is consistent with the hypothesis, but the lack of a dose–response effect or of an effect in the light phase weakens this evidence.

More compelling support from these authors came from identical infusions of polyclonal antibodies to synthetic human insulin with 60% cross-reactivity to rat insulin prior to the first nocturnal meal. Antibody infusions increased meal size significantly. This is strong evidence that endogenous insulin is necessary for the control of spontaneous meal size in the dark in rats.

Mechanisms of Action of Insulin

Peripheral Primacy

It is likely that vagal sensory fibers are among the very first to "see" changes in insulin secretion during a meal. Indeed, as cited earlier, vagal stimulation can enhance insulin secretion (Kaneto et al., 1967), and vagal fibers in the hepatic portal bed, liver, and pancreatic efferent and afferent blood flow would be affected by insulin release and indirectly by glucose incorporation into glycogen synthesis. Some have cited vagal tone in playing a role in hyperinsulinemia and obesity (autonomic tone hypothesis; Powley, 1977; Bray, 1985). My own work anticipated that the vagus might be an important, and perhaps redundant, route, signaling nutrient availability and eliciting insulin secretion in free-feeding animals. Thus, the vagus might participate as both an afferent and efferent route in enhancing satiety during prolonged increases in insulin release.

To investigate mediation of the satiating effect of insulin infusions, female rats underwent total abdominal vagotomy or a sham procedure, according to the technique of Martin et al. (1978), and were allowed to recover eating and body weight over a 3-week period. Animals were then observed for a 10-day baseline period of food intake and weight change. Then minipumps that delivered 2.4 U of regular insulin/rat/day were implanted. Food intake and body weight changes were again recorded during the insulin infusions. Insulin infusions reduced nocturnal 24-hour intake by approximately 16% (Table 8.1) in sham rats, but had no significant effect in vagotomized rats. The insulin infusions had no significant effect on the rate of weight gain in either group of rats (VanderWeele, 1993).

Glucose and Metabolic Utilization Hypothesis

While the above study strengthened my initial thinking about mode of action for insulin, it did not address the transduction of that signal. I did not believe that an insulin receptor on or innervated by vagal sensory terminals was directly limiting intake. Recall in the earlier cited studies with tolbutamide and phentolamine that rats reduced ingestion consequent to increased glucose clearance from the blood, provided hypoglycemia was not produced. Steadily increasing levels of insulin produce hypoglycemia but should provide a greater signal for a simple insulin receptor and continue to suppress intake of food; this does not occur. Instead, based on the work of others, I proposed a metabolic explanation.

Shimazu (1967) has shown that insulin or vagal stimulation can modulate glycogenic enzymes in the liver to turn on or off storage of carbohydrate. Thus, insulin is involved in carbohydrate storage in the liver as modulated and, perhaps, sensed by the vagus nerve. Raising peripheral insulin levels chronically or stimulating insulin release pharmacologically during a meal might enhance a vagal signal feeding back to the brain to inhibit further intake.

This hypothesis would relate most readily to Russek's hypothesis (1981) of liver-based regulation of short-term food intake, as well as to the ideas of Friedman and Tardoff (1986) and to my own that developed out of earlier work with Novin (Novin and VanderWeele, 1977; VanderWeele and Novin, 1981). It is my contention that pancreatic glucagon co-released with insulin during the initial stages of ingestion (preabsorptively) provides an early cue to limit food intake in a meal (see Chapter 7). As insulin and blood glucose levels rise, however, the liver shifts from glycogenolysis stimulated by glucagon to glycogenesis stimulated by insulin, and this metabolic shift sends a vagal inhibitory input to the brain to end the meal.

Recent, work (Campfield et al.,1985; Campfield and Smith, 1990) may support my contention. An elegant series of experiments indicate that initiation of a spontaneous meal in the rat occurs following a decline in blood glucose. The authors state that the glycemic fall may be precipitated by a brief spike in insulin; the work of Steffens and colleagues cited earlier, however, indicates insulin to be at or near its nadir prior to meal initiation

TABLE 8.1.

Mean Food Intake and Body Weight for Intact and Vagotomized Rats Before and During Insulin Infusions

	Intact Rats				Vagotomized Rats					
	Baseline		Pump		Baseline		Pump 1		Pump 2	
	N	D	N	D	N	D	N	D	N	D
Intake (g)	23.9 ± 1.2	11.7 ± 1.4	19.4* ± 1.0	10.6 ± 1.5	17.1 ± 1.4	12.9 ± 1.3	16.8 ± 1.5	12.2 ± 2.0	15.6 ± 1.6	12.9 ± 2.3
Body weight	297 ± 14		302 ± 13		279 ± 15		284 ± 12		300 ± 14	

Pump indicates 10 days of insulin infusion (2.4 U/rat/24 hours). Pump 1 and Pump 2 indicate two infusions administered sequentially. N, nocturnal; D, diurnal. Intake and body weight data are mean ± SE. Insulin infusions decreased nocturnal intake in intact rats (*$p<0.05$), but not in vagotomized rats. From VanderWeele (1993), with permission of the publisher.

(Steffens, 1975; Strubbe et al., 1977). I believe that it is more likely that an insufficiency of glucose relative to circulating insulin (whether spiking or not) accounts for the transient decline in glycemia signaling the start of a meal. Additionally, in the work of Campfield and Smith, the meal often did not begin following these small declines until restoration of glycemia had begun to occur, probably through glucagon or adrenaline release as the animal arouses, seeks, locates, and begins to ingest a suitable food. Thus, when glycogenolysis or gluconeogenesis occurs, food seeking and ingestion occur. Shortly after, these liver enzymatic processes are reversed by insulin released during ingestion, and eating stops. As discussed already, if insulin is absent, satiety from meal ingestion is impaired, and meal size increases.

Taste Effects of Insulin in the Nucleus Tractus Solitarius

One final mechanism remains for insulin's interaction with satiety that has some support; insulin may decrease evoked taste responses to sweet tastants in the brainstem. This view is essentially a hedonic hypothesis and would not be in conflict with the earlier mechanisms cited. Redundancy within the central nervous system, from the standpoint of either a number of structures of the brain involved in the regulation of any single behavior or more than one mechanism by which a behavioral outcome can be produced is not a new concept.

In a series of experiments, Giza and Scott (1983) investigated the multiunit responses of nucleus tractus solitarius (NTS) neurons to the four basic taste stimuli while animals experienced alterations in glycemia via different hormone or nutrient injections. The initial study administered glucose intravenously and noted a decline in the responsiveness of NTS neurons specifically to glucose placed on the tongue. Salt, bitter, and sour responses seemed relatively unaffected. Because the infusion of glucose probably released insulin, a later study infused insulin while NTS neurons were recorded (Giza et al., 1992). The administration of insulin also selectively reduced activity of NTS neurons to glucose administered to the tongue. In these experiments, glucose responsiveness is considered to be representative for sweet tastes, and this suggests that insulin reduces the hedonic attraction of sweet tastes. Perhaps this mechanism is involved in satiation and particularly in conditioned satiation to familiar sweet foods.

Summary

My work has emphasized that pancreatic insulin is one of a cascade of separate but interactive factors that act to terminate a meal and produce the physiological state of postprandial satiety. Ingestion of food produces insulin and glucagon release, which liberates glucose via liver glycogenolysis before absorption from the gut and the meal ends. As glucose and other macronutrients such as fat and insulin-sensitive amino acids begin to be absorbed in substantial quantities from the gut, insulin secretion rises and relative glucagon titers fall, producing enhanced effectiveness of insulin's satiating action and a restoration of euglycemia. At the same time, sweet-tasting foods may become less hedonically attractive, and the rat stops eating. For these reasons, the absence or impairment of insulin release, such as occurs in diabetes, decreases the satiating potency of ingested food and meal size increases.

Progress and Remaining Problems

Interaction of Insulin at Additional Sites and with Other Peptides

There are, of course, other authors and hypotheses relating insulin to satiety, as cited at one or two points earlier in this chapter. Certainly, a stimulus to my work was that of Woods, Porte and their colleagues. Since the early 1970s, these authors have advanced the view that circulating insulin gains access to the brain and functions there as a negative-feedback signal of adiposity (Woods et al., 1974, 1979). There is now very strong evidence for this (Schwartz et al., 1994; Woods, 1995).

Insulin apparently acts at sites within the medial hypothalamus to reduce food intake by enhancing the gain or efficacy of other prandial satiating hormones or signals (Woods and VanderWeele, 1989; Woods, 1995). For example, a synergism between centrally administered insulin and peripherally administered CCK has been reported in baboons (Figlewicz et al., 1986) and rats (Riedy et al., 1995). It is interesting that a similar synergism occurs between peripherally administered insulin and peripherally administered CCK (VanderWeele, 1984).

It is important to note that the central action of insulin and the peripheral one that I have discussed are not incompatible. Insulin could accumulate within the brain as a function of sustained, modest elevations in the periphery (Woods and VanderWeele, 1989; Steffens et al., 1988). This redundancy of the satiating effects of peripheral and central insulin could be an advantage for the control of meal size.

Effect of Insulin in Humans and Therapeutic Potential

Only one study has tested the satiating effect of insulin in humans (Woo et al., 1984). This study infused insulin with glucose into a peripheral vein in the forearm of human normal-weight volunteers. Amounts of insulin and glucose were matched so that plasma insulin was increased significantly before and during the test meal without producing a significant change in plasma glucose. The infused insulin and glucose did not change the size or duration of meals under these conditions.

Further experiments in humans are desirable. They may require prolonged infusions, such as are efficacious in rats, but prolonged administration of insulin in humans could lead to insulin resistance. Thus, at this time there appears to be no therapeutic potential for peripheral administration of insulin in the treatment of human hyperphagia except, of course, in insulin-dependent diabetes.

Summary and Conclusions

Criteria for a Satiety Signal

Gibbs et al. (1973) hypothesized that gut peptides (and perhaps pancreatic peptide hormones—my inclusive amendment) that are released peripherally during food intake may function in satiation and the termination of eating. This led them to publish a number of studies assessing CCK as a satiety signal (see Chapter 5). Additionally, that laboratory developed a set of criteria that must be met before a peptide or any other putative satiating signal can be proven to operate under physiological conditions. Geary (1990) recently reviewed and revised these criteria in the course of evaluating the evidence for glucagon as a satiating signal (see Chapter 7).I will use Geary's criteria to summarize the evidence for the satiating action of peripheral insulin reviewed in this chapter (Table 8.2).

The first criterion is satisfied. Insulin is released during a spontaneous meal.

The second criterion, that the proposed satiety signal, when administered exogenously, should produce a specific satiety effect, has also been satisfied. Insulin administered in doses not exceeding 2 U/day/rat or 1 or 2 mU/meal/rat reduced food intake (VanderWeele, 1993, 1994). This reduction in intake was specific to food because water intake was not affected. The reduced intake was due to a decrease in the size of spontaneous meals

TABLE 8.2.
Criteria for a Prandial Satiety Signal

1. The release of the hormone must change at an active site during a spontaneous meal
2. Exogenous administration of the hormone must produce a specific satiety effect
3. Agonism (or antagonism) of prandial changes or release of the hormone must have enhancing (or inhibiting) effects on satiety
4. The changes observed in the endogenous release of the hormone must be sufficient for its satiety effects
5. The hormone's satiety effect must appear in and have ecological relevance

Adapted from Geary (1990).

without a change in the number of meals, and it occurred in the absence of obvious signs of distress or malaise.

The third criterion that stimulation or inhibition of release of insulin should also reduce or increase meal size, respectively, has also been fulfilled. Phentolamine and tolbutamide increase insulin release during eating and decrease meal size (Barnett et al., 1988; VanderWeele et al., 1988). Decreasing insulin release by streptozotocin (VanderWeele, 1993) or diazeoxide (VanderWeele, 1994) increases meal size.

The fourth criterion appears to be satisfied by my most recent experiments (VanderWeele, 1994). Doses of only 1–2 mU of insulin, co-infused into the liver with a spontaneously taken meal, significantly reduced the size of that meal. I assert that this may be within the physiological range for the rat, given its insensitivity to porcine insulin.

The final criterion of Table 8.2 requires that satiety effects of the putative signal be demonstrable under "ecological conditions."A strict interpretation of this criterion has not been applied to any animal species, to my knowledge, at least not in a truly open environment with diets normally ingested by that species outside the laboratory. But it is relevant that we have demonstrated the satiating effect of insulin in free-feeding animals.

In summary, there is evidence that peripheral administration of insulin satisfies all of the criteria for a prandial satiety signal.One limitation of the evidence, particularly for criteria 2, 3, and 4, may be that most of the evidence comes from my laboratory. Until other laboratories perform such experiments, it is not possible to know how robust my results are.

Current Status of Peripheral Insulin as a Prandial Satiety Signal

The present chapter has presented a case for insulin as a prandial satiety signal, one I have been advocating for approximately 13 years. I believe that this case is a strong one; however, there are experiments that remain to be done that could clarify this issue. For example, the administration of rat insulin antibodies to free-feeding animals combined with assessment of spontaneous meal patterns would help to solidify the case. Assessing the role of labeled glucose uptake and incorporation into key tissue(s) and storage or metabolism after physiological and meal-related insulin infusions could clarify the site of action and cellular mechanism. Finally, measuring the interaction of insulin with other putative satiety peptides, hormones, and physiological stimuli relevant to body weight regulation is important.

I maintain the view that satiation during a meal is produced by a sequence of stimuli that occur as food is sampled, ingested, absorbed, metabolized, and stored. Although insulin plays a key role in many steps of this sequence, insulin's primary action is to alter intracellular metabolic function in the liver during a meal, and this produces a stimulus that is carried by vagal afferent nerves to the brain where it is processed into a negative-feedback signal for the control of meal size.

ACKNOWLEDGMENTS

The accumulation of data and full development of the ideas contained in this chapter would not have been possible without the aid and support of research development funds from Occidental College. Also, Dr. Kenneth A. Fox, an early student and collaborator, contributed both interaction and support at important times in the development of the theory. Dr. Donald Novin should also receive much credit for teaching, challenging my ideas and thinking, and, above all else, serving as an excellent role model for what a scientist should be. Finally, I would like to thank Drs. Gerard Smith and James Gibbs for the opportunity of presenting this work; their tutelage, support, and friendship while at the Bourne Laboratory on visits and sabbatical have continued to sustain and make worthwhile my research in the understanding of the role of insulin in the regulation of food intake.

REFERENCES

Anika, S. M., T. R. Houpt, and K. A. Houpt. Insulin as a satiety hormone. *Physiol. Behav.* 25:21–23, 1980.

Barnett, C. R., D. R. Talbert, and D. A. VanderWeele. Phentolamine stimulates insulin release and glucose clearance and suppresses food ingestion. *Physiol. Behav.* 44:685–689 1988.

Berthoud, H. R., and B. Jeanrenaud. Sham feeding-induced cephalic phase insulin release in the rat. *Am. J. Physiol.* 242:E280–E285, 1982.

Bray, G. A. Autonomic and endocrine factors in the regulation of food intake. *Brain Res. Bull.* 14:505–510, 1985.

Campfield, L. A., P. Brandon, and F. J. Smith. On-line continuous measurement of blood glucose and meal pattern in free-feeding rats: the role of glucose in meal initiation. *Brain Res. Bull.,* 14:605–616, 1985.

Campfield, L. A., and F. J. Smith. Systemic factors in the control of food intake. In Stricker, E. M. (ed.): *Handbook of Behavioral Neurobiology, Vol. 10, Neurobiology of Food and Fluid Intake.* New York: Plenum Press, 1990, pp. 183–206.

deCastro, J. M., and S. Balagura. Meal patterning in the strepotozotocin diabetic rat. *Physiol. Behav.* 15:259–263, 1975.

Deetz, L. E., P. J. Wangsness, J. F. Kavanaugh, and L. C. Griel, Jr. Effect of intraportal and continuous intrajugular administration of insulin on feeding in sheep. *J. Nutr.* 110:1983–1991, 1980

DeStefano, M. B., J. S. Stern, and T. W. Castonguay. Effect of chronic insulin administration on food intake and body weight in rats. *Physiol. Behav.* 50:801–806, 1991.

Figlewicz, D. P., D. B. West, L. J. Stein, S.C. Woods, and D. Porte, Jr. Insulin alters the sensitivity of baboons to CCK-induced single meal suppression. *Am. J. Physiol.* 250:R856–R860, 1986.

Friedman, M. I., and M. G. Tordoff. Fatty acid oxidation and glucose utilization interact to control food intake in rats. *Am. J. Physiol.* 251:R840–R845, 1986.

Geary, N. Pancreatic glucagon signals postprandial satiety. *Neurosci. Biobehav. Rev.* 14:323–338, 1990.

Gibbs, J., R. C. Young, and G. P. Smith. Cholecystokinin decreases food intake in rats. *J. Comp. Physiol. Psychol.* 84:488–495, 1973.

Giza, B. K., and T. R. Scott. Blood glucose selectively affects taste-evoked activity in the rat nucleus tractus solitarius. *Physiol. Behav.* 31:643–650, 1983.

Giza, B. K., T. R. Scott, and D. A. VanderWeele. Administration of satiety factors and gustatory responsiveness in the nucleus tractus solitarius of the rat. *Brain Res. Bull.* 28:637–639, 1992.

Grill, H. J., K. C. Berridge, K. C., and D. J. Ganster. Oral glucose is the prime elicitor of preabsorptive insulin secretion. *A. J. Physiol.* 246:R88–R95, 1984.

Kaneto, A., K. Kosaka, and K. Nakao. Effects of stimulation of the vagus nerve on insulin secretion. *Endocrinology* 80:530–536, 1967.

Larue-Achagiotis, C., and J. LeMagnen. Effect of long-term insulin on body weight and food intake intravenous versus intraperitoneal routes. *Appetite* 6:319–329, 1985.

Lindberg, N. O., P. C. Coburn, and E. M. Stricker. Increased feeding by rats after subdiabetogenic streptozotocin treatment: A role for insulin in satiety. *Behav. Neurosci.* 98:138–145, 1984.

Lovett, D., and D. A. Booth. Four effects of exogenous insulin on food intake. *Q. J. Exp. Psychol.* 22:406–419, 1970.

MacKay, E. M., J. W. Callaway, and R. H. Barnes. Hyperalimentation in normal animals produced by protamine insulin. *J. Nutr.* 20, 59–66, 1940.

Martin, J. R., D. Novin, and D. A. VanderWeele. Loss of glucagon suppression of feeding after vagotomy in rat. *Am. J. Physiol.* 234:E314–E318, 1978.

McIntyre, N., C. D. Holdsworth, and D. S. Turner. Intestinal factors in the control of insulin secretion. *J. Clin. Endocrinol.* 25:1317–1324, 1965.

Novin, D., and D. A. VanderWeele. Visceral involvement in feeding: There is more to regulation than the hypothalamus. In Sprague J. M., and A. N. Epstein (eds.); *Progress in Psychobiology and Physiological Psychology.* New York: Academic Press, 1977, pp. 193–241.

Oetting, R. L. and D. A. VanderWeele. Insulin suppresses intake without inducing illness in sham feeding rats. *Physiol. Behav.* 34:557–562, 1985.

Panksepp, J., A. Pollack, K. Krost, R. Meeker, and M. Ritter. Feeding in response to repeated protamine zinc insulin injections. *Physiol. Behav.* 14:487–493, 1975.

Panksepp, J. and M. Ritter. Mathematical analysis of energy regulatory patterns of normal and diabetic rats. *J. Comp. Physiol. Psychol.* 89:1019–1028, 1975.

Powley, T. L. The ventromedial hypothalamic syndrome, satiety and a cephalic phase hypothesis. *Psychol. Rev.* 84:89–126, 1977.

Riedy, C. A., M. Chavez, D. P. Figlewicz, and S. C. Woods. Central insulin enhances sensitivity to cholecystokinin. *Physiol. Behav.* 58:755–760, 1995.

Rowland, N. and S. Nicolaidis. Metering of intravenous versus oral nutrients and regulation of energy balance. *Am. J. Physiol.* 231:661–669, 1976.

Russek, M. Current status of the hepatostatic theory of food intake control, commentaries. *Appetite* 2:137–162, 1981.

Schwartz, M. W., D. P. Figlewicz, D. G. Baskin, S. C. Woods, and D. Porte, Jr. Insulin and the central regulation of energy balance: Update 1994. *Endocrinol. Rev.* 2:109–113, 1994.

Shimazu, T. Glycogen synthetase activity in liver; Regulation by autonomic nerves. *Science* 156:1256–1257, 1967.

Steffens, A. B. Influence of reversible obesity on eating behavior, blood glucose and insulin in the rat. *Am. J. Physiol.* 228:1738–1744, 1975.

Steffens, A. B., J. W. Scheurink, D. Porte, Jr., and S. C. Woods. Penetration of peripheral glucose and insulin into cerebrospinal fluids in rats. *Am. J. Physiol.* 255:200–204, 1988.

Strubbe, J. H., and A. B. Steffens. Rapid insulin release after ingestion of a meal in the unanesthetized rat. *Am. J. Physiol.* 229:1019–1022, 1975.

Strubbe, J. H., A. B. Steffens, and L. DeRuitter. Plasma insulin and the time pattern of feeding in the rat. *Physiol. Behav.* 18:81–86, 1977.

Surina-Baumgartner, D. M., W. Langhans, and N. Geary. Hepatic portal insulin antibody infusion increases, but insulin does not alter, spontaneous meal size in rats. *Am. J. Physiol.* 269:R978–R982, 1995.

VanderWeele, D. A. The alimentary canal, liver and vagus play a role in short-term feeding: But is the role regulation, correlation, glucostasis or spurious association? *Int. J. Obesity* 8(Suppl.1):51–63, 1984.

VanderWeele, D. A. Insulin and satiety. *Soc. Neurosci. Abstr.* 11:62, 1985.

VanderWeele, D. A. Insulin and satiety from feeding in pancreatic-normal and diabetic rats. *Physiol. Behav.* 54:477–485, 1993.

VanderWeele, D. A. Insulin is a prandial satiety hormone. *Physiol. Behav.* 56:619–622, 1994.

VanderWeele, D. A., J. A. Granja, and D. A. Deems. Discomfort or satiety; The spontaneous meal pattern may serve as a predictor. In (Hoebel B. G., and D. Novin (eds.); *The Neural Basis of Feeding and Reward.* Bunswick, ME: Haer Institute Press, 1982a, pp. 167–173.

VanderWeele, D. A., E. Haraczkiewicz, and T. B. Van Itallie. Elevated insulin and satiety in obese and normal-weight rats. *Appetite* 3:99–109, 1982b.

VanderWeele, D. A., E. Haraczkiewicz, and J. E. Vaselli. Tolbutamide affects food ingestion in a manner consistent with its glycemic effects in the rat. *Physiol. Behav.* 44:679–683, 1988.

VanderWeele, D. A., and D. Novin. On hepatic involvement in the short-term regulation of food ingestion. *Appetite* 2:153–156, 1981.

VanderWeele, D. A., R. Oetting-Deems, and R. B. Kanarek. Insulin modifies flavor aversions and preferences in real- and sham-feeding rats. *Am. J. Physiol.* 259:R823–R828, 1990.

VanderWeele, D. A., F. X. Pi-Sunyer, D. Novin, and M. J. Bush. Chronic insulin infusion suppresses food ingestion and body weight gain in rats. *Brain Res. Bull.* 5(Suppl. 4):7–11, 1980.

West, D. B., D. Fey, and S. C. Woods. Cholecystokinin persistently suppresses meal size but not food intake in free-feeding rats. *Am. J. Physiol.* 246:R776-R787, 1984.

Woo, R., H. R. Kissileff, F. X. Pi-Sunyer. Elevated postprandial insulin levels do not induce satiety in normal-weight humans. *Am. J. Physiol.* 247:R745–R749, 1984.

Woods, S. C. Insulin and the brain: A mutual dependency. In Fluharty, S., A. R. Morrison, J. Sprague, and E. Stellar (eds.): *Progress in Psychobiology and Physiological Psychology, Vol. 16.* New York: Academic Press, 1995, pp. 53–81.

Woods, S. C., E. Decke, and J. R. Vaselli. Metabolic hormones and regulation of body weight. *Psychol. Rev.* 81:26–43, 1974.

Woods, S. C., E. C. Lotter, L. D. McKay, and D. Porte, Jr. Chronic intracerebroventricular infusion of insulin reduces food intake and body weight of baboons. *Nature* 282:503–505, 1979.

Woods, S. C., and D. A. VanderWeele. Insulin and intake—A short summary. Letter from the president. *SSIBlings* 2:1–4, 1989.

9

Serotonin and the Structure of Satiation

KENNY J. SIMANSKY

There has been pharmacological evidence since at least 1960 (Soulairac and Soulairac, 1960) that implicates serotonin (5-hydroxytryptamine [5-HT]) in the regulation of food intake. This is not a surprising function given the anatomical localization of 5-HT in the gut (Gershon et al., 1989) and brain (Steinbusch, 1981; Jacobs and Azmitia, 1992), its actions on smooth muscle tone and digestive processes in the periphery (Tonini and De Ponti, 1995), and its diverse roles in the central modulation of motor activity, other complex behaviors, and mechanisms regulating the internal milieu (e.g., McCall, 1990; Wilkinson and Dourish, 1991). Data relating 5-HT specifically to feeding came primarily from studies in which drugs that enhanced serotonergic activity routinely decreased the consumption of food, whereas agents that diminished serotonergic activity often increased intake (Blundell, 1979). Depending on the experimental conditions used, however, these results led to different, often controversial, hypotheses regarding the precise role of 5-HT. It has been suggested, for example, that 5-HT governs the selection of macronutrients in the diet (Wurtman and Wurtman, 1988), influences responses to taste qualities of food (Cooper and Neill, 1987), and modulates gastric activity to reduce feeding (Booth et al., 1986).

Perhaps the most inclusive hypothesis is that serotonergic activity increases the rate of satiation and prolongs the state of satiety (Blundell, 1986). This proposal came initially from the observation that the serotonergic agonist fenfluramine reduced the size of meals without delaying the onset of feeding by rats (Blundell et al., 1976). Blundell emphasized the importance of analyzing the alterations in feeding behavior, not simply the changes in food intake, produced by serotonergic agents. Using this strategy, later work generally supported a role for 5-HT in satiety (for reviews, see Blundell, 1979, 1986; Blundell and McArthur, 1981). Inferences about the specific mechanisms involved, however, were limited by relative ignorance of the diversity of 5-HT receptor subtypes and by the paucity of agents for selectively manipulating their function. Also, the original investigations used systemic administration of drugs that readily entered the central nervous system. Thus, questions of the anatomical sites of action for serotonergic drugs in feeding remained untested.

Three major classes of serotonergic receptors (5-HT_1-like, 5-HT_2, and 5-HT_3) were recognized in the nomenclature established by Bradley et al. in 1986. On the basis of data from operational (i.e., functional and ligand-binding), transductional (i.e., intracellular signaling), and structural (i.e., molecular biological) studies, seven classes containing at least 16 separate subtypes have now been described in peripheral and central tissues of various mammalian species (for review, see Hoyer et al., 1994). In animal models, new agonists and antagonists have been employed for more selectively defining the effects of altering synaptic levels of 5-HT, and especially of stimulating the 1A, 1B, 1-like, 2A, 2C, and 3 subtypes on feeding behavior (see Table 9.1, below). Some studies have explored the ingestive actions of administering 5-HT and related compounds into the brain, while renewed interest in the pharmacology of 5-HT in the gut has led to a parallel assessment of peripheral mechanisms. Increasingly, sophisticated behavioral analyses have characterized the alterations in feeding behavior and meal patterns underlying the anorectic actions of serotonergic agents.

Numerous reviews have described the pharmacological mechanisms by which these drugs reduce food intake (e.g., Garattini et al., 1986; Cooper, 1992; Dourish, 1992; Simansky, 1996b). The present work focuses on the view that 5-HT influences the process of satiation by examining the consequences of altering serotonergic function on the structure of feeding behavior. To assess results from studies using different experimental strategies, the review evaluates serotonergic actions on the distribution of meals and of feeding during meals, on modifications in the pattern of behaviors during and after meals, and on related work using instrumental paradigms. In addition to the general relationships between pharmacological mechanisms and the organization of feeding, the roles of peripheral and central sites in mediating serotonergic effects are considered.

Serotonergic Modulation of the Microstructure of Meals

Drugs that Enhance Serotonergic Neurotransmission

FENFLURAMINE

Drugs that indirectly increase synaptic levels of endogenous 5-HT (releasing agents, selective serotonin reuptake inhibitors, metabolic precursors) or that directly stimulate postsynaptic 5-HT receptors have been used to test the effects of enhancing serotonergic neurotransmission on the organization of feeding. Fenfluramine, which releases 5-HT and blocks its reuptake, has been the prototypical agent for probing the role of endogenous 5-HT in regulating food intake (for reviews, see Garattini et al., 1986, 1992; Cooper, 1992). Early experiments used the racemic (*d,l*-) form of this agent. Subsequently, the dextroisomer of fenfluramine was shown to be more potent as an anorectic drug than the levoisomer in reducing food intake and also more selective in altering serotonergic (compared with catecholaminergic) function (Garattini et al., 1986). Thus, *d*-fenfluramine replaced the racemic form in most studies of ingestive behavior during the past 10 years. Qualitatively, *d,l*- and *d*-fenfluramine have produced the same effects on meal parameters.

Fenfluramine reduces meal size and the rate of eating. Blundell and his colleagues first examined the effects of fenfluramine on the organization of feeding in a series of studies that incorporated automated and observational techniques. In the initial study, they used an eatometer (Kissileff, 1970) that recorded the time when each pellet was removed from a trough for determining the changes produced by *d,l*-fenfluramine in the distribution of responses for food (Blundell et al., 1976). In free-feeding rats, fenfluramine reduced the size of meals without changing their frequency. After food deprivation, fenfluramine did not alter the latency until feeding began, but caused a dose-related decrease in the duration of feeding. These data suggested that fenfluramine inhibited processes that maintained feeding or enhanced those that terminated this behavior.

Blundell and Latham (1978, 1980) extended the behavioral analysis of the anorectic actions of fenfluramine in food-deprived rats by characterizing the microstructure of ingestion within bouts of feeding. Observers continuously monitored the behavior of the rats during a 1-hour test or watched videotapes and rated the occurrence of eating, drinking, grooming, locomotor activity, resting, and other responses. As in the study using an eatometer, fenfluramine decreased the amount of food eaten without altering the latency to initiate the meal. In addition, the microstructural analysis revealed that fenfluramine reduced the rate of eating as calculated from the quotient of the food consumed and the time the rats were observed eating. The rats did not compensate for the decreased rate by increasing the duration of feeding or by changing their nonfeeding activity. In contrast, anorectic doses of the indirectly acting, catecholaminergic agonist amphetamine delayed the start of feeding, slightly increased its rate, increased nonfeeding activity, and disrupted the pattern of behavior. The actions of fenfluramine also contrasted with the slower, but prolonged, feeding with normal intake displayed by rats injected with the dopaminergic antagonists pimozide and α-flupenthixol. Blundell and Latham (1980) concluded that fenfluramine acted in a behaviorally specific manner to diminish food consumption by inhibiting the rate of eating and by facilitating the process of satiation.

The effect of fenfluramine to decrease meal size has been confirmed in lean (Grinker et al., 1980; Burton et al., 1981; Davies et al., 1983; Grignaschi et al., 1992) and genetically obese (Grinker et al., 1980) rats, in lean and obese humans (see Blundell and Hill, 1992), and in baboons (Foltin and Fischman, 1989). Generally, rats and primates do not compensate for the reduced size of meals by initiating the next meal sooner or by increasing the frequency of meals or bouts of feeding (Blundell and Latham, 1978; Davies et al., 1983; Fletcher and Burton, 1986a; Grignaschi and Samanin, 1992a). Indeed, fenfluramine has been reported to decrease meal and bout frequency in some studies using rats (e.g., Blundell and Latham, 1980; Grinker et al., 1980), to decrease snacking between meals in humans (Blundell and Hill, 1992), and to decrease meals obtained by baboons in an operant paradigm using a multicomponent chained schedule in which satisfying the contingency in the first component (procurement) led to food availability in the second component (meal) (Foltin and Fischman, 1989). Thus, across species, fenfluramine has consistently altered eating patterns in a manner suggesting that this drug augments the satiating effect of food. In humans, this inference has been corroborated using a visual analog scale with fenfluramine, causing substantial increases in self-ratings of feelings of fullness (Blundell and Hill, 1987) and persistent decreases in the postprandial motivation to eat (Hill and Blundell, 1990).

Fenfluramine's action to decrease the rate of eating has also been demonstrated repeatedly in studies of rats that were eating pellets under ad libitum conditions or after varying lengths of food deprivation using eatometers (Cooper and Francis, 1979; Burton et al., 1981; Fletcher and Burton, 1986a; Grignaschi et al., 1992, 1993; Grignaschi and Samanin, 1992a), analysis of videotapes (Burton et al., 1981), or direct observation (Cooper and Francis, 1980; Cooper and Sweeney, 1980). Similarly, fenfluramine slowed eating in humans (Rogers and Blundell, 1979; Kyriakides and Silverstone, 1979; McGuirk et al., 1991) and decreased the rate of operant responding by baboons in the meal component of the chained schedule described above (Foltin and Fischman, 1989).

In the study with baboons, fenfluramine decreased the rate of responding in the second meal but not in the first meal. As noted above, the decrease in rate occurred during the meal component, but not during the procurement phase of the schedule. Furthermore, the decrease became more pronounced as the meal progressed. These data suggested that fenfluramine interacted with stimuli produced by feeding to inhibit responding in a behaviorally specific manner.

This experiment with baboons measured the rate of an operant lever pull to obtain food rather than the rate at which the food was either taken or ingested. When the rate of intake was estimated by determining the distribution of intervals between successive pellet removals from an eatometer in rats, fenfluramine reduced the estimated rate of intake throughout discrete bouts of feeding (Burton et al., 1981). Fenfluramine also slowed eating throughout a test meal in humans, when the rate of eating was calculated, more strictly, by measuring food intake (bread, meats) and the time actually spent chewing (Rogers and Blundell, 1979).

Reductions in eating rate may reveal a normal consequence of satiation or more direct alteration of motor function. The ability of fenfluramine to slow eating beginning with the onset of the meal in rats and humans suggests that satiety signals from postingestive stimuli are not necessary for this drug to suppress the rate of eating. One interpretation of these data is that fenfluramine prematurely activates physiological processes that control motor function to normally retard feeding as a meal progresses. Evidence for decreased rates of ingestion during later portions of meals has been reported for rats (Blundell, 1986; Davis and Smith, 1988; Davis et al., 1993), baboons (Foltin and Fischman, 1989), and humans (Rogers and Blundell, 1979; Spiegel et al., 1993—but see Burton et al., 1981; Hsiao and Spencer, 1983). In related work, rats ate test meals more slowly if they were partially satiated in one study (Cooper and Francis, 1993), but not in another (Blundell et al., 1985).

Discrepancies about changes in the rate of ingestion during a meal probably reflect methodological differences across studies. Spiegel et al. (1993) manipulated rates of intake in humans by varying the bite size of solid food provided during a test meal. When larger pieces of food were eaten, the initial rates were faster and eating decelerated quickly over the course of the meal. In contrast, eating small pieces produced relatively low initial rates that declined minimally. Regardless of the differing initial rates engendered by varying the bite size, subjects finished the meal at a common pace. These results lead to the speculation that, in drug studies, fenfluramine clamps intake closer to a terminal rate that is characteristic for the conditions of the test.

Cooper and Francis (1979) reported that *d,l*-fenfluramine decreased food intake and the eating rate of pellets, but not of powdered food. Furthermore, *d,l*-fenfluramine inhibited consumption of coarse food, but not of fine crumbs (Baker and Booth, 1990). Baker and Booth (1990) concluded that fenfluramine disrupts motor function, with greater consequences for ingestion of larger pieces of food (see also Cooper, 1992). Rats, however, eat pellets more quickly than powder (Cooper and Francis, 1979), and fenfluramine's ability to fix eating at a slow terminal rate would be more detectable when initial rate of intake is fast.

The effects of fenfluramine on oral motor function have been proposed to reflect changes in responsiveness to the gustatory qualities of food. Asin et al. (1992) characterized the effects of *d*-fenfluramine in rats on the microstructure of licking a solution of 0.4 M (13.7%) sucrose after 4 hours of food deprivation. The analysis grouped licks into bursts, in which the interlick intervals (ILI) were less than 250 msec, and clusters, with ILIs less than 500 msec (Davis and Smith, 1992; see also Chapter 4). Fenfluramine decreased the number of licks within bursts and clusters, decreased the rate of licking within bursts, and increased the proportion of responses with ILIs between 251 and 500 msec. The microstructural changes in the sizes of bursts and clusters were qualitatively identical to those produced in nondrugged rats by decreasing the concentration of sucrose (Davis and Smith, 1992).

Asin et al. (1992) concluded that the sucrose tasted less palatable to rats that had been injected with fenfluramine. Higgins et al. (1994), however, observed that *d*-fenfluramine produced dose-related decreases across several concentrations of saccharin and reduced peak intake, without shifting the concentration-intake curve. They concluded that fenfluramine did not alter the responsiveness of the rats to the palatability of the diet. Fletcher (1988a) also argued against a role for fenfluramine in modulating palatability, because the drug did not change the preference of water-deprived rats for saccharin solution compared with water. He suggested that fenfluramine slowed consummatory responses because it interfered with the ability of rats to maintain high rates of responding.

At least partly, the data of Asin et al. (1992) appeared to support this interpretation because, during real feeding, *d*-fenfluramine (2.5 mg/kg, but not lower doses) selectively decreased the speed of the most rapid licking (within bursts). These data are opposite to the *increased* intraburst rate of licking observed when less palatable solutions of sucrose are offered to nondrugged rats (Davis and Smith, 1992). The greater proportion of responses occurring with ILIs between 251 and 500 msec after fenfluramine and the decreased efficiency of licking (ml/lick) might also be consistent with reduced motor performance. Thus, rather than altering palatability, fenfluramine may reduce burst size and cluster size by limiting the ability of the rats to sustain long trains of responses at high rates. These issues can be assessed further by determining the action of fenfluramine on the licking of other solutions that display different microstructural patterns from sucrose (e.g., maltose; Davis and Smith, 1992).

Fenfluramine reduces food intake independently of effects on the rate of eating. A primary action of fenfluramine to slow feeding could cause the decrease in food intake produced by this drug (Burton et al., 1981). To investigate this possibility, Grignaschi et al. (1992) assessed the effects of repeated, single daily injections of fenfluramine on the relationships among meal parameters. The drug reduced the size and duration of meals during 5 days of treatment but inhibited the rate of eating only on the first day; meal frequency was unaltered. These data using pharmacological tolerance to manipulate the microstructure of feeding clearly implied that fenfluramine acted directly to reduce meal size without affecting the rate of eating.

Kaplan et al. (1997) specifically tested this hypothesis in rats that were delivered 12.5% glucose at a fixed rate through an intraoral cannula. *d*-Fenfluramine suppressed intake in this test. Furthermore, the drug reduced meal size by the same amount whether the glucose was infused continuously or with an interruption imposed during the meal. When rats licked at a spout for glucose, halving the size of drops obtained for each lick approximately doubled the number of licks during the meal after injection of either vehicle or drug. Fenfluramine (0.75 mg/kg) reduced glucose consumption by 43% without changing the licking rate within bursts.

Overall, the data argued compellingly that at least moderate doses of fenfluramine reduce food intake by reducing meal size—presumably by enhancing satiation—and not by effects on oral motor movements of ingestion or other aspects of behavioral performance. As noted below, however, actions of serotonergic drugs on the rate of eating still require further evaluation.

Other Indirectly Acting 5-HT Agonists

Manipulating serotonergic activity with other indirectly acting agents causes predictable changes in meal parameters. Systemic administration of the selective serotonin reuptake inhibitors (SSRIs) ORG 6582 (Blundell and Latham, 1978) and fluoxetine (Clifton et al., 1989; Lee and Clifton, 1992) decreased meal size and eating rate without altering the frequency of meals in rats given ad libitum access to pellets. The SSRIs fluoxetine, ORG 6582, femoxetine, and sertraline (Blundell and Latham, 1978; Simansky and Vaidya, 1990) and the 5-HT precursor 5-hydroxy-L-tryptophan (L-5-HTP) (Blundell and Latham, 1979; Fletcher and Burton, 1986a) also slowed the rate at which rats ate pellets and sweet liquid diets after food deprivation. As with 2.5 mg/kg fenfluramine, fluoxetine decreased the intraburst rate of licking, the sizes of bursts and clusters, and licking efficiency and increased the proportion of ILIs between 251 and 500 msec when rats drank 0.4 M sucrose (Asin et al., 1992). Also as with fenfluramine, fluoxetine decreased ingestion of saccharin across a range of concentrations without shifting the psychophysical function for intake (Leander, 1987).

Although these data are consistent with the view that SSRIs decreased the rate of feeding by a primary action on motor function, tolerance developed to the inhibitory action of sertraline on the rate of eating, but not to decreases in food intake or meal size (Grignaschi et al., 1992). Furthermore, Latham and Blundell (1979) found that acute injection of the 5-HT precursor, L-tryptophan decreased meal size, but not the rate of eating or number of meals.

These data argue that the decreased rate of eating observed after indirectly enhancing serotonergic neurotransmission is not necessary for the reduction of meal size. More direct tests of the effect of precursors and SSRIs on meal size by the methods used by Kaplan et al. (1997) need to be performed, and a convincing argument that 5-HT inhibits feeding by modulating responses to gustatory stimuli remains to be established.

Receptors Mediating Actions of Indirect 5-HT Agonists on Meal Parameters

Separate 5-HT receptors mediate reductions in eating rate and meal size produced by fenfluramine. The 5-HT antagonist metergoline has been shown repeatedly to block the ability of fenfluramine to reduce food intake in rodents and humans (e.g., Blundell and Latham, 1980; Goodall and Silverstone, 1988; Neill and Cooper, 1989a; Grignaschi and Samanin, 1992a). Metergoline is a relatively nonselective drug with moderate affinities at 5-HT_1 receptors and very high affinities at 5-HT_2 (and 5-HT_7) subtypes (Zifa and Fillion, 1992; Hoyer et al., 1994). Neill and Cooper (1989a) extended the analysis by demonstrating that, besides metergoline, the nonselective 5-HT antagonist methiothepin, the 5-$HT_{2A/2C}$ antagonist ritanserin, and the 5-$HT_{1A/1B}$ antagonist cyanopindolol antagonized the reduction of food intake produced by *d*-fenfluramine in rats (Table 9.1). In contrast, the hypophagia was unaffected by the 5-HT_{2A} antagonists ketanserin and ICI 169 369, the peripherally acting 5-HT_2 antagonist xylamidine, the 5-$HT_{3/4}$ antagonist tropisetron (Neill and Cooper, 1989a), and the 5-HT_{1A} antagonist WAY 100,635 (Vickers et al., 1996). Thus, it appeared that 1B and central 2C serotonergic receptors mediated fenfluramine's action to reduce food consumption.

In a study using more limited pharmacological analysis, Grignaschi and Samanin (1992a) examined the mechanisms for fenfluramine's microstructural actions in rats that were feeding freely from an eatometer. Metergoline antagonized the effects of *d*-fenfluramine to reduce eating rate, meal size, and total intake; in contrast, a single dose of the 1A/1B blocker cyanopindolol selectively antagonized the reductions of meal size and intake, whereas a dose of the 2A/2C blocker ritanserin antagonized only the slower rate of eating. These results suggested that separate subtypes of 5-HT receptors mediate the effects of fenfluramine on eating rate and meal size. The data also reinforced the concept that, even without changes in the rate of consumption, enhancing the inhibitory control of meal size can reduce food intake.

The failure of ritanserin to reverse the fenfluramine-induced decrease in food intake agrees with some other work (Samanin et al., 1989; Lawton and Blundell, 1993) but conflicts with the findings of Neill and Cooper (1989a) (and with work in humans; Goodall et al., 1993).

Methodological differences among the studies in rats suggested that the taste or other properties of the test diet may dictate partly the relative importance of 5-HT_{1B} and 5-HT_{2C} receptors in the action of fenfluramine to reduce total food intake. Samanin et al. (1989), however, reported that ritanserin attenuated fenfluramine-induced anorexia only at doses of antagonist (1–2 mg/kg) that reduced central concentrations of *d*-norfenfluramine, an active metabolite of the parent compound. In contrast, a dose of metergoline that completely reversed the anorectic effect of fenfluramine occupied a smaller proportion of cortical 5-HT_{2A} receptors than ritanserin without altering central concentrations of fenfluramine or *d*-norfenfluramine. Samanin et al. (1989) concluded that ritanserin acts pharmacokinetically, not pharmacodynamically, to reduce the actions of fenfluramine. They also used these data to implicate 1B sites, and to discount 5-HT_2 subtypes, in mediating fenfluramine's effects.

Although these results excluded serotonergic 2A sites in frontal cortex, and probably elsewhere, from a role in fenfluramine's effect to reduce food intake, ritanserin also has very high affinity for 5-HT_{2C} sites (Hoyer et al., 1994). Lower doses of ritanserin were used to antagonize the reduction in eating rate produced by *d*-fenfluramine (Grignaschi and Samanin, 1992a) than to decrease the concentration of *d*-norfenfluramine in the brain (Samanin et al., 1989). Therefore, the 5-HT_{2C} receptor remains a viable candidate for mediating the control of eating rate. This conclusion is particularly interesting because, unlike *d*-fenfluramine, *d*-norfenfluramine has moderate affinity for 5-HT_{2C} serotonergic receptors in rodents, guinea pigs, and primates (Mennini et al., 1991; Caccia et al., 1993). Brain concentrations of this metabolite associated with hypophagia are adequate to directly stimulate 5-HT_{2C} receptors but not other subtypes of the serotonergic 1, 2,

Table 9.1.
Properties of 5-HT Receptor Subtypes Probed in Feeding Studies

	$5\text{-}HT_1$ Class				$5\text{-}HT_2$ Class		$5\text{-}HT_3$ Class
	1A Subtype	1B Subtype	1D Subtype	1-Like Subtype	2A Subtype	2C Subtype	
Transduction system	G-protein coupled Inhibits adenylyl cyclase Stimulates opening of potassium channel (1A subtype)				G-protein coupled Stimulates phospholipase C		Ligand-gated cation channel
Comments	Cell body autoreceptor; also postsynaptic	Found in rodents; terminal autoreceptor; also postsynaptic	$1D_b$ is homolog of rodent 1B	Group of receptors "1D-like"	Classic $5\text{-}HT_2$	Formerly $5\text{-}HT_{1C}$	Subtypes may exist across species
Prototypical agonists	8-OH-DPAT 5-CT	5-CT CP-93,129	5-CT L 694247 Sumatriptan	5-CT	α-Me-5-HT DOI	α-Me-5-HT DOI	2-Me-5-HT mCPBG
Prototypical antagonists	WAY 100,635 Cyanopindolol[a] Spiperone Methiothepin	Cyanopindolol[a] GR 127935 Methiothepin	GR 127935 Metergoline[a] Methiothepin	Methiothepin	Ketanserin Ritanserin Spiperone Metergoline	SB200646A Ritanserin Mesulergine Metergoline	Tropisetron (ICS 205-930) MDL-72222

Drugs used as agonist probes for feeding	8-OH-DPAT Buspirone Gepirone Ipsapirone RU 24969 Eltoprazine	CP-93,129 CP-94,253 Eltoprazine mCPP RU 24969 TFMPP	Not characterized	5-CT	α-Me-5-HT DOI Quipazine	α-Me-5-HT DOI mCPP Quipazine TFMPP	Not characterized
Drugs used as antagonist probes for feeding	Cyanopindolol[a] Metergoline[a] Methysergide[a] WAY 100,135[a]	Cyanopindolol[a] Metergoline[a] Methysergide[a]	Not characterized	Methysergide[a]	Ketanserin ICI 169 369 Cyproheptadine Ritanserin Metergoline Methysergide Mianserin Mesulergine Xylamidine	SB200646A Cyproheptadine Ritanserin Metergoline Methysergide Mianserin Mesulergine Xylamidine	Tropisetron (ICS 205–930)

Nomenclature based on the International Union of Pharmacology Classification of Receptors for 5-Hydroxytryptamine (serotonin) (see Hoyer et al., 1994). Four other classes of 5-HT receptors and a number of subtypes are not listed (e.g., 1E, 1p, 2B). Some of the 5-HT_1-like subtypes may prove ultimately to belong to one of the other classes. The 5-HT_{2B} subtype or its transcripts (mRNA) have been detected in the periphery and brain; expressed 2B clones are linked in vitro to phospholipase C. α-Me-5-HT and 5-HT are more potent than 5-CT at this receptor, which is very sensitive to blockade by metergoline but insensitive to ketanserin in rats (see Simansky, 1996a). Ketanserin binds appreciably to 2A and to 2C receptors but it has 100–1,000-fold higher binding affinity for the former and has displayed great 2A vs. 2C functional selectivity in other feeding studies (e.g., Kennett and Curzon, 1991). Agonists and antagonists listed for feeding are those that are described in this review; others have been used, but those studies were not cited (see Cooper, 1992; Dourish, 1992). Drugs are listed according to the purposes for which they were used; all of the agents display varying affinities and efficacies for multiple 5-HT and other receptors. For example, metergoline and ritanserin also have moderate to high affinity for 5-HT_7 receptors (see Hoyer et al., 1994). 5-CT, 5-carboxamidotryptamine; CP-93,129, 3-(1,2,5,6-tetrahydropyrid-4-yl)- pyrrolo[3,2-b]pyrid-5-one; CP-94,253, 3-(1,2,5,6-tetrahydro-4-pyridyl)-5-propoxypyrrolo [3,2-b]pyridine; DOI, 1-(2,5-dimethoxy-4-iodophenyl)-2-aminopropane; GR 127935, *N*-[4-methoxy-3-(4-methyl-1-piperazinyl)phenyl]-2′-methyl-4′-(5-methyl-1,2,4-oxadiazol-3-yl)[1,1-biphenyl]-4-carboxamide; L 694247, 2-{5-[3-(-4-methylsulphonylamino)- benzyl-1,2,4-oxadiazol-5-yl]-1H-indole-3-yl}ethylamine; mCPBG, *meta*-chlorophenylbiguanide; mCPP, 1-(3-chlorophenyl)piperazine (or, *meta*-chlorophenylpiperazine); α-Me-5-HT, α-methyl-5-hydroxytryptamine; 2-Me-5-HT, 2-methyl-5-hydroxytryptamine; 8-OHDPAT, 8-hydroxy-di-*N*-propyl aminotetralin; RU 24969, 5-methoxy-3-(1,2,3,6-tetrahydropyridin-4-yl)-1H-indole; TFMPP, 1-(3-trifluoro-methyl-phenyl)-piperazine; WAY 100,135, *N-tert*-butyl-3-[4-(2-methoxyphenyl)piperazin-1-yl]-2-phenylpropionamide; WAY 100,635, *N*-{2-[4-(2-methoxyphenyl)-1-piperazinyl]ethyl}-*N*-(2-pyridinyl)cyclohexanecarboxamide trihydrochloride.

[a]Partial agonist at this subtype. Partial agonist properties of GR 127935 remain to be defined.

or 3 classes (Mennini et al., 1991). Thus, *d*-fenfluramine may act partly via biotransformation to a directly acting 5-HT_{2C} agonist to modulate feeding behavior (see Gibson et al., 1993).

The receptors mediating microstructural effects of selective serotonin reuptake inhibitors require further analysis. Lee and Clifton (1992) reported that metergoline (1 mg/kg) attenuated the reductions in food intake and meal size produced by fluoxetine (5 mg/kg) in free-feeding rats. Metergoline, itself, slowed eating and produced an additive reduction in rate with fluoxetine.

The ability of metergoline to restore meal size and food consumption mimics its action to normalize the smaller meals and intake produced by fenfluramine. These data appeared consistent with a role for serotonergic receptors, and for endogenous 5-HT, in mediating the effect of indirectly acting 5-HT agonists to decrease meal size. However, given that metergoline is such a potent 5-HT_{2C} antagonist, its failure to block the fluoxetine-induced decrease in eating rate was unexpected. Lee and Clifton (1992) proposed that either metergoline acted as a partial agonist to slow eating or metergoline and fluoxetine were each stimulating an unspecified nonserotonergic mechanism controling feeding rate.

Metergoline does display partial or full agonist properties in some functional and transductional assays of the activation of 5-HT_1 subtypes (Hoyer et al., 1994), but it is a potent and silent (pure) antagonist or an *inverse agonist* (intrinsic effect opposite to 5-HT) at 2C receptors (Westphal and Sanders-Bush, 1994). Furthermore, metergoline was effective in blocking the action of fenfluramine to slow eating (Grignaschi and Samanin, 1992a).

These data suggest that non-5-HT_{2C} serotonergic receptors or nonserotonergic receptors may mediate the retardation of eating produced by fluoxetine. It is important to note that metergoline failed, more generally, to antagonize even the hypophagia induced by fluoxetine in several other studies that used large doses (10–20 mg/kg) of the SSRI (Wong et al., 1988; Grignaschi and Samanin, 1992b; Lightowler et al., 1996). Furthermore, reduction of 5-HT concentrations in the whole brain by 70% produced by the neurotoxin 5,7-dihydroxytryptamine (5,7-DHT) (Grignaschi and Samanin, 1992b) and by 90% after the 5-HT synthesis inhibitor *p*-chlorophenylalanine (PCPA) did not reverse fluoxetine (20–30 mg/kg) (Lightowler et al., 1996). Based on pharmacokinetic data and characteristics of fluoxetine and its metabolites in blocking 5-HT and catecholamine reuptake, in vitro, Caccia et al. (1992) suggested that fluoxetine inhibited food intake by indirectly enhancing dopaminergic neurotransmission. Depleting central dopamine with the neurotoxin 6-hydroxydopamine, however, also failed to antagonize fluoxetine-induced hypopohagia (Grignaschi and Samanin, 1992b). Furthermore, in vivo, hypophagic doses of fluoxetine selectively block 5-HT and not dopamine reuptake (Fuller et al., 1994). Thus, the specific roles of serotonergic and other mechanisms in fluoxetine's ability to reduce food intake remain to be determined. It would not be surprising if low and high doses of this SSRI involve different receptor mechanisms (metergoline sensitive and insensitive) and several loci (forebrain, brainstem, peripheral; see Central 5-HT and Satiety).

Serotonergic mechanisms have been implicated in the actions of other SSRIs to inhibit feeding. The anorectic potencies of a series of six of these agents (including fluoxetine) correlated with their potencies in blocking 5-HT reuptake in vivo, but not in vitro (Angel et al., 1988). Depletion of 5-HT in the brain with 5,7-DHT attenuated the anorectic action of sertraline (Lucki et al., 1988; Cervo et al., 1991). In studies with antagonists, metergoline blocked or attenuated decreased food consumption produced by sertraline (Lucki et al., 1988; Garattini et al., 1992; although, see also Grignaschi and Samanin, 1993) and by another SSRI, SL 81.0385 (Angel et al., 1988). Sertraline was also inhibited by the nonselective 5-$HT_{1/2}$ antagonist methysergide, but not by the 5-HT_{2A} antagonist ketanserin or by the peripherally acting 5-$HT_{2A/2C}$ antagonist xylamidine (Lucki et al., 1988). Thus, as with fenfluramine, 5-HT_1 or 5-HT_{2C} receptors may also mediate decreases in food consumption produced by sertraline. However, in the only attempt to block microstructural adjustments produced by sertraline, metergoline failed to antagonize the action of this SSRI in feeding (Grignaschi and Samanin, 1993). Thus, the receptor mechanisms mediating ingestive effects of SSRIs require further investigation.

Directly Acting 5-HT Agonists

Directly acting 5-HT agonists have been used for probing the functions of 5-HT receptors in the microstructure of feeding. Simansky and Vaidya (1990) analyzed videotapes to compare the actions of sertraline in feeding with effects of the 5-$HT_{2C/1B}$ agonist mCPP, the 5-$HT_{1A/1B}$ agonist RU 24969, and the 5-$HT_{2A/2C}$ agonist DOI (see Table 9.1) in rats consuming sweetened milk. Sertraline reduced the amount of time that rats fed and their rate of eating as calculated from the intake and the actual time drinking at the spout. mCPP reduced the rate of eating but not the duration of feeding, whereas RU 24969 and DOI decreased the duration but not the rate. Sertraline, mCPP, and RU 24969 did not alter the continuity of feeding, but, after DOI, the rats frequently interrupted licking to engage in other behavior.

These three agonists bind to multiple subtypes of serotonergic receptors (Hoyer et al., 1994). The relative binding affinities of the drugs for the various subtypes differ, however, and feeding studies have dissociated the pharmacological profiles for the actions of mCPP, RU 24969, and DOI to reduce food intake. Specifically, 5-HT_{1B} receptors were implicated in the hypophagic effect of RU 24969 (Bendotti and Samanin, 1987; Kennett and Curzon, 1988), 1B and especially 2C receptors in this effect of mCPP (Kennett and Curzon, 1988, 1991), and 2A and possibly 2C receptors in the action of DOI (Schechter and Simansky, 1988; Aulakh et al., 1992).

It is interesting that mCPP and DOI each stimulate 2C receptors but produce different effects on feeding behavior. The greater efficacy of mCPP than DOI at 2C receptors (Conn and Sanders-Bush, 1987; Sanders-Bush and Breeding, 1991) and the lack of efficacy of mCPP (but not DOI) in several models of 2A activity (Conn and Sanders-Bush, 1987; Simansky and Schechter, 1988) may contribute to the behavioral distinctions between these drugs. Perhaps more significantly, mCPP, but not DOI, concurrently stimulates 2C and 1B receptors. Clifton et al. (1993) found that mCPP reliably decreased the rate at which rats ate pellets from an eatometer; a higher dose of this drug than that used by Simansky and Vaidya (1990) reduced both rate and duration of feeding. mCPP also increased the ratio of the first intermeal interval to the size of the preceding meal (satiety ratio). These data prompt the hypothesis that only the higher doses of mCPP activate 1B receptors adequately to affect feeding, whereas all doses within the usual range tested engage 2C sites.

mCPP also suppressed drinking of isotonic and hypertonic sodium chloride solutions, but not water (Neill and Cooper, 1989c). Clifton et al. (1993) observed a reduction in water intake, but suggested that it was secondary to changes in feeding. An alternative, or complementary, hypothesis is that gustatory stimuli (from drinking sodium chloride or from food being eaten) enable 5-HT_{2C} stimulation to slow eating or drinking.

Once again, the evidence suggests that separate subtypes of 5-HT receptors mediate different aspects of the behavioral reorganization of feeding in response to drugs. Stimulating 5-HT_{2C} receptors appears to inhibit the rate of eating; 5-HT_{1B} receptors mediate the duration of feeding; and activating 5-HT_{2A} receptors fragments feeding into discontinous episodes.

Drugs that Diminish Serotonergic Neurotransmission

The hypothesis that endogenous 5-HT serves physiological roles in the inhibitory control of feeding demands that manipulations that interfere with serotonergic activity increase food intake. During recent years, an increasing number of studies have used drugs that stimulate autoreceptors on serotonergic neurons in the brainstem, and agents that block postsynaptic serotonergic receptors, to test whether reducing 5-HT function produces hyperphagia.

Autoreceptor Agonists

Systemic administration of low doses of the directly acting 5-HT_{1A} agonist 8-OHDPAT or of other agonists at this subtype (e.g., buspirone, gepirone, ipsapirone) increases food intake by free-feeding rats. This hyperphagic action is associated with the ability of 1A agonists to reduce the firing of serotonergic neurons by stimulating somatodendritic autoreceptors on 5-HT perikarya within the raphe nuclei of the midbrain and pons. The best evidence for this is that the hyperphagia can be produced by direct injection of 1A agonists into the raphe (for review, see Dourish, 1992).

The increased food intake produced by systemic 8-OHDPAT is associated with an increase in the duration of feeding. Several challenges to the behavioral specificity of this hyperphagia have been mounted, but a variety of experimental approaches have supported the hypothesis that 8-OHDPAT and related drugs act via 1A sites (Hartley and Fletcher, 1994) to selectively enhance feeding, probably by disinhibiting satiation (Dourish, 1992). Detailed microstructural analyses of the parameters of feeding after peripheral or central injections of hyperphagic doses of 1A agonists remain to be performed.

Receptor Antagonists

Administration of 5-HT receptor antagonists provides another means of probing the physiological role of the endogenous neurotransmitter in feeding. Compared with injections of 5-HT_{1A} agonists, antagonists are more selective pharmacologically in interrupting serotonergic neurotransmission at subsets of postsynaptic receptors. Systemically administered antagonists are less selective anatomically by interrupting 5-HT activity, centrally and peripherally, at neuronal synapses and at smooth muscle and endocrine targets.

Early reports that cyproheptadine, a 5-HT and histamine antagonist, increased food intake and weight gain in humans and laboratory animals suggested that serotonergic activity served a physiological role in inhibiting feeding (see Blundell, 1979). Even a cursory review of the literature, however, revealed numerous studies in which 5-HT antagonists failed to increase food consumption. These investigations generally used experimental conditions, such as food deprivation before the test meal, that were not optimal for demonstrating increases in food intake because baseline intakes were already large.

Fletcher (1988b) tested the effects of four 5-HT antagonists on food intake in rats that were given a wet mash diet either after ad libitum access to pellets or after a 30-minute prefeeding with mash. In presatiated rats, methysergide, metergoline, and ritanserin increased mash consumption during the 2-hour test; cyproheptadine did not. In contrast, none of the antagonists produced hyperphagia in the rats that were fed ad libitum but not presatiated, and high doses of methysergide, ritanserin, and cyproheptadine decreased intake. These data demonstrated the importance of the baseline level of intake in determining the effects of 5-HT antagonists on feeding. Fletcher (1988b) concluded that these drugs disinhibited the state of satiety rather than slowing the process of satiation during the meal. Fantino et al. (1990), however, suggested that antiserotonergic drugs specifically reduce satiation because cyproheptadine decreased the negative effects of a 50-g glucose load on the affective responses of fasted human subjects to sucrose (negative alimentary alliesthesia) without altering hunger ratings.

Each of the drugs that increased food intake is very potent in inhibiting cellular signaling at 5-HT_{2C} receptors in vitro; cyproheptadine is 10-fold less potent (Hoyer et al., 1994). Because ritanserin binds very poorly to 5-HT_{1B} receptors (Zifa and Fillion, 1992), the results supported 5-HT_{2C} blockade as the mechanism for the orexigenic effect of these antagonists. Other reports that mianserin, cyproheptadine, 1-naphthylpiperazine, metergoline, methysergide, methiothepin, mianserin, mesulergine, and ritanserin produced increased intake, but the 1A antagonists WAY 100,135 and WAY 100,635, the 2A antagonist ketanserin, the 5-HT_3 antagonist MDL 72222, or the 5-$HT_{3/4}$ antagonist tropisetron did not, suggested that blocking 5-HT_{2C}, or 5-HT_{1B} and 5-HT_{2C}, receptors decreased the satiating potency of ingested food (Kennett and Curzon, 1988; Dourish et al., 1989; Simansky, 1991; Hartley and Fletcher, 1994; Vickers et al., 1996). The hyperphagic effects of methysergide may involve, at least partly, feeding mediated by neuropeptide Y, because methysergide increased the concentrations of this peptide in the arcuate and paraventricular nuclei of the hypothalamus (Dryden et al., 1993).

Despite this accumulating evidence that 5-HT antagonists produce hyperphagia, very few studies have examined the actions of these drugs on the microstructure of feeding. On occasion, antagonists have been reported to increase parameters of feeding, such as bout size or duration (e.g., Fletcher and Burton, 1986a,b), but more often no changes are noted (e.g., Blundell and Latham, 1980) or the antagonists even mimic partial agonist actions (e.g., Grignaschi and Samanin, 1992a; Lee and Clifton, 1992). The nonuniform effects of antagonists have been noted in free-feeding and food-restricted rats. None of these studies, however, attempted to optimize the conditions for demonstrating hyperphagia because the major purpose of the investigations was to assess actions of agonists, not antagonists, in feeding.

The macronutrient composition of the diet or its taste may influence the hyperphagic effects of 5-HT antagonists. Metergoline increased the intake of a high-carbohydrate diet in rats that were given a choice of several diets that varied in their macronutient content (Stallone and Nicolaidis, 1989; Leibowitz, 1993; Leibowitz et al., 1993). Leibowitz et al. (1993) administered metergoline systemically, immediately before onset of dark, to rats that had free access to separate sources of carbohydrate, protein and fat. During the first nocturnal hour, metergoline increased the size of carbohydrate meals, elevated the rate of feeding carbohydrates, and decreased the interval until the next carbohydrate meal. In humans, metergoline also increased the intake of sweet-tasting food without altering that of non-sweet food (Goodall and Silverstone, 1988).

Thus, metergoline selectively reduced the satiating effect of carbohydrates. The relevance of this finding for central serotonergic control of the structure of satiation is discussed below (see Medial Hypothalamic Control of Macronutrient-Specific Satiety).

Summary and Discussion

The data consistently raise the theme that increasing serotonergic activity decreases food intake by enhancing the satiating potency of food. Additionally, 5-HT stimulation can inhibit the rate of eating. This latter behavioral effect diminishes the amount of food consumed before the process of satiation terminates the meal. In this manner, the slowing of eating contributes to the reduction in total food intake produced by serotonergic drugs. Nonetheless, pharmacological studies of receptor mechanisms and tolerance demonstrated, and direct manipulations of food delivery established, that meal size can be decreased independently of changes in eating rate. Thus, the physiological significance of the retardation of eating by 5-HT agonists remains in question.

Some have argued that the reduced rate of intake upon stimulation by drugs reflects an alteration in the output of a brainstem pattern generator controlling oral movements (e.g., Hsiao and Spencer, 1983; Baker and Booth, 1990). This action could reflect either toxicity or a physiologically relevant change. For example, injection of lithium chloride or adulterating diet with quinine reduced the rate of feeding under circumstances where prefeeding rats did not change this measure (Blundell et al., 1985). An alternative, and more optimistic, view suggests that serotonergic inhibition of the rate of eating reflects a normal modification associated with satiation or with changes in some other factor, such as taste (Rogers and Blundell, 1979; Davis et al., 1993).

The pharmacological data argue that postsynaptic 1B receptors in rodents (and the homologue 1D subtype in other mammals) mediate serotonergic enhancement of the satiating effect of food and that the 2C subtype mediates inhibition of eating rate. Some investigators have proposed already that both 5-HT_{1B} and 5-HT_{2C} receptors are involved in the reduction of total food intake produced by serotonergic drugs (e.g., Kennett and Curzon, 1988; Neill and Cooper, 1989a; Clifton et al., 1993; Kitchener and Dourish, 1994; Simansky, 1996b). Others have suggested that the 1B subtype predominates in serving this purpose during normal feeding (Grignaschi and Samanin, 1992a). We determined recently that the relatively selective 5-HT_{1B} agonist CP-94,253 reduced the duration of meals in rats (Lee and Simansky, 1995). Nonetheless, the data using antagonists suggest that selectively antagonizing the 5-HT_{2C} receptor subtype disinhibits feeding. This observation would appear to conflict with the view that 5-HT_{2C} receptors control oral motor function in feeding and that this function is not critical for regulating the size of meals.

Two recent developments promise to shed more light on the role of 5-HT_{2C} receptors in feeding and satiation. First, Kennett et al. (1994) demonstrated that the selective 5-$HT_{2B/2C}$ antagonist SB 200646A blocked the hypophagic effect of mCPP in rats. This antagonist has 100-fold lower affinity for 2A receptors than for the 2B and 2C sites. It is important to determine whether SB 200646A will release feeding under appropriate testing conditions.

Second, Tecott et al. (1995) reported that mutant mice lacking 5-HT_{2C} receptors did not respond to the hypophagic effect of a large dose of mCPP. The 5-HT_{2C} "knockout" mice weighed more as adults than their wild-type littermates. Furthermore, these mice did not increase their weight when pair-fed with their controls. These data suggest that the transgenic mice may have a deficit in 5-HT_{2C}–mediated satiety. Surprisingly, in comparison, mice with 5-HT_{1B} knockouts have not been reported to display hyperphagia or obesity (Ramboz et al., 1996), although their feeding behavior has yet to be analyzed carefully. It should be emphasized that gene deletions force the nervous system to develop in the absence of receptors that may themselves influence the formation of circuitry that controls feeding and numerous other functions. The nervous system of transgenic mice therefore might include atypical wiring to preserve such critical regulatory functions. The implication that the transgenic 5-HT_{2C} knockout mouse may nonetheless have a deficit in regulating food intake is important and demands detailed analysis of the behavioral structure of the meal in these animals. The vulnerability of the 1B knockout to hyperphagic effects of 2C antagonists is crucial to determine. The converse experiment awaits development of selective 1B blockers.

Serotonergic Modulation of the Topography of Feeding Behavior

Pharmacological agents used to probe physiological mechanisms underlying satiation should reduce food intake while preserving the general structure of normal feeding behavior. This fundamental tenet of the current behavioral pharmacology of ingestion is echoed throughout this book and has been applied in questioning the role of 5-HT in the process of satiation.

Analyses of the pattern, or topography, of feeding and other periprandial behaviors after administration of serotonergic drugs have been based on the model described by Antin et al. They reported that, at the end of a meal, rats stopped eating, groomed and explored for a period of time, and then rested or slept (Antin et al., 1975). When rats with open gastric cannulas sham fed after 17 hours of food deprivation, they did not display this behavioral sequence of satiation. This observation implied that gastric or postgastric signals are necessary to initiate or enable the processes leading to satiety. The finding that peripheral injection of the gut peptide cholecystokinin (CCK) stopped sham feeding and elicited the behavioral sequence demonstrated that exogenous agents could replace the missing critical stimuli (see Chapter 5). From this work, many investigators have assumed that drugs that augment satiation during real feeding should accelerate the progression of the sequence from feeding, through the termination of feeding and the increase in nonfeeding activity, and finally to the onset of resting.

Fenfluramine

Blundell and McArthur (1981) first applied this type of analysis to characterizing actions of serotonergic anorectics in feeding. They videotaped rats that were given free access to two isocaloric powdered diets containing either 5 or 45 percent casein. After *d,l*-fenfluramine, the rats ate more of the 45 percent casein diet, decreased the duration of the meal, decreased the time spent grooming, and increased resting. The increase in resting occurred in its normal place at the end of the sequence of satiation rather than interspersed among feeding and other activities. Blundell and McArthur emphasized the normal, orderly sequence of behaviors, the hastening of the end of feeding, and the quicker onset of resting.

In subsequent studies in which animals were tested after food deprivation, rather than with free access to food, fenfluramine interfered with the normal pattern of behavior. Thus, in several experiments using liquid or mash diets, and the *d*-isomer or the racemate, fenfluramine disrupted the order and altered the frequency of the behavioral components of the sequence of satiation (Montgomery and Willner, 1988; Willner et al., 1990; McGuirk et al., 1992a). For example, Montgomery and Willner (1988) used a time-sampling method adapted from Antin et al. (1975) in which they recorded the behaviors of the rats at specified intervals during the testing period. For test diets, rats were given either sucrose solution after 4 hours of deprivation or wet mash after 20 hours of deprivation. The concentration of the sucrose solution varied, as did the doses of fenfluramine, in a systematic fashion. Nondrugged controls displayed a normal sequence of satiation after consuming 1.0 M sucrose or mash, but had persistent, high levels of nonfeeding activity with relatively little resting after lower concentrations of sucrose. The racemate and the *d*-isomer of fenfluramine each suppressed resting regardless of the testing conditions used and at doses lower than required for decreasing food intake. Higher doses of drug not only inhibited resting, but also increased nonfeeding activity throughout the period of observation. These data indicated that fenfluramine might have reduced rather than enhanced the influence of gustatory, osmotic, or caloric stimuli on periprandial behavior. Recently, however, Vickers et al. (1996) reported that *d*-fenfluramine accelerated the sequence of satiation in a normal manner when *nondeprived* rats were tested with mash. Metergoline attenuated the hypophagia and the quicker termination of feeding, but not the advanced onset of resting.

5-HT Reuptake Inhibitors

Acute administration of selective inhibitors of 5-HT reuptake increased the rate of the orderly transition from feeding to resting and increased the incidence of resting in nondeprived and in deprived rats. Fluoxetine, sertraline, femoxetine, and paroxetine each accelerated the onset of satiety, and this action occurred under a number of different testing conditions (Clifton et al., 1989; Simansky and Vaidya, 1990; McGuirk et al., 1992b). Simansky and Vaidya (1990), for example, showed that sertraline decreased milk intake in a dose-related fashion that correlated highly with the decreased incidence of feeding observed by the time-sampling method (Fig. 9.1, top). Sertraline also systematically decreased the latency until the

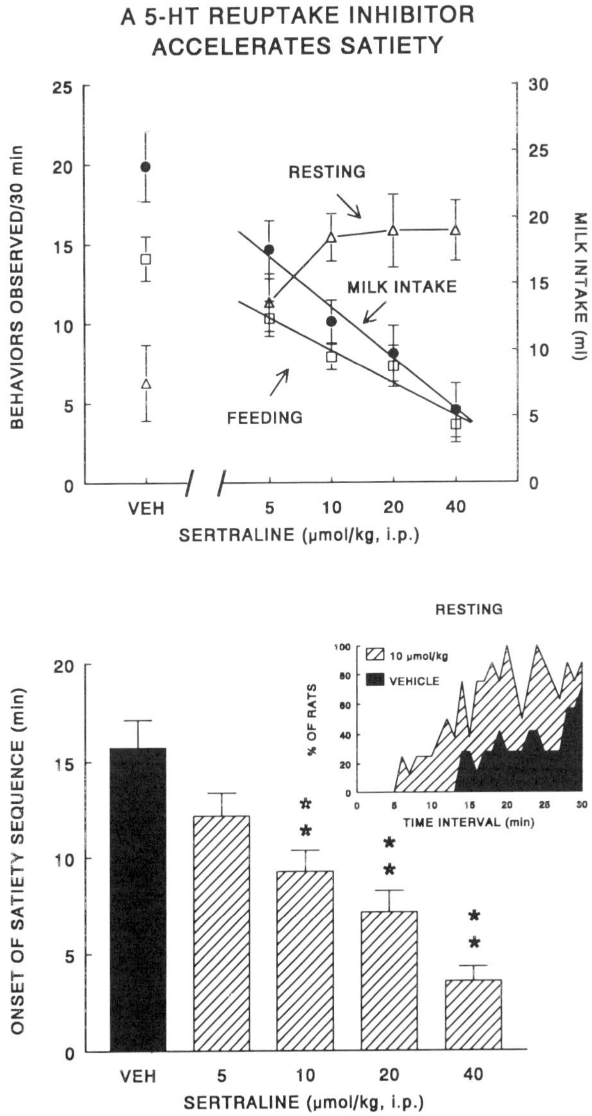

FIGURE 9.1. A 5-HT Reuptake Inhibitor Accelerates Satiety. Actions of the 5-HT reuptake inhibitor sertraline to reduce the consumption of sweetened milk and alter the periprandial behavior of rats during a 30-minute testing period after 18-hour food deprivation. Behaviors were recorded once each minute during the test. **Top:** Doses of 5–40 mmol/kg reduced milk intake and the number of observations of feeding compared with controls injected with vehicle; those doses also increased the frequency of resting. In contrast to the asymptotic data for resting, sertraline produced a monotonically increasing function across the entire dose range for the combination of resting and standing (not shown). **Bottom:** Reduction by sertraline of the time until rats displayed behaviors other than feeding for three successive periods. Rats given sertraline groomed and explored sooner but with the same frequency as controls (not shown). Sertraline also accelerated the onset of resting after nonfeeding activity (e.g., see inset comparing controls and rats given 10 mmol/kg). $^{**}p < 0.01$ compared with vehicle. Adapted from Simansky and Vaidya, (1990), with permission of the publisher.

onset of the postfeeding sequence of satiety-related behaviors without altering their incidence and reduced the time until rats began resting (Fig. 9.1, bottom).

Lee and Clifton (1992) found that fluoxetine also reduced the duration of drinking milk, and this effect was antagonized by metergoline; they did not characterize the sequence of satiation. Clifton et al. (1989) did demonstrate that fluoxetine reduced food intake in free-feeding rats and decreased the time until they began resting after the meal.

The data obtained by observational analysis distinguished the ability of reuptake inhibitors to enhance the rate of postprandial behavioral satiation from the action of fenfluramine to interfere with this sequence after food deprivation. Willner et al. (1990) used operant analysis of the behavior of rats maintained on random-interval schedules with different densities of food reinforcement to compare actions of fenfluramine and fluoxetine. Fluoxetine, but not fenfluramine, altered responding in a manner typical of that produced by reducing food deprivation in nondrugged controls. The dissociation of the ingestive effects of these two drugs may reflect differences in their major neurochemical mechanisms.

Chronic administration of fenfluramine and fluoxetine markedly changes their effects on feeding and satiation. Thus, repeated daily administration of these agents caused pronounced tolerance to the anorectic effect of fenfluramine, but little tolerance to fluoxetine's reduction of food consumption (McGuirk et al., 1992a). After repeated administration, fenfluramine accelerated the onset of resting, but fluoxetine no longer produced this effect. The dissociation of the effects to reduce intake and to increase resting led McGuirk et al. (1992a) to suggest that separate mechanisms might underlie serotonergic decreases in feeding and in the time until the onset of satiety (cf. metergoline blockade of decreased feeding but not enhanced resting with fenfluramine; Vickers et al., 1996; see above). Furthermore, they suggested that drugs might enhance satiation, but stimulate concurrently competing behaviors that interfere with resting. Thus, they cautioned that "the absence of a temporal shift in the behavioral satiety sequence cannot be used to infer the absence of drug effects on satiety processes (p. 406)." Moreover, they implied that appropriate neuropharmacological manipulations might block competing responses and unmask normal satiety-related behaviors.

The presence of a normal behavioral sequence that is shifted in time would appear to provide one important piece of evidence that a drug enhanced satiation. Resting is an important marker in that sequence given that normal feeding characteristically elicits the progression of feeding and nonfeeding activities that is terminated by a phase of resting, that adulterating food with substances, such as quinine, interrupts feeding but does not lead to resting, that rats with open gastric cannulas continue feeding without proceeding normally to resting, and that closing the gastric cannulas restores the sequence of satiation and resting (Antin et al., 1975; Blundell et al., 1985). Nonetheless, McGuirk et al. (1992a) contributed an important caveat to the analysis of drug actions in ingestion and further differentiated the actions of fenfluramine and fluoxetine in modulating feeding.

Direct-Acting 5-HT Agonists

A small, but increasing, number of studies have probed the role of different subtypes of 5-HT receptors in the topography of feeding. In one, the direct-acting agonist eltoprazine decreased the amount and duration of sucrose ingestion and accelerated the onset of resting at the expense of grooming (McGuirk et al., 1992b). Biochemical data suggest that eltoprazine is a partial agonist at 5-HT_{1A} and 5-HT_{1B} receptors and a weak antagonist at 5-HT_{2C} sites (Schipper et al., 1990). Against the background implicating 5-HT_{1B} and 5-HT_{2C} receptors in anorectic effects of fenfluramine and other agents, these data suggested that 5-HT_{1B} sites mediate the behavioral transition from feeding to satiety in rats.

Kitchener and Dourish (1994) extended the analysis of direct-acting 5-HT agonists in satiety. They compared the effects of mCPP, RU 24969, DOI, and the 5-$HT_{1B/2C}$ agonist TFMPP on the behavioral sequence of satiation in rats that were given pellets after 17 hours of food deprivation. At some doses, the 5-$HT_{1B/2C}$ agonists TFMPP and mCPP produced behavioral effects that mimicked those elicited by prefeeding. In contrast, the 5-HT_{1B} agonist RU 24969 increased nonfeeding activity without accelerating resting, and the 5-$HT_{2A/2C}$ agonist DOI fragmented behavior. Kitchener and Dourish suggested that combined stimulation of the 1B and 2C subtypes elicits satiety. The distinction between the ability of eltoprazine to accelerate resting and RU 24969 to increase activity without enhancing resting is important because both drugs are agonists at 1A and 1B receptors. High doses of eltoprazine did, however, increase nonfeeding activity before the onset of resting, although, even then, rats rested sooner. The locomotor excitation produced by RU 24969 was blocked by interfering with dopaminergic activity; the hypophagia was not. The hyperactivity was potentiated rather than blocked by metergoline (e.g., Bendotti and Samanin, 1987).

In one recent study, the more selective 5-HT_{1B} agonist CP-94,253 (Koe et al., 1992) shortened meal duration and accelerated the transition through the sequence of satiation (Halford and Blundell, 1996). Lee and Simansky (1995) found that this agent reduced the duration of feeding without significantly increasing nonfeeding activity. This 5-HT_{1B} agonist did not advance or increase resting in our experiments. CP-94,253 reduced the number, but not the size, of bursts and clusters of licking when rats consumed 10% sucrose solution. This outcome is typical of an increase in postingestive feedback, or enhanced conditioned oropharyngeal inhibitory stimulation associated with that feedback, but not of an alteration in the hedonic value of food (Davis et al., 1993). Together, the data imply that 5-HT_{1B} stimulation decreases the size of meals by enhancing negative feedback from food, but that activating this subtype is inadequate for reliably eliciting the complete expression of behavioral satiation.

The difference between the behavioral profiles of mCPP and DOI using either microstructural analysis (Simansky and Vaidya, 1990) or time-sampling observations (Kitchener and Dourish, 1994) in real feeding was intriguing because of their overlapping, but dissimilar, pharmacological profiles (see above). The effects of these two agents were compared further during sham feeding because this procedure yields sustained high rates of ingestive responses without interference from postingestive satiation signals (Fig. 9.2). Previous studies had demonstrated that a number of indirectly acting agonists and quipazine reduced sham feeding of sucrose (Neill and Cooper, 1988, 1989b), but neither DOI nor mCPP was tested and none of the agents was characterized by time-sampling analysis. As can be seen in Figure 9.2, mCPP reduced intake during sham feeding, but not the incidence of observations of feeding. In contrast, DOI produced an identical decrease in intake by reducing the incidence of feeding and promoting exploratory behaviors and standing. These behaviors were interspersed throughout episodes of feeding. Neither drug elicited the sequence of satiation. Thus, the ability of mCPP to enhance the rate of satiation appears to require the presence of gastric or postgastric signals produced by feeding, whereas the disruptive activity pattern associated with DOI occurs independently of those stimuli.

Summary and Discussion

The data show that SSRIs characteristically alter periprandial behavior in a manner suggesting that these drugs enhance the rate of satiation. The behavioral differences between fenfluramine and the reuptake inhibitors in deprived rats question whether deprivation dissociates the neurochemical mechanisms of action of fenfluramine and reuptake inhibitors in feeding. Still, the extent to which serotonergic receptors mediate the effects of fenfluramine and SSRIs on the pattern of feeding behavior remains to be established. The ability of direct-acting agonists that concurrently stimulate

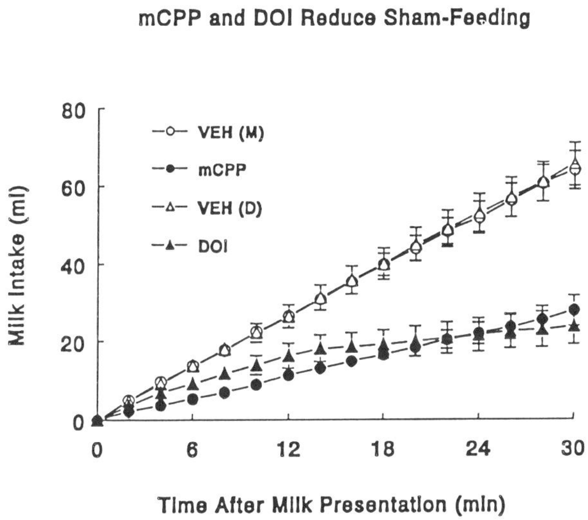

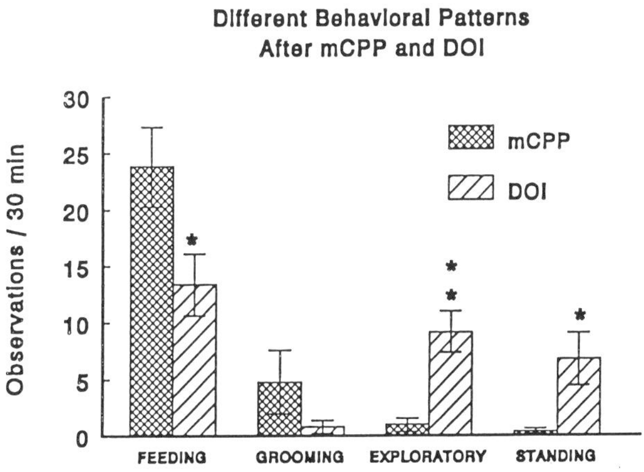

FIGURE 9.2. Actions of mCPP and DOI on Sham Feeding. Actions of mCPP (5.0 mmol/kg, ip) and DOI (3.0 mmol/kg, ip) to reduce milk consumption and alter behavior during gastric sham feeding after 18-hour food deprivation. **Top:** Cumulative milk intakes after injection of mCPP or its vehicle in five rats or after DOI or its vehicle in a separate group of five rats. mCPP and DOI reduced 30-minute intakes by 57% and 64%, respectively (both p values < 0.01). **Bottom:** Distribution of behaviors determined by time-sampling analysis once each minute during the sham-feeding sessions described at top. After administration of vehicle (not shown), rats were observed feeding continuously, except one rat in the Veh (M) group groomed during two intervals and one rat was recorded licking once in the Veh (D) group. mCPP did not alter significantly the distribution of behaviors displayed. DOI reduced observations of feeding and increased exploratory behavior (rearing, locomotion, sniffing, licking) and standing compared with baseline after vehicle (p values < 0.01). Rats did not rest after either vehicle or drug. Asterisks indicate difference from corresponding mean after mCPP: $^*p < 0.05$; $^{**}p < 0.01$.

5-HT_{1B} and 5-HT_{2C} receptors to accelerate satiation clearly begs the question whether these subtypes mediate the actions of the reuptake inhibitors. This analysis may prove problematic given the difficulties noted previously in blocking even the overall hypophagic actions of fluoxetine and other SSRIs with metergoline.

Peripheral 5-HT and Satiety

The localization of high concentrations of 5-HT within the gastrointestinal system, its actions to modulate enteric neuronal activity and smooth muscle tone, and the ability of 5-HT to reach the blood from the gut make 5-HT an important candidate for a peripheral mediator of preabsorptive satiation. Because fenfluramine and the other agents discussed to this point readily penetrate the blood–brain barrier, these drugs can influence feeding behavior by stimulating either peripheral or central 5-HT receptors, or both. Accordingly, systemic injection of 5-HT, which penetrates poorly into the parenchyma of the central nervous system, has been used to probe the effects of selectively activating peripheral serotonergic receptors on feeding and satiation.

The data have been reviewed in detail elsewhere (Simansky, 1996a) and have established that (1) given either intraperitoneally or subcutaneously, 5-HT reduces food intake in rats; (2) the hypophagic action of peripheral, exogenous, 5-HT occurs at doses that do not impair sensorimotor activity, reduce exploration of an open field, produce a conditioned taste aversion, or decrease water intake; (3) under some conditions, peripheral 5-HT increases water consumption, but this dipsogenic action can be dissociated from the hypophagic effect by their different time courses, by various pharmacological antagonists and inhibition of angiotensin synthesis, and by vulnerability to peripheral and central neural lesions; (4) on the basis of studies with a series of 5-HT antagonists, the hypophagic effect of 5-HT is mediated by two classes of 5-HT receptors—a 5-HT_1-like subtype not 1A, 1B, 1D, 1E, 1F, or 1p and a 5-HT_2 subtype ($5\text{-HT}_{3/4}$ subtypes are not involved); and (5) the receptors for peripheral serotonergic hypophagia can be probed selectively by the 5-HT_1 agonist 5-CT (5-carboxamidotryptamine) and the 5-HT_2 agonist α-Me-5-HT (Table 9.1).

The hypophagic action of α-Me-5-HT is eliminated by the peripherally acting 5-HT_2 antagonist xylamidine (Simansky et al., 1989–1990), although this agent does not interfere with the effect of the 5-HT_2 agonist DOI, which does enter the brain (Schechter and Simansky, 1988). Based on its structure, 5-CT presumably acts outside the brain (at least partly via the area postrema; Simansky and Adipudi, 1996); further analysis of this assertion is in progress.

Microstructural and time-sampling analyses have been used for characterizing the behaviorally specific effects of peripheral 5-HT on feeding. Using a standard eatometer device, Fletcher and Burton (1986b) demonstrated that 5-HT reduced food intake by decreasing the size and duration of bouts of feeding without altering the frequency of bouts or eating rate during a 60-minute test after 20 hours of food deprivation. After 5-HT, rats displayed a normal sequence of behaviors during satiation, but the transition from feeding to resting occurred faster (Edwards and Stevens, 1991; Simansky et al., 1992). 5-HT also reduced consumption of sweet liquid diets during sham feeding (Neill and Cooper, 1989b; Simansky et al., 1992), but peripheral serotonergic stimulation did not elicit behavioral evidence of satiety under these conditions (Simansky et al., 1992).

Overall, the results suggest that peripherally administered 5-HT has a satiating effect and accelerates the onset of satiety in rats. Although 5-HT produced an immediate, dose-related, and persistent reduction of milk consumption during sham feeding (Simansky et al., 1992), it did not elicit the satiety sequence. Apparently, the satiating effect of 5-HT requires synergistic action with a gastric or postgastric satiating mechanism to elicit satiety. More refined experimental analysis is required to determine whether the decreased intake during sham feeding produced by 5-HT was due to a change in the rate of licking or to a change in its pattern, e.g., an increase in interbout intervals. Such studies would be informative because it is possible that 5-HT reduces the rate of feeding when high baseline rates are sustained during sham feeding, but not during real feeding of solid (Fletcher and Burton, 1986b) or liquid diet.

Although 5-CT and α-Me-5-HT reduce food intake by different pharmacological mechanisms, they apparently do so via a common behavioral process. Thus, videoanalysis of the effects of these analogues on feeding behavior demonstrated that doses of 5-CT and a-Me-5-HT that reduced milk intake by 35%–40% decreased the time that rats fed by the same proportion (Fig. 9.3). The effects of the analogues on intake and on the time spent feeding were additive. In contrast, only the combination of the analogues slowed the rate of feeding and then only by approximately 20%. Clearly, decreases in the time rather than the rate of feeding constitute the primary mechanism for the anorectic effect of the analogues. Further analysis is necessary to determine how feeding is distributed during real and sham feeding and whether the analogues differentially influence the behavioral components of the sequence of satiation. A comparison of the re-

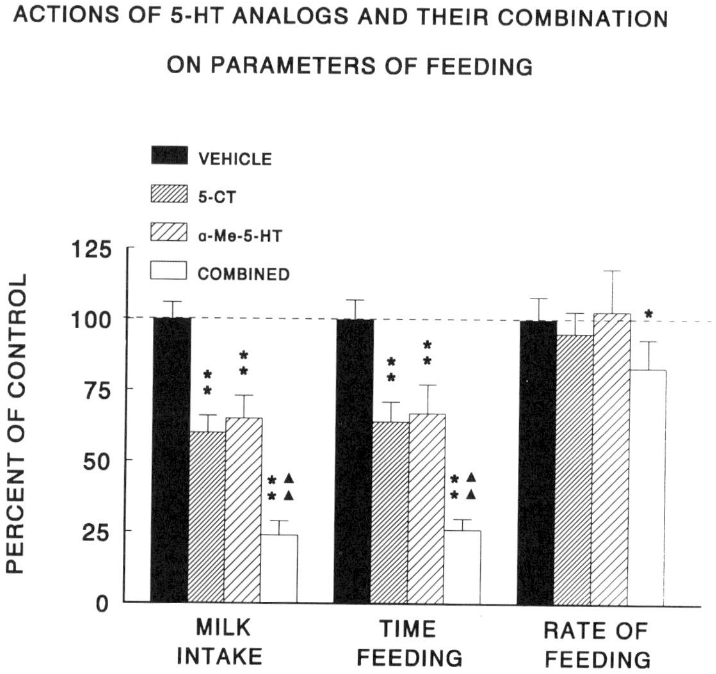

FIGURE 9.3. Actions of 5-HT Analogs and Their Combination on Parameters of Feeding. Effects of intraperitoneal administration of 5-CT (0.03 mmol/kg), α-Me-5-HT (4.0 mmol/kg), or their combination on milk intake, the amount of time spent feeding, and the actual (i.e., local) rate of feeding by eight rats during a 30-minute test with sweetened milk diet after 18-hour food deprivation. Behavior of the rats was analyzed from videotapes. Rats were tested in a random order under each condition. Baselines after vehicle were 31.1 ± 2.0 ml, 16.5 ± 1.2 min, and 1.94 ± 0.16 ml/min, respectively. Asterisks indicate difference from vehicle: $^{*}p < 0.05$; $^{**}p < 0.01$. Triangles, different from 5-CT alone and from α-Me-5-HT alone, $p < 0.01$.

sponses to the peripherally acting 5-$HT_{2A/2C}$ agonist α-Me-5-HT and the centrally acting drug DOI is also in order.

The above account distinguishes the behavioral actions of peripheral 5-HT in feeding from those of systemically administered fenfluramine or L-5-HTP. Specifically, after food deprivation, peripheral 5-HT accelerates the normal transition through the phases of the behavioral sequence of satiation without affecting the local rate of feeding, whereas fenfluramine disrupts the pattern of responses and slows eating rate. λ-5-HTP also inhibits the rate of eating by rats after food deprivation and in nondeprived rats (Blundell and Latham, 1979; Fletcher and Burton, 1986a). In one study, the peripheral decarboxylase inhibitor MK-486 failed to antagonize the action of L-5-HTP to slow eating, which suggested that this precursor of

5-HT acted centrally to decrease rate (Blundell and Latham, 1979). Injecting 5-HT directly into the hypothalamic paraventricular nucleus also decreases rate (Shor-Posner et al., 1986). Fletcher and Burton (1986a), however, reported that xylamidine partially attenuated the action of L-5-HTP to decrease rate and implicated a peripheral mechanism. It is possible that a certain intensity or pattern of peripheral serotonergic receptor stimulation will reduce feeding rate as suggested by the effects of combining 5-CT and α-Me-5-HT (Fig. 9.3). Under conditions that remain to be specified, L-5-HTP may, via its conversion to 5-HT, produce the appropriate pattern of stimulation. This hypothesis predicts that higher doses of 5-HT than those tested previously will also decrease feeding rate. Nonetheless, it remains evident that a reduction in the time spent feeding, and not rate, is most responsible for peripheral 5-HT's action to reduce food intake in deprived rats. Given that different mechanisms appear to be recruited by fenfluramine as a function of deprivation state, it remains essential to characterize the actions of peripheral 5-HT in free-feeding rats.

The actions of peripheral 5-HT also differ from those of CCK and bombesin. In contrast with the peptides (e.g., see Antin et al., 1975; Martin and Gibbs, 1980), 5-HT does not elicit satiety in sham-feeding rats. The mechanisms underlying actions of 5-HT and CCK differ insofar as abdominal vagotomy blocks the inhibitory effect of CCK, but not of 5-HT, on food intake (e.g., Smith et al., 1981; Fletcher and Burton, 1985; Eberle-Wang et al., 1993; Simansky, 1996a). Furthermore, the CCK_A antagonist devazepide blocks the hypophagic effect of CCK, but not that of 5-HT (Eberle-Wang and Simansky, 1992). Thus, *peripheral* 5-HT apparently does not recruit endogenous CCK to reduce food intake.

Devazepide has been reported to inhibit the hypophagic effect of fenfluramine (Cooper et al., 1990a; Grignaschi et al., 1993). Given the dissociation of the actions of peripheral 5-HT and CCK in satiety, these data might suggest that *central* serotonergic stimulation releases central CCK to modulate satiation. In other studies, however, blocking CCK_A receptors did not antagonize the actions of either fenfluramine (Li and Rowland, 1994) or fluoxetine (Cooper et al., 1990b) to reduce food intake. Thus, before exploring for serotonergic recruitment of CCKergic pathways in the brain, it will be necessary to determine (1) more precisely the conditions under which devazepide reverses the effect of fenfluramine on meal size and (2) whether devazepide antagonizes hypophagia produced by fluoxetine or other 5-HT reuptake inhibitors under any testing conditions.

Better evidence exists that 5-HT mediates the satiating effect of peripheral CCK (see Chapter 5). In two studies, the $5\text{-HT}_{1/2}$ antagonist metergoline reversed the hypophagic action of systemically administered

CCK-8 (Stallone et al., 1989; Grignaschi et al., 1993). Poeschla et al. (1993) demonstrated further that the reduction in food intake produced by CCK-8 was blocked by the 5-$HT_{2A/2B/2C}$ antagonist mianserin, but not by the 5-HT_{2A} antagonist ketanserin or by three 5-HT_3 antagonists. Pharmacological inhibition of the synthesis of 5-HT (Esfehani et al., 1995) and of the firing of central serotonergic neurons (Poeschla et al., 1992) also attenuated the effect of peripheral CCK-8 on food intake. Thus, a body of data suggests that peripheral CCK-8 recruits endogenous 5-HT to elicit satiety, that the critical serotonergic pathways involved in this process are located in the brain, and that 5-HT_{2C} receptors may ultimately mediate this serotonergic function although ritanserin reportedly did not block CCK's action (Samanin and Grignaschi, 1996). Identification of the loci and roles of 5-HT_{2C} (and 5-HT_{1B}) receptors in central serotonergic modulation of the satiating effect of CCK requires further investigation.

Central 5-HT and Satiety

Serotonergic pathways in brain innervate structures that have been implicated in regulating food intake at all levels of the neuraxis. Nonetheless, the specific sites involved in this inhibitory function remain to be identified. Attempts to define more precisely these central loci have focused on the hypothalamus. This emphasis derived somewhat from the hypothalamocentric view of ingestive control that long dominated this field, but more importantly from a systematic series of studies in which serotonergic drugs inhibited feeding when injected directly into various hypothalamic nuclei (for review, see Leibowitz, 1993). In those investigations, microinjection of 5-HT into the paraventricular nucleus of the hypothalamus (PVN) reduced food intake that was elicited by food deprivation, by the onset of the dark photoperiod during free access to food, or by injection of norepinephrine into the PVN (Shor-Posner et al., 1986; Weiss et al., 1986). Injecting fenfluramine, fluoxetine, and various other serotonergic agents into the PVN mimicked, and antagonists prevented, the inhibitory action of 5-HT in the PVN.

The effects of injecting 35 nmol of 5-HT unilaterally into the PVN on the microstructure of feeding were analyzed using an eatometer (Shor-Posner et al., 1986). The rats were maintained and tested with a single diet of sweetened powdered chow, and the parameters of feeding were measured beginning with the onset of the dark photoperiod. 5-HT reduced food intake, the duration of the first meal after darkness, and the rate of intake, but not the latency to begin eating. 5-HT also increased the intermeal interval until the second meal, but did not alter the frequency of meals. *d,l*-Norfenfluramine reduced meal size and the rate of intake, although rats actually ate for a longer period during the first meal after injection of this indirect agonist. The decreased meal size and eating rate resembled data reported after systemically administered 5-HT agonists, but differed from the results obtained after selectively stimulating peripheral serotonergic receptors. Accordingly, it might be hypothesized that the PVN plays a critical role in mediating hypophagic actions of systemically administered serotonergic drugs that enter the brain.

The receptor subtypes reponsible for the satiety actions of 5-HT in the medial hypothalamus remain to be identified. Leibowitz (1993) proposed that 5-HT_{1B} or 5-HT_{2C} receptors serve this function. Consistent with this suggestion, intra-PVN injections of the highly selective 5-HT_{1B} agonist CP-93,129, the less selective 5-HT_{1B} agonist RU 24969, and the 5-$HT_{1B/2C}$ agonist TFMPP reduced food intake (Hutson et al., 1988; Macor et al., 1990; Samanin and Grignaschi, 1996; although not Fletcher et al., 1992). Local injection of the 5-$HT_{1A/1B}$ and β-adrenergic receptor antagonist (−)propranolol blocked these hypophagic effects (Samanin and Grignaschi, 1996). Moderate concentrations of 5-HT_{1B} receptors exist in the dorsomedial (DMH) and ventromedial (VMH) hypothalamus, where these receptors are located primarily postsynaptically (Frankfurt et al., 1993). Furthermore, studies using in vivo microdialysis have demonstrated that extracellular concentrations of 5-HT and its acid metabolite (5-HIAA) increase in the region between the PVN and VMH during meals in spontaneously feeding rats (Orosco and Nicolaidis, 1992). Together, these data suggest that 5-HT is released within the medial hypothalamus during feeding, where the neurotransmitter stimulates 5-HT_{1B} receptors to mediate satiation.

This discussion has focused on the role of the hypothalamus in central serotonergic inhibition of feeding. Nonetheless, microdialysis failed to demonstrate a relationship between fenfluramine-induced release of 5-HT in either the perifornical hypothalamus or PVN and the hypophagic action of this drug (Oluyomi et al., 1994; Raiteri et al., 1995). Furthermore, local injection of 5-HT_{1B} antagonist into the PVN did not inhibit the hypophagic effects of *systemically* administered *d*-fenfluramine (Grignaschi et al., 1995). Moreover, lesions of the PVN did not inhibit hypophagia induced by systemically administered 5-HT agonists (Fletcher et al., 1993). Thus, activating 5-HT receptors in the PVN may be sufficient, but not necessary, to reduce food intake. Clearly, sites outside the PVN, including those involving central 5-HT_{1B} receptors and peripheral loci, are important for serotonergic modulation of satiation.

Medial Hypothalamic Control of Macronutrient-Specific Satiety

Although the medial hypothalamus may not be required for a general inhibitory effect of serotonergic neurotransmission in feeding, this region of the brain may play a more specific role in regulating satiation for carbohydrates. In numerous studies from Leibowitz' laboratory, injection of 5-HT into the hypothalamus decreased the consumption of a carbohydrate-rich diet so that rats generally ingested a smaller proportion of their total caloric intake in the form of this macronutrient (for review, see Leibowitz, 1993). Local serotonergic stimulation with 5-HT, *d*-norfenfluramine, or fluoxetine produced this effect on dietary selection after microinjection into the PVN, DMH, VMH, or suprachiasmatic nucleus (SCN), but not other hypothalamic sites. The proposed influence of serotonergic stimulation to suppress carbohydrate intake occurred selectively at the onset of the dark photoperiod and *at doses that frequently did not change total caloric intake.* Analysis of the structure of the early meals showed that intra-PVN injection of a dose of 5-HT as small as 2.5 nmol decreased the actual and proportional intakes of a carbohydrate-rich diet, decreased the time spent feeding and the rate of feeding on this diet, and increased the satiety ratio for the carbohydrate diet (Leibowitz et al., 1993). In comparison, earlier studies testing the effects of 5-HT on total food intake used doses that were 10-fold higher to inhibit feeding. Thus, paradigms using dietary selection may provide a particularly sensitive method for detecting the satiating effects of localized manipulations of serotonergic activity in the brain. Furthermore, systemically administered fluoxetine reduced carbohydrate intake and increased the extracellular concentration of 5-HT in the PVN as measured by microdialysis (Paez and Leibowitz, 1993). Thus, it has been proposed that a medial hypothalamic serotonergic system modulates satiation that is specific for carbohydrate; this process is linked to circadian controls of macronutrient intake (Leibowitz et al., 1989; Leibowitz, 1993).

Further Behavioral Analysis of a Role for 5-HT in Satiety

Procedures other than analyzing feeding behavior during a meal have been used for assessing whether serotonergic activity modulates satiety. The most common alternative involved testing the actions of serotonergic drugs on performance to obtain food in a runway. In those studies, rats typically were food-restricted and tested on 15 successive trials in which they traversed a runway to gain access to food in a goalbox for 2–3 minutes.

Fenfluramine consistently increased the latency for rats to leave the startbox at the beginning of the trial, decreased the speed at which rats ran to the goalbox, and decreased the amount of food eaten at the end of the trials (Thurlby and Samanin, 1981; Thurlby et al., 1983, 1985; Kirkham and Blundell, 1986; Neill et al., 1990; Samanin et al., 1991). These effects occurred beginning with the initial trials and mimicked the behavior of nondrugged controls on later trials after they satiated normally during the runway test (Thurlby et al., 1983). The early retardation of performance after fenfluramine also matched that observed in the first block of trials in nondrugged controls that were satiated by a 30-minute meal before testing (Thurlby and Samanin, 1981). Thus, these data were taken as further evidence that fenfluramine induces satiety (Thurlby and Samanin, 1981; Thurlby et al., 1983; Kirkham and Blundell, 1986).

Very limited information exists about the pharmacological mechanism for serotonergic modulation of performance in the runway or in other instrumental tasks. Carli and Samanin (1992) compared the effects of mCPP, *d*-fenfluramine, quipazine, and lysergic acid diethylamide (LSD) on responding in a five-choice serial reaction time task reinforced by food reward. mCPP and fenfluramine altered responding in a manner similar to that produced by prefeeding. In contrast, quipazine and LSD produced attentional disturbances. As with DOI, quipazine produces anorexia at least partly via 5-HT_{2A} receptors (Shukla et al., 1990). Thus, results across a spectrum of behavioral paradigms argues against a role for 5-HT_{2A} receptors in satiety and validate mCPP as a useful direct-acting agonist for studying serotonergic mechanisms in satiety.

mCPP and fenfluramine also produce common actions in the runway test (Thurlby and Samanin, 1981). This observation suggests that 5-HT_{1B} or 5-HT_{2C} receptors mediate satiation in this task. In studies with antagonists, metergoline but not ritanserin or xylamidine inhibited the effects of fenfluramine (Thurlby et al., 1985; Neill et al., 1990). Furthermore, only metergoline enhanced performance in the runway (Neill et al., 1990). These results would appear to implicate activation of 5-HT_1 receptors rather than 2C sites in decreasing running speed and food intake. However, the single dose of ritanserin tested (0.5 mg/kg) was only half of that used for metergoline. The latter drug is 100-fold more potent than ritanserin in blocking signal transduction at 5-HT_{2C} receptors in vitro (Hoyer et al., 1994). Thus, although this dose of ritanserin was adequate to antagonize the slower rate of eating produced by fenfluramine (Grignaschi and Samanin, 1992a), larger doses may be required for blocking these behavioral actions of fenfluramine in runway tasks. Unfortunately, using ritanserin is problematic because doses of 1–2 mg/kg interfere with the disposition of fenfluramine (Samanin et al., 1989), but other 5-HT_{2C} antagonists may avoid this pharmacokinetic effect.

The specific mechanism may depend also on the testing conditions because metergoline failed to block fenfluramine-induced retardation of performance when the drug was combined with prefeeding (Samanin et al., 1991). These data suggest that a metergoline-insensitive serotonergic receptor, or an entirely nonserotonergic mechanism, is responsible for fenfluramine's behavioral action in the runway when this indirect agonist is administered to fed rather than deprived rats. Metergoline, however, was administered 6.5 hours before the runway test. In a previous study, the antagonist was given 2.5 hours before the test and did block fenfluramine in rats that were not prefed (Neill et al., 1990). Thus, shorter pretreatment times should be assessed for antagonizing fenfluramine's satiating action in prefed rats in the runway. Metergoline did prevent the action of fenfluramine to delay gastric emptying in prefed rats (Samanin et al., 1991). Apparently, increased gastric retention of food is unnecessary for the satiety-like action of fenfluramine in the runway.

The fact that gastric mechanisms are not necessary for fenfluramine-induced satiety in the runway does not mitigate against peripheral serotonergic mechanisms influencing behavior in this task. Fletcher (1987) administered 5-HT systemically to food-deprived rats before testing them in a standard 15-trial assay. 5-HT reduced running speed and food intake. Unlike fenfluramine, however, 5-HT produced these effects only in the later blocks of trials. Furthermore, after injection of 5-HT, rats displayed a normal relationship between running speed and the amount of food eaten at the end of a trial. These data suggest that 5-HT did not induce satiety by itself but, instead, enhanced the satiating effects of food consumed during the early trials. This conclusion is identical to the inference drawn from the observation that 5-HT accelerates satiation during real feeding but does not elicit satiety in sham-feeding rats (Simansky et al., 1992).

Conclusions

The results of the last two decades of research provide compelling evidence that serotonin mediates inhibitory controls of feeding in mammals. Across diverse paradigms, drugs that enhance activity at postsynaptic serotonergic receptors decrease the amount of food consumed before a meal is terminated. Given the explosion of 5-HT receptor subtypes that have been identified, it is somewhat surprising how well the data converge on roles for postsynaptic 5-HT_{1B} and 5-HT_{2C} receptors in mediating satiation. Furthermore, it appears that these two receptor systems exert cooperative actions in satiation, and their concurrent activation is required for the full behavioral expression of satiety.

An important caveat exists in that the apparent consistency of the results, especially with direct agonists, has been achieved largely with the same few pharmacological agents. These drugs, of course, vary in their relative affinities and efficacies at the various serotonergic and nonserotonergic receptors. The possibility exists that the results to date reveal uniformity in the properties of specific drugs across paradigms, not in the behavioral roles of the receptor subtypes with which they have become associated. Increasingly, however, this issue is being addressed by using newer and more selective agents. Of particular importance, the evidence that 5-HT antagonists (and receptor gene deletions) increase food intake, probably by disinhibiting satiety, is the most convincing support for a physiological role for serotonergic pathways in regulating meals. The emergence of more selective antagonists for blocking 5-HT_{2C} and 5-$HT_{1B/1D}$ receptors will establish more definitively the roles of these subtypes in controlling meal size. Humans, nonhuman primates, lagomorphs, and other nonrodent species express a subtype of 5-HT_{1D} receptor instead of 5-HT_{1B} receptors. Thus, it is a pressing issue to use species other than rodents for future studies because of the clinical relevance of 5-HT_{1D} function for developing novel therapeutic agents in humans.

The mechanisms for the satiety action of SSRIs must be elucidated. The extent, if any, to which 5-HT_{1B} and 5-HT_{2C} receptors mediate the effect of these drugs to enhance satiation is virtually untested. It is quite possible that serotonergic receptors other than those in the 5-HT_1, 5-HT_2, or 5-HT_3 classes subserve their actions. It has been suggested that 5-HT_7 receptors may transduce serotonergic actions of fenfluramine in satiation (Raiteri et al., 1995). This class of receptors should be considered also when analyzing mechanisms for the hypophagic effects associated with SSRIs, although the difficulty of blocking the satiating action of these agents with metergoline might argue against a role for 5-HT_7 receptors.

The pharmacological profiles of peripheral serotonergic receptors must also be defined more precisely because these sites may mediate satiating actions of fenfluramine and SSRIs. Also, peripheral receptors may offer novel therapeutic targets outside the central nervous system.

In the brain, future investigation needs to identify the critical loci where serotonergic drugs act to influence feeding, satiation, and oral motor responses. Although research has focused on hypothalamic mechanisms, anatomically broader studies of 5-HT function in ingestion may yield better insights about the role of this important neurotransmitter in satiation, in particular, and in brain function more generally. For example, recent evidence that inhibiting 5-HT synthesis failed to alter fenfluramine-induced hypophagia (Gibson et al., 1993) might be explained by inadequate depletions of 5-HT more caudally in the brainstem or in peripheral sites.

The sequence of responses leading to satiety is a relatively fixed action pattern of behavior. In cats, 5-HT neurons within the dorsal raphe and in more caudal serotonergic cell groups of the deep brainstem increase their firing in a very regular fashion during repetitive motor activities, including chewing, biting, lapping, and grooming (Jacobs and Azmitia, 1992; Veasey et al., 1995; Fornal et al., 1996). An excitatory role for 5-HT in these behaviors fits into the broader interpretation that the 5-HT cell groups innervating the forebrain and brainstem are involved in modulating motor activity and motor sequences (Veasey et al., 1995; Fornal et al., 1996). The possibility that excessive 5-HT_{2C} stimulation actually reduces oral motor output to slow feeding needs to be explored.

Serotonergic pathways innervate structures within the hindbrain that process visceral and gustatory input and promote the electroencephalographic shift toward reduced vigilance (e.g., Steinbusch, 1981; Thor and Helke, 1987; Nosjean et al., 1987)—a characteristic of satiety (Antin et al., 1975; Danguir et al., 1979). Some of the evidence that was reviewed above suggested that stimuli associated with the eating or digestion of food enhance or enable the ability of serotonergic drugs to produce satiety. Thus, serotonergic drugs may amplify physiological signals in the hindbrain that lead to decreased arousal. Overall, serotonergic systems appear to mediate motor sequences involved in feeding and to set the gain that determines the rate of transition from feeding to satiety, while other neurotransmitter and hormonal systems exert principal control to trigger this fixed action pattern.

ACKNOWLEDGEMENTS

I thank Wayne Kachelries for his invaluable assistance in making the figures and in preparing the manuscript. I am especially grateful to my former students Drs. Kim Eberle-Wang and Lee Schechter for their insightful research and collegial discussions about the role of 5-HT in satiation. Preparation of this chapter was supported by USPHS grant MH41987 from the National Institute of Mental Health.

REFERENCES

Angel, I., M.-A. Taranger, Y. Claustre, B. Scatton, and S. Z. Langer. Anorectic activities of serotonin uptake inhibitors: Correlation with their potencies at inhibiting serotonin uptake *in vivo* and ^{3}H-mazindol binding in vitro. *Life Sci.* 43:651–658, 1988.

Antin, J., J. Gibbs, J. Holt, R. C. Young, and G. P. Smith. Cholecystokinin elicits the complete behavioral sequence of satiety in rats. *J. Comp. Physiol. Psychol.* 89:784–790, 1975.

Asin, K. E., J. D. Davis, and L. Bednarz. Differential effects of serotonergic and catecholaminergic drugs on ingestive behavior. *Psychopharmacology,* 109:415–421, 1992.

Aulakh, C. S., J. L. Hill, H. T. Yoney, and D. L. Murphy. Evidence for involvement of 5-HT_{1c} and 5-HT_2 receptors in the food intake suppressant effects of 1-(2,5-dimethoxy-4-indolphenyl)-2-aminopropane (DOI). *Psychopharmacology* 109:444–448, 1992.

Baker, B. J., and D. A. Booth. Effects of *dl*-fenfluramine on dextrin and casein intakes influenced by textural preferences. *Behav. Neurosci.* 104:153–159, 1990.

Bendotti, C., and R. Samanin. The role of putative 5-HT_{1A} and 5-HT_{1B} receptors in the control of feeding in rats. *Life Sci.* 41:635–642, 1987.

Blundell, J. E. Serotonin and feeding. In Essman, W. B. (ed.): *Serotonin in Health and Disease.* New York: Spectrum, 1979, pp. 403–450.

Blundell, J. E. Serotonin manipulations and the structure of feeding behaviour. *Appetite* 7(Suppl.):39–56, 1986.

Blundell, J. E., and A. J. Hill. Serotoninergic modulation of the pattern of eating and the profile of hunger-satiety in humans. *Int. J. Obesity* 11:141–155, 1987.

Blundell, J. E., and A. J. Hill. Dexfenfluramine and appetite in humans. *Int. J. Obesity* 16(Suppl. 3): S51–S59, 1992.

Blundell, J. E., and C. J. Latham. Pharmacological manipulation of feeding behavior: Possible influences of serotonin and dopamine on food intake. In Garattini, S., and R. Samanin, (eds.): *Central Mechanisms of Anorectic Drugs.* New York: Raven Press, 1978, pp. 83–109.

Blundell, J. E., and C. J. Latham. Serotonergic influences on food intake: Effect of 5-hydroxytryptophan on parameters of feeding behaviour in deprived and free-feeding rats. *Pharmacol. Biochem. Behavior.* 11:431–437, 1979.

Blundell, J. E., and C. J. Latham. Characterization of adjustments to the structure of feeding behavior following pharmacological treatment: Effects of amphetamine and fenfluramine and the antagonism produced by pimozide and methergoline. *Pharmacol. Biochem. Behavior.* 12:717–722, 1980.

Blundell, J. E., C. J. Latham, and M. B. Leshem. Differences between the anorexic actions of amphetamine and fenfluramine-possible effects on hunger and satiety. *J. Pharmacy Pharmacol.* 28:471–477, 1976.

Blundell, J. E., and R. A. McArthur. Behavioral flux and feeding: Continuous monitoring of food intake and food selection, and the video-recording of appetitive and satiety sequences for the analysis of drug action. In Garattini, S., and R. Samanin (eds.): *Anorectic Agents: Mechanisms of Action and Tolerance.* New York: Raven Press, 1981, pp. 19–42.

Blundell, J. E., P. J. Rogers, and A. J. Hill. Behavioral structure and mechanisms of anorexia: Calibration of natural and abnormal inhibition of eating. *Brain Res. Bull.* 15:371–376, 1985.

Booth, D. A., E. L. Gibson, and B. J. Baker. Gastromotor mechanism of fenfluramine anorexia. *Appetite* 7(Suppl.):57–69, 1986.

Bradley, P. B., G. Engel, W. Feniuk, J. R. Fozard, P. P. A. Humphrey, D. N.

Middlemiss, E. J. Mylecharane, B. P. Richardson, and P. R. Saxena. Proposals for the classification and nomenclature of functional receptors for 5-hydroxytryptamine. *Neuropharmacology* 25:563–576, 1986.

Burton, M. J., S. J. Cooper, and D. A. Popplewell. The effect of fenfluramine on the microstructure of feeding and drinking in the rat. *Br. J. Pharmacol.* 72:621–633, 1981.

Caccia, S., M. Anelli, C. Fracasso, E. Frittoli, P. Giorcelli, M. Gobbi, C. Taddei, S. Garattini, and T. Mennini. Anorectic effect and brain concentrations of D-fenfluramine in the marmoset: Relationship to the in vivo and in vitro effects on serotonergic mechanisms. *Naunyn-Schmiedeberg Arch. Pharmacol.* 347:306–312, 1993.

Caccia, S., A. Bizzi, G. Coltro, C. Fracasso, E. Frittoli, T. Mennini, and S. Garattini. Anorectic activity of fluoxetine and norfluoxetine in rats: Relationship between brain concentrations and in vitro potencies on monoaminergic mechanisms. *J. Pharmacy Pharmacol.* 44:250–254, 1992.

Carli, M., and R. Samanin. Serotonin receptor agonists and serotonergic anorectic drugs affect rats' performance differently in a five-choice serial reaction time task. *Psychopharmacology* 106:228–234, 1992.

Cervo, L., G. Grignaschi, C. Rossi, and R. Samanin. Role of central serotonergic neurons in the effect of sertraline in rats in the forced swimming test. *Eur. J. Pharmacol.* 196:217–222, 1991.

Clifton, P. G., A. M. Barnfield, and G. Curzon. Effects of food deprivation and mCPP treatment on the microstructure of ingestive behaviour of male and female rats. *J. Psychopharmacol.* 7:257–264, 1993.

Clifton, P. G., A. M. C. Barnfield, and L. Philcox. A behavioural profile of fluoxetine-induced anorexia. *Psychopharmacology* 97:89–95, 1989.

Conn, P. J., and E. Sanders-Bush. Relative efficacies of piperazines at the phosphoinositide hydrolysis-linked serotonergic (5-HT-2 and 5-HT-1c) receptors. *J. Pharmacol. Exp. Thera.* 242:552–557, 1987.

Cooper, S. J. 5-HT and ingestive behaviour. In Marsden, C. A., and D. J. Heal, (eds.): *Central Serotonin Receptors and Psychotropic Drugs*. London: Blackwell Scientific Publications, 1992, pp. 260–291.

Cooper, S. J., C. T. Dourish, and D. J. Barber. Reversal of the anorectic effect of (+)-fenfluramine in the rat by the selective cholecystokinin receptor antagonist MK-329. *Br. J. Pharmacol.* 99:65–70, 1990a.

Cooper, S. J., C. T. Dourish, and D. J. Barber. Fluoxetine reduces food intake by a cholecystokinin-independent mechanism. *Pharmacol. Biochem. Behavi.* 35:51–54, 1990b.

Cooper, S. J., and R. L. Francis. Feeding parameters with two food textures after chlordiazepoxide administration, alone or in combination with *d*-amphetamine of fenfluramine. *Psychopharmacology* 62:253–259, 1979.

Cooper, S. J., and R. L. Francis, Interactions of chlordiazepoxide and anorectic agents on rate and duration parameters of feeding in the rat. *Psychopharmacology* 69:261–265, 1980.

Cooper, S. J., and R. L. Francis. A microstructural analysis of the effects of presatiation on feeding behavior in the rat. *Physiol. Behav.* 53:413–416, 1993.

Cooper, S. J., and J. C. Neill. 5-HT receptors and the sweet tooth. *Trends Pharmacol. Sci.* 8:199–200, 1987.

Cooper, S. J., and K. F. Sweeney. Effects of spiperone alone and in combination with anorectic agents on feeding parameters in the rat. *Neuropharmacology* 19:997–1003, 1980.

Danguir, J., S. Nicolaidis, and H. Gerard. Relations between feeding and sleep patterns in the rat. *J. Comp. Physiol. Psychol.* 93:820–830, 1979.

Davies, R. F., J. Rossi III, J. Panksepp, N. J. Bean, and A. J. Zolovick. Fenfluramine anorexia: A peripheral locus of action. *Physiol. Behav.* 30:723–730, 1983.

Davis, J. D., and G. P. Smith. Analysis of lick rate measures the positive and negative feedback effects of carbohydrates on eating. *Appetite* 11:229–238, 1988.

Davis, J. D., and G. P. Smith. Analysis of the microstructure of the rhythmic tongue movements of rats ingesting maltose and sucrose solutions. *Behav. Neurosci.* 106:217–228, 1992.

Davis, J. D., G. P. Smith, and J. Meissner. Postpyloric stimuli are necessary for the normal control of meal size in real feeding and sham feeding rats. *Am. J. Physiol.* 34:R888–R895, 1993.

Dourish, C. T. 5-HT receptor subtypes and feeding behaviour. In P. B. Bradley, S. L. Handley, S. J. Cooper, B. J. Key, N. M. Barnes, and J. H. Coote. (eds.): *Serotonin, CNS Receptors and Brain Function.* London: Pergamon Press, Ltd., 1992, pp. 179–202.

Dourish, C. T., M. L. Clark, A. Fletcher, and S. D. Iversen. Evidence that blockade of postsynaptic 5-HT_1 receptors elicits feeding in satiated rats. *Psychopharmacology* 97:54–58, 1989.

Dryden, S., H. D. McCarthy, U. H. Malabu, M. Ware, and G. Williams. Increased neuropeptide Y concentrations in specific hypothalamic nuclei of the rat following treatment with methysergide: Evidence that NPY may mediate serotonin's effects on food intake. *Peptides* 14:791–796, 1993.

Eberle-Wang, K., P. Levitt, and K. J. Simansky. Abdominal vagotomy dissociates the anorectic mechanisms for peripheral serotonin and cholecystokinin. *Am. J. Physiol.* 265:R602–R608, 1993.

Eberle-Wang, K., and K. J. Simansky. The CCK-A receptor antagonist, devazepide, blocks the anorectic action of CCK but not peripheral serotonin in rats. *Pharmacol. Biochem. Behav.* 43:943–947, 1992.

Edwards, S., and R. Stevens. Peripherally administered 5-hydroxytryptamine elicits the full behavioural sequence of satiety. *Physiol. Behav.* 50:1075–1077, 1991.

Esfehani, N., I. Bednar, G. A. Qureshi, and P. Södersten. Inhibition of serotonin synthesis attenuates inhibition of ingestive behavior by CCK-8. *Pharmacol. Biochem. Behav.* 51:9–12, 1995.

Fantino, M., L. Brondel, A. H. Swiergiel, and O. Lebec. Reduction of negative alliesthesia for sweet gustatory stimuli by cyproheptadine, a serotonin antagonist. *Life Sci.* 46:1381–1387, 1990.

Fletcher, P. J. The anorectic action of peripheral 5-HT examined in the runway: Evidence for an action on satiation. *Psychopharmacology* 93:498–501, 1987.

Fletcher, P. J. The effects of *d*-fenfluramine on saccharin intake and preference, and on food and water intake. *Pharmacol. Biochem. Behav.* 29:687–691, 1988a.

Fletcher, P. J. Increased food intake in satiated rats induced by the 5-HT antagonists methysergide, metergoline and ritanserin. *Psychopharmacology* 96:237–242, 1988b.

Fletcher, P. J., and M. J. Burton. The anorectic action of peripherally administered 5-HT is enhanced by vagotomy. *Physiol. Behav.* 34:861–866, 1985.

Fletcher, P. J., and M. J. Burton. Dissociation of the anorectic actions of 5-HTP and fenfluramine. *Psychopharmacology* 89:216–220, 1986a.

Fletcher, P. J., and M. J. Burton. Microstructural analysis of the anorectic action of peripherally administered 5-HT. *Pharmacol. Biochem. Behav.* 24:1133–1136, 1986b.

Fletcher, P. J., P. J. Currie, J. W. Chambers, and D. V. Coscina. Radiofrequency lesions of the PVN fail to modify the effects of serotonergic drugs on food intake. *Brain Res.* 630:1–9, 1993.

Fletcher, P. J., H. M. Zhi, M. H. Zack, and D. V. A. Coscina. A comparison of the effects of the 5-HT_1 agonists TFMPP and RU 24969 on feeding following peripheral or medial hypothalamic injection. *Brain Res.* 580:265–272, 1992.

Foltin, R. W., and M. W. Fischman. Food intake in baboons: effects of d-amphetamine and fenfluramine. *Pharmacol. Biochem. Behav.* 31:585–592, 1989.

Fornal, C. A., C. W. Metzler, F. Marrosu, L. E. Ribiero-do-Valle, and B. L. Jacobs. A subgroup of dorsal raphe serotonergic neurons in the cat is strongly activated during oral-buccal movements. *Brain Res.* 716:123–133, 1996.

Frankfurt, M., S. D. Mendelson, C. R. McKittrick, and B. S. McEwen. Alterations of serotonin receptor binding in the hypothalamus following acute denervation. *Brain Res.* 601:349–352, 1993.

Fuller, R. W., S. K. Hemrick-Luecke, and H. D. Snoddy. Fluoxetine at anorectic doses does not have properties of a dopamine uptake inhibitor. *J. Neural Transm.* 96:165–177, 1994.

Garattini, S., A. Bizzi, S. Caccia, and T. Mennini. Progress report on the anorectic effects of dexfenfluramine, fluoxetine and sertraline. *Int. J. Obesity* 16(Suppl. 3): S43–S50, 1992.

Garattini, S., T. Mennini, C. Bendotti, R. Invernizzi. Neurochemical mechanism of action of drugs which modify feeding via the serotoninergic system. *Appetite* 7(Suppl.):15–38, 1986.

Gershon, M. D., G. Mawe, and T. Branchek. 5-Hydroxytryptamine and enteric neurons. In Fozard, J. R. (ed.): *The Peripheral Actions of 5-HT.* New York: Oxford University Press, 1989, pp. 247–273.

Gibson, E. L., A. J. Kennedy, and G. Curzon. *d*-Fenfluramine– and *d*-norfenfluramine–induced hypophagia: Differential mechanisms and involvement of postsynaptic 5-HT receptors. *Eur. J. Pharmacol.* 242:83–90, 1993.

Goodall, E. M., P. J. Cowen, M. Franklin, and T. Silverstone. Ritanserin attenuates anorectic, endocrine and thermic responses to *d*-fenfluramine in human volunteers. *Psychopharmacology* 112:461–466, 1993.

Goodall, E., and T. Silverstone. Differential effect of *d*-fenfluramine and metergoline on food intake in human subjects. *Appetite* 11:215–228, 1988.

Grignaschi, G., B. Mantelli, C. Fracasso, M. Anelli, S. Caccia, and R. Samanin. Reciprocal interaction of 5-hydroxytryptamine and cholecystokinin in the control of feeding patterns in rats. *Br. J. Pharmacol.* 109:491–494, 1993.

Grignaschi, G., J. C. Neill, A. Petrini, S. Garattini, and R. Samamin. Feeding pattern studies suggest that *d*-fenfluramine and sertraline specifically enhance the state of satiety in rats. *Eur. J. Pharmacol.* 211:137–142, 1992.

Grignaschi, G., and R. Samanin. Role of 5-HT receptors in the effect of *d*-fenfluramine on feeding patterns in the rat. *Eur. J. Pharmacol.* 212:287–289, 1992a.

Grignaschi, G., and R. Samanin. Role of serotonin and catecholamines in brain in the feeding suppressant effect of fluoxetine. *Neuropharmacology* 31:445–449, 1992b.

Grignaschi, G., and R. Samanin. Role of serotonin receptors in the effect of sertraline on feeding behavior. *Psychopharmacology* 110:203–208, 1993.

Grignaschi, G., F. Sironi, and R. Samanin. The 5-HT_{1B} receptor mediates the effect of *d*-fenfluramine on eating caused by intra-hypothalamic injection of neuropeptide Y. *Eur. J. Pharmacol.* 274:221–224, 1995.

Grinker, J. A., A. Drewnowski, M. Enns, and H. Kissileff. Effects of *d*-amphetamine and fenfluramine on feeding patterns and activity of obese and lean zucker rats. *Pharmacol. Biochem. Behav.* 12:265–275, 1980.

Halford, J. C. G., and J. E. Blundell. The 5-HT_{1B}-receptor agonist CP-94,253 reduces food intake and preserves the behavioral satiety sequence. *Physiol. Behav.* 60:933–939, 1996.

Hartley, J. E., and A. Fletcher. The effects of WAY-100135 and 8-hydroxy-2-(di-*n*-propylamino)tetralin on feeding in the rat. *Eur. J. Pharmacol.* 252:329–332, 1994.

Higgins, G. A., D. M. Tomkins, C. X. Poulos, and E. M. Sellers. Effect of dexfenfluramine on saccharin drinking: Behavioural and pharmacological studies. *Pharmacol. Biochem. Behav.* 47:307–315, 1994.

Hill, A. J., and J. E. Blundell. Sensitivity of the appetite control system in obese subjects to nutritional and serotoninergic challenges. *Int. J. Obesity* 14:219–233, 1990.

Hoyer, D., D. E. Clarke, J. R. Fozard, P. R. Hartig, G. R. Martin, E. J. Mylecharane, P. R. Saxena, and P. P. A. Humphrey. VII. International Union of Pharmacology Classification of receptors for 5-hydroxytryptamine (serotonin). *Pharmacol. Rev.* 46:157–203, 1994.

Hsiao, S., and R. Spencer. Analysis of licking responses in rats: effects of cholecystokinin and bombesin. *Behav. Neurosci.* 97:234–245, 1983.

Hutson, P. H., T. P. Donohoe, and G. Curzon. Infusion of the 5-hydroxytryptamine agonists RU 24969 and TFMPP into the paraventricular nucleus of the hypothalamus causes hypophagia. *Psychopharmacology* 95:550–552, 1988.

Jacobs, B. L., and E. C. Azmitia. Structure and function of the brain serotonin system. *Physiol. Rev.* 72:165–229, 1992.

Kaplan, J. M., J. Donahey, J.-P. Baird, K. Simansky, and H. J. Grill. *d*-Fenfluramine anorexia: Dissociation of ingestion rate, meal duration, and meal size effects. *Pharmacol. Biochem. Behav.* (1997 in press).

Kennett, G. A., G. and Curzon. Evidence that hypophagia induced by mCPP and TFMPP requires 5-HT_{1C} and 5-HT_{1B} receptors; hypophagia induced by RU 24969 only requires 5-HT_{1B} receptors. *Psychopharmacology* 96:93–100, 1988.

Kennett, G. A., and G. Curzon. Potencies of antagonists indicate that 5-HT_{1C} receptors mediate 1–3(chlorophenyl)piperazine-induced hypophagia. *Br. J. Pharmacol.* 103:2016–2020, 1991.

Kennett, G. A., M. D. Wood, A. Glen, S. Grewal, I. Forbes, A. Gadre, and T. P. Blackburn. In vivo properties of SB 200646A, a 5-$HT_{2C/2B}$ receptor antagonist. *Bri. J. Pharmacol.* 111:797–802, 1994.

Kirkham, T. C., and J. E. Blundell. Effect of naloxone and naltrexone on the development of satiation measured in the runway: Comparisons with *d*-amphetamine and *d*-fenfluramine. *Pharmacol. Biochem. Behav.* 25:123–128, 1986.

Kissileff, H. R. Free feeding in normal and "recovered lateral" rats monitored by pellet-detecting eatometer. *Physiol. Behav.* 5:163–173, 1970.

Kitchener, S. J., and C. T. Dourish. An examination of the behavioural specificity of hypophagia induced by 5-$HT_{1B,}$ 5-HT_{1C} and 5-HT_2 receptor agonists using the post-prandial satiety sequence in rats. *Psychopharmacology* 113:369–377, 1994.

Koe, B. K., J. A. Nielsen, J. E. Macor, and J. Heym. Biochemical and behavioral studies of the 5-HT_{1B} receptor agonist, CP-94,253. *Drug Dev. Res.* 26:241–250, 1992.

Kyriakides, M., and T. Silverstone. Comparison of the effects of *d*-amphetamine and fenfluramine on hunger and food intake in man. *Neuropharmacology* 18:1007–1008, 1979.

Latham, C. J., and J. E. Blundell. Evidence for the effect of tryptorhan on the pattern of food consumption in free feeding and food deprived rats. *Life Sci.* 24:1971–1978, 1979.

Lawton, C. L., and J. E. Blundell. 5-HT and carbohydrate suppression: Effects of 5-HT antagonists on the action of *d*-fenfluramine and DOI. *Pharmacol. Biochem. Behav.* 46:349–360, 1993.

Leander, J. D. Fluoxetine suppresses palatability-induced ingestion. *Psychopharmacology* 91:285–287, 1987.

Lee, M. D., and P. G. Clifton. Partial reversal of fluoxetine anorexia by the 5-HT antagonist metergoline. *Psychopharmacology* 107:359–364, 1992.

Lee, M., and K. J. Simansky. Reduction of food intake and alteration of licking distribution by the 5-HT_{1B} agonist CP-94,253 in rats. *Soc. Neurosci. Abstr.* 21:1460, 1995

Leibowitz, S. F. Hypothalamic serotonin in relation to appetite for macronutrients and eating disorders. In Vanhoutte, P. M., P. R. Saxena, R. Paoletti, N. Brunello, and A. S. Jackson (eds.): *Serotonin: From Cell Biology to Pharmacology and Therapeutics*. Boston, MA: Kluwer Academic Publishers, 1993, pp. 383–391.

Leibowitz, S. F., J. T. Alexander, W. K. Cheung, and G. F. Weiss. Effects of serotonin and the serotonin blocker metergoline on meal patterns and macronutrient selection. *Pharmacol. Biochem. Behav.* 45:185–194, 1993.

Leibowitz, S. F., G. F. Weiss, U. A. Walsh, and D. Viswanath. Medial hypothalamic nuclei: Role in circadian patterns of feeding and macronutrient selection. *Brain Res.* 503:132–140, 1989.

Li, B., and N. E. Rowland. Cholecystokinin- and dexfenfluramine-induced anorexia compared using devazepide and c-*fos* expression in the rat brain. *Regul. Pept.* 50:223–233, 1994.

Lightowler, S., M. Wood, T. Brown, A. Glen, T. Blackburn, I. Tulloch, and G. Kennett. An investigation of the mechanism responsible for fluoxetine-induced hypophagia in rats. *Eur. J. Pharmacol.* 296:137–143, 1996.

Lucki, I., M. S. Kreider, and K. J. Simansky. Reduction of feeding behavior by the serotonin uptake inhibitor sertraline. *Psychopharmacology* 96:289–295, 1988.

Macor, J. E., C. A. Burkhart, J. H. Heym, J. L. Ives, L. A. Lebel, M. E. Newman, J. A. Nielsen, K. Ryan, D. W. Schulz, L. K. Torgersen, and B. K. Koe. 3-(1,2,5,6-Tetrahydropyrid-4-yl)pyrrolo[3,2-b]pyrid-5-one: A potent and selective serotonin (5-HT_{1B}) agonist and rotationally restricted phenolic analogue of 5-methoxy-3- (1,2,5,6-tetrahydropyridi4-yl)indole. *J. Med. Chem.* 33:2087–2093, 1990.

Martin, C. F., and J. Gibbs, Bombesin elicits satiety in sham feeding rats. *Peptides* 1:131–134, 1980.

McCall, R. B. Role of neurotransmitters in the central regulation of the cardiovascular system. *Prog. Drug Res.* 35:25–84, 1990.

McGuirk, J., E. Goodall, T. Silverstone, and P. Willner. Differential effects of *d*-fenfluramine and *d*-amphetamine on the microstructure of human eating behavior. *Behav. Pharmacol.* 2:113–119, 1991.

McGuirk, J., R. Muscat, and P. Willner. Effects of chronically administered fluoxetine and fenfluramine on food intake, body weight and the behavioural satiety sequence. *Psychopharmacology* 106:401–407, 1992a.

McGuirk, J., R. Muscat, and P. Willner. Effects of the 5-HT uptake inhibitors, femoxetine and paroxetine, and a 5-$HT_{1A/B}$ agonist, eltoprazine, on the behavioural satiety sequence. *Pharmacol. Biochem. Behav.* 41:801–805, 1992b.

Mennini, T., A. Bizzi, S. Caccia, A. Codegoni, C. Fracasso, E. Frittoli, G. Guiso, I. M. Padura, C. Taddei, A. Uslenghi, and S. Garattini. Comparative studies on the anorectic activity of d-fenfluramine in mice, rats, and guinea pigs. *Naunyn-Schmiedeberg Arch. Pharmacol.* 343:483–490, 1991.

Montgomery, A. M. J., and P. Willner. Fenfluramine disrupts the behavioral satiety sequence in rats. *Psychopharmacology* 94:397–401, 1988.

Neill, J. C., C. Bendotti, and R. Samanin. Studies on the role of 5-HT receptors in satiation and the effect of *d*-fenfluramine in the runway test. *Eur. J. Pharmacol.* 190:105–112, 1990.

Neill, J. C., and S. J. Cooper. Evidence for serotonergic modulation of sucrose sham-feeding in the gastric-fistulated rat. *Physiol. Behav.* 44:453–459, 1988.

Neill, J. C., and S. J. Cooper. Evidence that *d*-fenfluramine anorexia is mediated by 5-HT_1 receptors. *Psychopharmacology* 97:213–218, 1989a.

Neill, J. C., and S. J. Cooper. Effects of 5-hydroxytryptamine and *d*-fenfluramine on sham feeding and sham drinking in the gastric-fistulated rat. *Physiol. Behav.* 46:949–953, 1989b.

Neill, J. C., and S. J. Cooper. Selective reduction by serotonergic agents of hypertonic saline consumption in rats: evidence for possible 5-HT_{1C} receptor mediation. *Psychopharmacology* 99:196–201, 1989c.

Nosjean, A., M. Alruison, and R. F. Laguzzi. Increase in paradoxical sleep after destruction of serotoninergic innervation in the nucleus tractus solitarius of the rat. *Neuroscience* 23:469–481, 1987.

Oluyomi, A. O., E. L. Gibson, A. M. C. Barnfield, and G. Curzon. *d*-Fenfluramine and *d*-norfenfluramine hypophagias do not require increased hypothalamic 5-hydroxytryptamine release. *Eur. J. Pharmacol.* 264:111–115, 1994.

Orosco, M., and S. Nicolaidis. Spontaneous feeding-related monoaminergic changes in the rostromedial hypothalamus revealed by microdialysis. *Physiol. Behav.* 52:1015–1019, 1992.

Paez, X., and S. F. Leibowitz. Changes in extracellular PVN monoamines and macronutrient intake after idazoxan or fluoxetine injection. *Pharmacol. Biochem. Behav.* 46:933–941, 1993.

Poeschla, B., J. Gibbs, K. J. Simansky, D. Greenberg, and G. P. Smith. Cholecystokinin-induced satiety depends on activation of 5-HT1C receptors. *Am. J. Physiol.* 264:R62–R64, 1993.

Poeschla, B., J. Gibbs, K. J. Simansky, and G. P. Smith. The 5-HT_{1A} agonist 8-OH-DPAT attenuates the satiating action of cholecystokinin. *Pharmacol. Biochem. Behav.* 42:541–543, 1992.

Raiteri, M., G. Bonanno, and F. Vallebuona. In vitro and in vivo effects of *d*-fenfluramine: No apparent relation between 5-hydroxytryptamine release and hypophagia. *J. Pharmacol. Exp. Ther.* 273:643–649, 1995.

Ramboz, S., F. Saudou, D. A. Amara, C. Belzung, L. Segu, R. Misslin, M.-C. Buhot, and R. Hen. 5-HT_{1B} receptor knock out—Behavioral consequences. *Behav. Brain Res.* 73:305–312, 1996.

Rogers, P. J., and J. E. Blundell. Effect of anorexic drugs on food intake and the micro-structure of eating in human subjects. *Psychopharmacology* 66:159–165, 1979.

Samanin, R., and G. Grignaschi. Role of 5-hydroxytryptamine receptor subtypoes in satiety and animal models of eating disorders. In Cooper, S. J., and P. G. Clifton, (eds.): *Drug Receptor Subtypes and Ingestive Behaviour.* London: Academic Press, 1996, pp. 39–58.

Samanin, R., T. Mennini, C. Bendotti, D. Barone, C. Caccia, and S. Garattini. Evidence that central 5-HT_2 receptors do not play an important role in the anorectic activity of *d*-fenfluramine in the rat. *Neuropharmacology* 28:465–469, 1989.

Samanin, R., J. C. Neill, G. Grignaschi, I. M. Padura, A. Bizzi, and S. Garattini. Role of 5-HT receptors in the effect of *d*-fenfluramine on gastric emptying and feeding behaviour as examined in the runway test. *Eur. J. Pharmacol.* 197:69–73, 1991.

Sanders-Bush, E., and M. Breeding. Choroid plexus epithelial cells in primary culture: A model of 5-HT_{1C} receptor activation by hallucinogenic drugs. *Psychopharmacology* 105:340–346, 1991.

Schechter, L. E., and K. J. Simansky. 1-(2,5-Dimethoxy-4-iodophenyl)-2-aminopropane (DOI) exerts an anorexic action that is blocked by 5-HT_2 antagonists in rats. *Psychopharmacology* 94:342–346, 1988.

Schipper, J., M. Th. M. Thulp, and H. Sjibesma. Neurochemical profile of eltoprazine. *Drug Metab. Drug Interact.* 8:85–114, 1990.

Shor-Posner, G., J. A. Grinker, C. Marinescu, O. Brown, and S.F. Leibowitz. Hypothalamic serotonin in the control of meal patterns and macronutrient selection. *Brain Res. Bull.* 17:663–671, 1986.

Shukla, R., D. MacKenzie-Taylor, and R. H. Rech. Evidence for 5-HT_2 receptor mediation in quipazine anorexia. *Psychopharmacology* 100:115–118, 1990.

Simansky, K. J. Peripheral 5-carboxamidotryptamine (5-CT) elicits drinking by stimulating 5-HT_1-like serotonergic receptors in rats. *Pharmacol. Biochemi. Behav.* 38:459–462, 1991.

Simansky, K. J. 5-HT receptor subtypes influencing feeding and drinking: Focus on the periphery. In Cooper, S. J., and P. G. Clifton, (eds.): *Drug Receptor Subtypes and Ingestive Behaviour.* London: Academic Press, 1996a, pp. 59–97.

Simansky, K. J. Serotonergic control of the organization of feeding and satiety. *Behav. Brain Res.* 73:37–42, 1996b.

Simansky, K. J., and V. J. Adipudi. Ablating the area postrema (AP) attenuates the decrease in food intake produced by 5-carboxamidotryptamine (5-CT) but not by α-methyl-5-hydroxytryptamine (α-ME-5-HT). *Soc. Neurosci. Abstr.* 22: 1414, 1996.

Simansky, K. J., J. Jakubow, F. C. Sisk, A. H. Vaidya, and K. Eberle-Wang. Peripheral serotonin is an incomplete signal for eliciting satiety in sham-feeding rats. *Pharmacol. Biochem. Behav.* 43:847–854, 1992.

Simansky, K. J., and L. E. Schechter. Properties of some 1-arylpiperazines as antagonists of stereotyped behaviors mediated by central serotonergic receptors in rodents. *J. Pharmacol. Exp. Ther.* 247:1073–1081, 1988.

Simansky, K. J., F. C. Sisk, A. H. Vaidya, and K. Eberle-Wang. Peripherally administered α-methyl-5-hydroxytryptamine and 5-carboxamidotryptamine reduce food intake via different mechanisms in rats. *Behav. Pharmacol.* 1:241–246, 1989–1990.

Simansky, K. J., and A. H. Vaidya. Behavioral mechanisms for the anorectic action of the serotonin (5-HT) uptake inhibitor sertraline in rats: comparison with directly acting 5-HT agonists. *Brain Res. Bull.* 25:953–960, 1990.

Smith, G. P., C. Jerome, B. Cushin, R. Eterno, and K. J. Simansky. Abdominal vagotomy blocks the satiety effect of cholecystokinin in the rat. *Science* 213:1036–1037, 1981.

Soulairac, A., and M. L. Soulairac. Action de la reserpine, de la serotonine et de l'iproniazide sur le comportement alimentaire du rat. *C. R. Soc. Biol. Filliales (Paris)* 154:510–513, 1960.

Spiegel, T. A., J. M. Kaplan, A. Tomassini, and E. Stellar. Bite size, ingestion rate, and meal size in lean and obese women. *Appetite* 21:131–145, 1993.

Stallone, D., and S. Nicolaidis. Increased food intake and carbohydrate preference in the rat following treatment with the serotonin antagonist metergoline. *Neurosci. Lett.* 102:319–324, 1989.

Stallone, D., S. Nicolaidis, and J. Gibbs. Cholecystokinin-induced anorexia depends on serotoninergic function. *Am. J. Physiol.* 256:R1138–R1141, 1989.

Steinbusch, H. W. M. Distribution of serotonin-immunoreactivity in the central nervous system of the rat—Cell bodies and terminals. *Neuroscience* 6:557–618, 1981.

Tecott, L. H., L. M. Sun, S. F. Akana, A. M. Strack, D. H. Lowenstein, M. F. Dallman, and D. Julius. Eating disorder and epilepsy in mice lacking 5-HT_{2C} serotonin receptors. *Nature* 374:542–548, 1995.

Thor, K. B., and C. J. Helke. Serotonin and substance P–containing projections to the nucleus tractus solitarii of the rat. *J. Comp. Neurol.* 265:275–293, 1987.

Thurlby, P. L., S. Garattini, and R. Samanin. Effects of serotonin antagonists on the performance of a simple food acquisition task in rats treated with fenfluramine isomers. *Pharmacol. Res. Commun.* 17:1129–1139, 1985.

Thurlby, P. L., V. E. Grimm, and R. Samanin. Feeding and satiation observed in the runway: The effects of *d*-amphetamine and *d*-fenfluramine compared. *Pharmacol. Biochem. Behav.* 18:841–846, 1983.

Thurlby, P. L., and R. Samanin. Effects of anorectic drugs and prior feeding on food-rewarded runway behavior. *Pharmacol. Biochem. Behav.* 14:799–804, 1981.

Tonini, M., and F. De Ponti. Serotonin modulation of gastrointestinal motility. In Gaginella, T. S., and J. J. Galligan (eds.): *Serotonin and Gastrointestinal Function.* Boca Raton, FL: CRC Press, 1995, pp. 53–84.

Veasey, S. C., C. A. Fornal, C. W. Metzler, and B. L. Jacobs. Response of serotonergic caudal raphe neurons in relation to specific motor activities in freely moving cats. *J. Neurosci.* 15:5346–5359, 1995.

Vickers, S. P., P. G. Clifton, and C. T. Dourish. Behavioural evidence that *d*-fenfluramine–induced anorexia in the rat is not mediated by the 5-HT_{1A} receptor subtype. *Psychopharmacology* 125:168–175, 1996.

Weiss, G. F., P. Papadakos, K. Knudson, and S. F. Leibowitz. Medial hypothalamic serotonin: Effects on deprivation and norepinephrine-induced eating. *Pharmacol. Biochem. Behav.* 25:1223–1230, 1986.

Westphal, R. S., and E. Sanders-Bush. Reciprocal binding properties of 5-hydroxytryptamine type 2C receptor agonists and inverse agonists. *Mol. Pharmacol.* 46:937–942, 1994.

Wilkinson, L. O., and C. T. Dourish. Serotonin and animal behavior. In Peroutka, S. J. (ed.): *Serotonin Receptor Subtypes: Basic and Clinical Aspects.* New York: John Wiley & Sons, 1991, pp. 147–210.

Willner, P., J. McGuirk, J., G. Phillips, and R. Muscat. Behavioural analysis of the anorectic effects of fluoxetine and fenfluramine. *Psychopharmacology* 102: 273–277, 1990.

Wong, D. T., L. R. Reid, and P. G. Threlkeld. Suppression of food intake in rats by fluoxetine: Comparison of enantiomers and effects of serotonin antagonists. *Pharmacol. Biochem. Behav.* 31:475–479, 1988.

Wurtman, R. J., and J. J. Wurtman. Do carbohydrates affect food intake via neurotransmitter activity? *Appetite* 11 (Suppl.):42–47, 1988.

Zifa, E., and G. Fillion. 5-Hydroxytryptamine receptors. *Pharmacol. Rev.* 44:401–458, 1992.

10

Satiation of Alcohol Intake

PAUL J. KULKOSKY

Viewpoint

In one of the earliest papers in alcohol research, Richter (1926) concluded that voluntary alcohol consumption was limited by the abilities of animals and humans to metabolize ethyl alcohol and to utilize it as a source of calories. When liver alcohol dehydrogenase (LADH) was demonstrated to be the rate-limiting enzyme in the oxidative metabolism of alcohol (Cornell et al., 1979), it was possible to test Richter's hypothesis. A review of the literature in 1985 revealed that variations in alcohol intake among individuals and species were highly correlated with the activity of LADH (Kulkosky, 1985a). For example, when intake of alcohol was maximized by adding alcohol gradually to a sweet saline vehicle, rats and hamsters consumed alcohol in proportion to metabolic capacities. As a result, blood alcohol was limited to levels that did not produce gross intoxication or physical dependence. These findings confirmed Richter' hypothesis and led us to view the satiation of alcohol intake as an adaptive response that optimizes reinforcement of alcohol and minimizes its aversive consequences (Kulkosky, 1985a).

We also saw the satiation of alcohol intake as consistent with Adolph's ideas (1980, 1981) about the short-term limitation of intake of nutrients, sodium, water, oxygen, and heat. Adolph noted that the limitation of intake that occurred with these substances was related to clearances of the substances that was a direct function of extant deficit and the conductances of handling the substances from the environment to the cells. Adolph argued that intake was limited by the "saturation" of the process of clearance, i.e., when the capacity to clear the specific substances was approached. We considered this explanation to generalize to the satiation of intake of ethanol solutions. Thus, we believe that alcohol intake is limited by Adolph's "trick of process-satiation" (Adolph, 1981, p. R362).

Satiety ensures beneficial consequences of consumption and prevents *nimiety*, the aversive experience of glutting caused by intake exceeding clearance capacity (Kulkosky, 1985b). This capacity is determined by the maximal ability to digest, store, and utilize ingesta. Nimiety, a rarely used word (Wolf, 1958), characterizes what is avoided by the limitation of alcohol and other intakes (Woods, 1991). Nimiety is an interoceptive perception of surfeit that completes the hedonic continuum of intake-related experience: appetite→satiety→nimiety. From this perspective, the function of the satiation of alcohol intake is to permit intake to reach satisfaction, i.e., satiety, and to avoid the distress of intoxication, i.e., nimiety.

Structure of Alcohol Satiation

Preabsorptive Satiation and Neuropeptides

Intakes typically stop before ingesta reach peak levels in circulation and cells (Adolph, 1980; Grossman, 1960). Adolph (1980) referred to this preabsorptive satiation as "alimentary satiety" to contrast it with "tissue satiety" that was produced by the full impact of ingestion on cells. A bout of ethanol intake follows this pattern. After overnight water deprivation, ethanol intake stops in about 10 minutes (Kulkosky et al., 1986; Glazner et al., 1988), but blood ethanol levels do not peak until about 20–80 minutes later.

When Smith, Gibbs, and their colleagues clearly demonstrated satiating effects on feeding of the brain–gut peptides cholecystokinin (CCK) and bombesin (BN) (Gibbs et al., 1973, 1979; Smith and Gibbs, 1992; see also Chapters 5 and 6), we investigated the possibility that these peptides would also inhibit intake of alcohol. This idea was supported by the following evidence: (1) the dose-related self-report of satiety (reduction of the disposition to eat) by humans receiving "disguised" alcoholic beverages (Booth, 1981); (2) the compendious literature showing that food deprivation increases, and repletion decreases, alcohol intake (reviewed by Kulkosky, 1985a); (3) the reports of increased ingestion of alcohol in rats drinking with gastric fistulas open compared with drinking with gastric fistulas closed (Rowland and Barnett, 1991); (4) increased alcohol consumption produced by bypassing oral intake and the cephalic phase of insulin secretion (Amit and Stern, 1969); and (5) the rapid 15-fold increase in plasma CCK in rats after intragastric instillation of ethanol (Liddle, et al., 1984). Thus, it appeared reasonable that CCK and perhaps other peptides released from the gut by ingested alcohol served as signals of "alimentary satiety" for ethanol as well as for food. Therefore, we tested the effects of CCK and BN on alcohol intake, preference, and blood levels.

The Ethogram of Ethanol Satiation

Initial studies showed that CCK and BN inhibited alcohol intake in rats at doses that did not inhibit water intake, and this effect was structurally specific, requiring a bioactive molecule (reviewed in Kulkosky, 1985a). We characterized the behavioral sequence or *ethogram* of ethanol satiation with time sampling, observational methods developed by Antin and colleagues (1975) for the study of feeding. Drinking of alcohol rapidly declined in the first 15 minutes of access, and BN produced a decrease in drinking only at 5–15 minutes after presentation (Glazner et al., 1988). BN did not affect the initial rapid drinking rate. Standing, feeding, and grooming increased after alcohol drinking decreased, and, lastly, resting increased to end the sequence (Fig. 10.1). A similar pattern of behaviors was observed in a study of CCK's behavioral effects on alcohol consumption (Kulkosky et al., 1986). This ethogram of satiation produced by BN and CCK resembles that described by Antin et al. (1975) in food-deprived rats with access to liquid diet. The effects of BN and CCK satisfy a criterion for a signal of satiation because the basic pattern of behavior occurs, but intake ends sooner (Booth, 1981; Kulkosky, 1985b; Smith, 1983). Although blood ethanol levels were reduced by BN and CCK, this was due to the decreased intake. The peptides did not artifactually lower blood ethanol levels by interfering with rates of elimination of ethanol.

Motivation of Alcohol Satiation: Role of CCK and BN

Water Deprivation Procedure

Our first studies of the satiating effects of CCK and BN on alcohol intake employed water deprivation to motivate substantial ethanol intake. This procedure may confound thirst and hunger in the interpretation of the inhibitory actions of the peptides on intake. Similar inhibitory actions, however, were observed whether solid food was present or removed and water intake was unaffected during access to alcohol. Thus, it appeared unlikely that the inhibitory effect on alcohol drinking was caused by reduced chow intake or decreased thirst.

To distinguish the effects of peptides on water and alcohol intake better, we first employed a simultaneous choice of 2% ethanol and water after

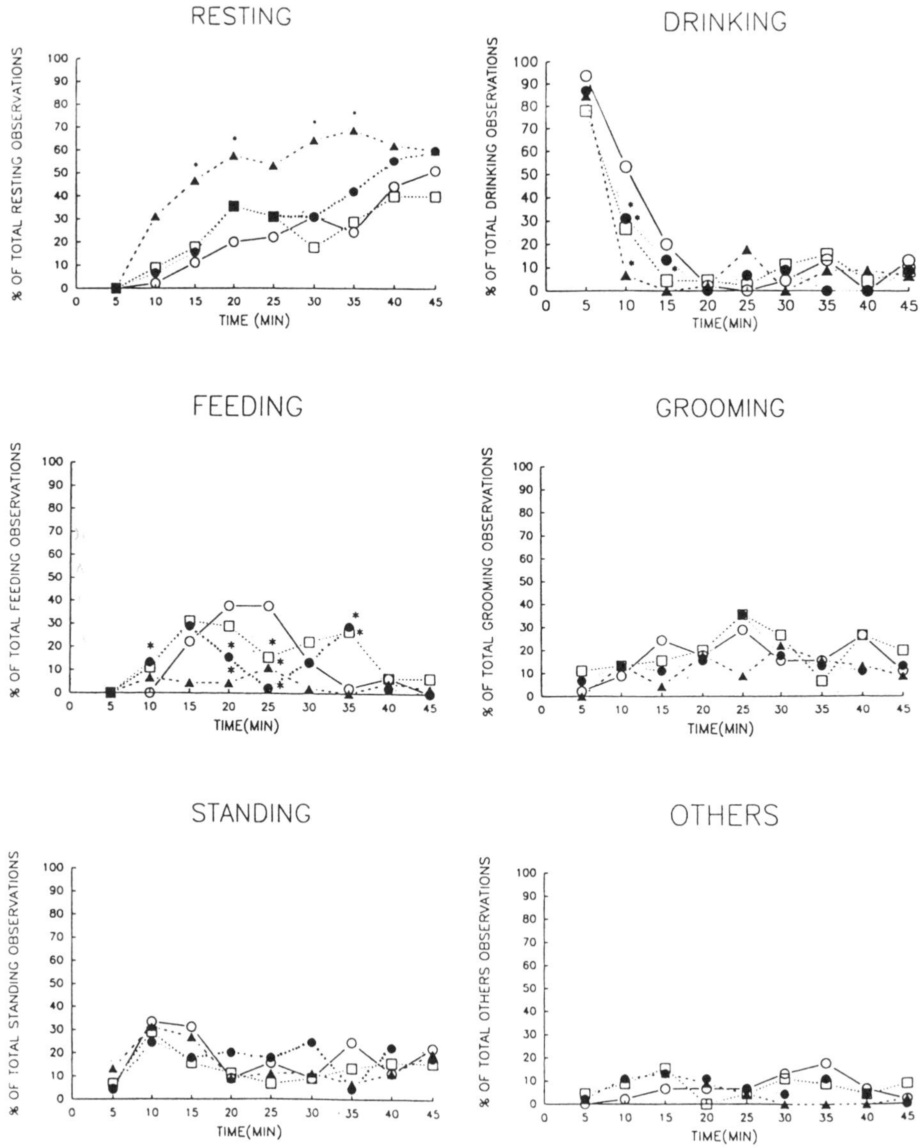

FIGURE 10.1. Mean percentage of total observations made in categories of resting, drinking, feeding, grooming, standing, and "others." Rats (N = 9) were water deprived for 23 hours and given access to one bottle of 5% w/v ethanol for 45 min with chow ad libitum. Intraperitoneal injection of saline (○——○) or bombesin (1μg/kg ●····●, 2 μg/kg □····□, 4 μg/kg ▲---▲) was given 1 minute prior to ethanol presentation. $p < 0.05$ compared to saline. From Glazner et al. (1988) with permission of the publisher.

water deprivation. We found that as CCK suppressed alcohol intake in this design, water intake increased, resulting in stable total fluid intake after 2 μg/kg of CCK (Kulkosky et al., 1989c). This demonstrated that the satiating effect of CCK under these conditions was specific to alcohol and could not be attributed to general inhibition of ingestive behaviors, such as is produced by sedatives, emetics, or bitter tastes.

Nutritive Expectancy Procedure

A more complex preference design for testing the satiating effect of peptides was introduced by Fedorchak and Bolles (1988). They reported that consumption of a flavor previously paired with alcohol was reduced in hungry rats by CCK when the flavor was presented separately from alcohol. These authors argued that the satiating effect of CCK depends on a "nutritive expectancy" of calories to be derived from ethanol and not on a simultaneous interaction of the peptide and alcohol.

We adapted their design to study the effects of CCK and BN on alcohol preference in rats that were not water deprived, but were food restricted to reduce body weight. A choice of 4% ethanol and water was presented to rats that had been adapted to alcohol consumption or had been maintained on water alone. CCK (4 μg/kg, ip) produced results in this nutritive expectancy design (Fig. 10.2) that were nearly identical to those obtained with BN (4 μg/kg) (Kulkosky et al., 1989c, 1992). In each case, the putative satiating peptide reduced alcohol intake, not water intake, but only in rats with previous experience with alcohol.

We interpreted these results as supporting the idea of formation of a natural association of alcohol's sensory cues with its physiological consequences. This "nutritive expectancy" of repletion from alcohol is a naturally conditioned interoceptive perception (Booth, 1985) cued by CCK or BN. The peptides provide effective information on the consequences of ingestion best in experienced animals. Our results and those of Fedorchak and Bolles (1988) and Bolles (1985) argue for the importance of acquired expectancy and adaptive responses to ingestive experiences for the satiating effects of CCK and BN on alcohol preference.

Limited Access Procedure

To provide a test of satiating peptides in rats that were not deprived of water or food, we adapted a design used by Toth et al. (1990b). Their "limited access procedure" (Macdonall and Marcucella, 1979; Linseman,

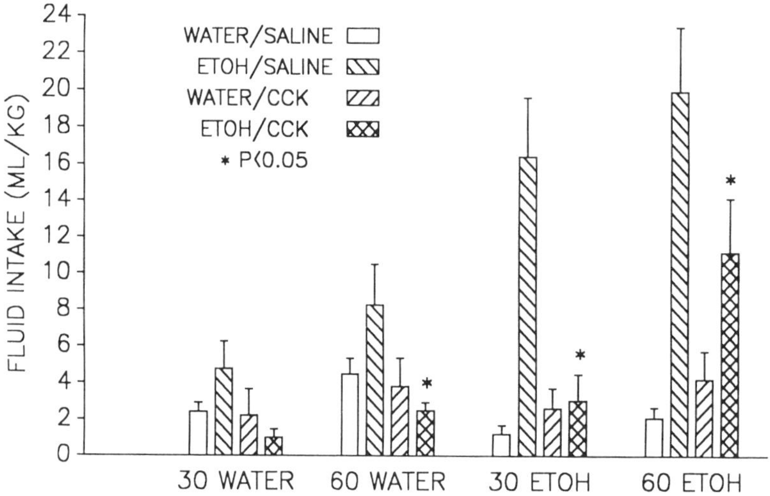

FIGURE 10.2. Mean (± SE) fluid intakes of rats with prior experience with ethanol (ETOH, N = 12) or water (N = 12) at 30 and 60 minutes after presentation. Food-restricted rats received a two-bottle choice of 4% w/v ethanol and water after ip injection of saline or CCK-8 (4 μg/kg). Asterisks denote a significant difference in intake of the fluid after CCK versus saline control. From Kulkosky et al. (1989c), with permission of the publishers.

1987) gives nondeprived rats a choice of 6% w/v ethanol and water for 40 minutes daily in experimental cages in the dark phase of the lighting cycle. They found that CCK reduced alcohol intake, but not water intake, in this procedure.

We extended their results by examining the phase of lighting, repeated experience with alcohol and peptide injection, sex, and dose of CCK and BN in the limited access procedure. CCK suppressed alcohol intake more potently in the dark phase (Fig. 10.3), the effect increased with experience with alcohol and CCK and with dose of CCK, but there was no effect of sex. Water intake remained unaffected or increased after certain doses of CCK (Kulkosky et al., 1991). In contrast, none of these factors changed the satiating potency of BN and BN had no effect on water intake (Kulkosky et al., 1992).

These results confirm that the satiating effects of CCK and BN on ethanol intake do not depend on motivation of drinking by water or food deprivation, are specific to alcohol, and are not the results of nonspecific debilitation or hypodipsia. Furthermore, the results with CCK accord with conclusions from nutritive expectancy experiments because experience with alcohol and CCK enhanced the satiating potency of CCK.

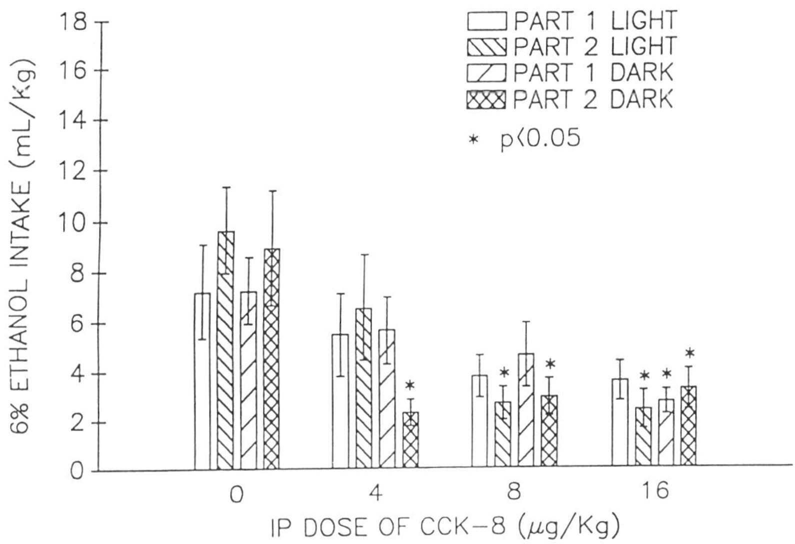

FIGURE 10.3. Mean (± SE) 6% w/v ethanol intake of nondeprived rats during the limited access procedure after ip injections of CCK-8. Rats were tested during the light (N = 12) or dark (N = 12) phase of the lighting cycle in two successive replications (parts 1 and 2) and had a two-bottle choice of 6% ethanol and water for 40 minutes in test cages. Asterisks denote a significant difference in intake after CCK versus saline (0) control. From Kulkosky et al. (1991), with permission of the publisher.

Temporal Parameters of Satiating Effects

Although water deprivation is not necessary for the satiating effects of CCK and BN, the most potent satiating effects of CCK and BN are obtained when alcohol intake is motivated by water deprivation and there is temporal contiguity of peptide injection and access to alcohol. A temporal criterion of a signal of satiation is that its action be prompt, begin within the duration of a meal, and last long enough to bridge the time between alimentary and tissue satieties. We examined the temporal properties of CCK and BN by varying the interval between peptide injection and alcohol presentation from −20 to −10 to 0 minutes. CCK was most potent when injected immediately prior to access to alcohol and had no reliable effect when access was delayed by 20 minutes (Fig. 10.4) (Kulkosky et al., 1989c). Similar results were obtained with bombesin (Kulkosky et al., 1987).

These results are consistent with the short half-lives of these peptides in the circulation and with the necessity of a relatively short interval for synergistic interactions between the peptides and pregastric and gastric

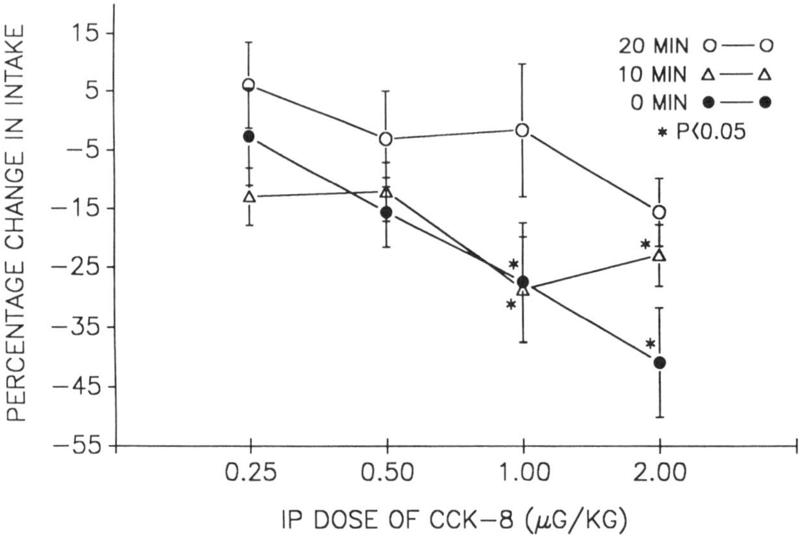

FIGURE 10.4. Mean (± SE) percentage change of 5% w/v ethanol intake, relative to saline, after ip injection of CCK-8. Rats (N = 6 per group) were 23-hour water deprived and injected 20, 10, or 0 minutes prior to one-bottle ethanol access for 30 minutes. From Kulkosky et al. (1989 c), with permission of the publisher.

satiating mechanisms demonstrated during the ingestion of food, some of which are also probably active during ingestion of alcohol (Antin et al., 1977, 1978; Campbell and Davis, 1974; see also Chapter 2).

Thus, CCK and BN could act within a drinking session as prompt, short-acting signals of "alimentary satiety" especially if prior experience enhances the potency of these "signs" of postabsorptive consequences of alcohol ingestion. These must, however, be other long-term satiating factors, such as blood concentration or metabolites of ethanol, that mediate the prolonged interval between the end of one bout of alcohol ingestion and the beginning of the next.

Stimulus Specificity of Satiating Effects of Ceruletide

We performed a study of the stimulus specificity of the satiating effect of ceruletide, an amphibian decapeptide with all of the biological actions of CCK (Kulkosky, Lozinsky, Antista, and Chiu, unpublished observations). Ceruletide had been shown to decrease the intake of alcohol, but not water, in water-deprived rats (Kulkosky and Chiu, 1986; Gewiss et al., 1991). First we determined the concentrations of alcohol (5% w/v), quinine sulfate (0.0075% w/v), and acetone (2% w/v) that depressed fluid intake

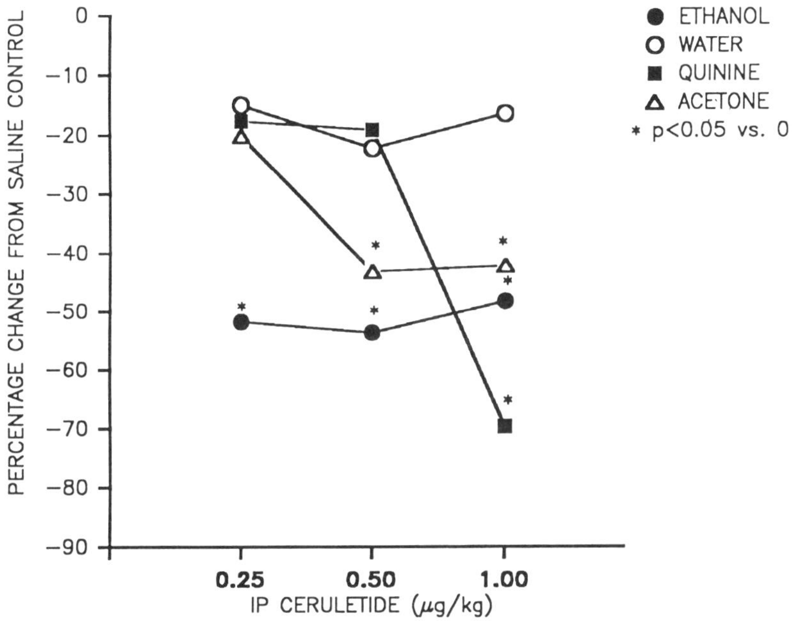

FIGURE 10.5. Mean percentage change of intake of 5% w/v ethanol, water, 0.0075% w/v quinine sulfate, and 2% w/v acetone, relative to saline, after ip injection of ceruletide. Rats (N = 6 per group) were 23-hour water deprived and injected immediately prior to one-bottle fluid access for 30 minutes. Unpublished observations of P. Kulkosky, G. Lozinsky, J. Antista, and N. Chiu.

equally in comparison with water. We chose acetone and quinine because they have some of the stimuli that ethanol has: Acetone has a bitter taste and intoxicating sedative effects, while quinine has only a bitter taste. Ceruletide (0.25–1.0 θg/kg) was injected prior to fluid access. No dose of ceruletide affected intake of water, in agreement with previous findings (Smith et al., 1984; Kulkosky and Chiu, 1986; Gewiss et al., 1991). In contrast, ethanol intake was markedly suppressed by all doses of ceruletide, acetone intake was decreased by the two larger doses, and quinine intake was decreased only by the largest dose (Fig. 10.5). These data suggest the following stimulus specificity for the potency of the satiating effect of ceruletide: Most potent for a bitter, intoxicating, caloric nutrient-containing substance (ethanol); less potent for a bitter, intoxicating, but noncaloric substance (acetone); least potent for a bitter, nonintoxicating and noncaloric substance (quinine). Thus, each stimulus property of ethanol apparently affects the satiating potency of ceruletide. To investigate this inter-pretation, a range of concentrations of each fluid stimulus should be tested across a range of doses of ceruletide.

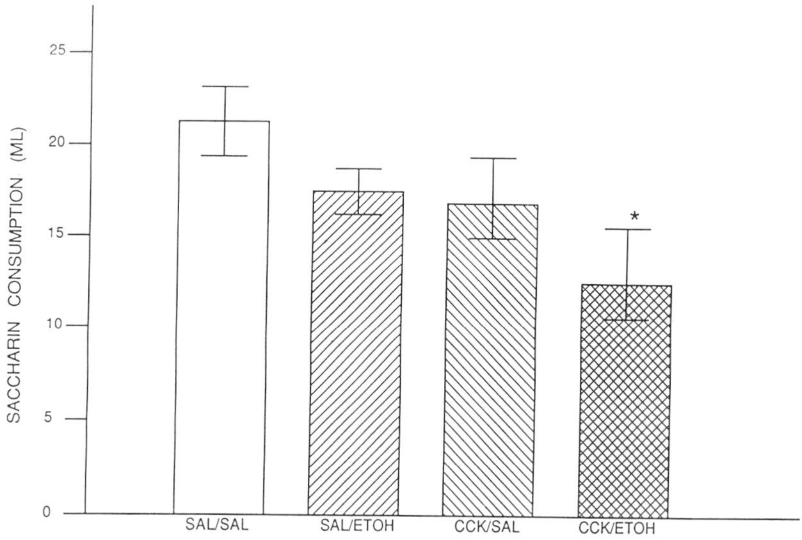

FIGURE 10.6. Mean (± SE) consumption of 0.125% w/v sodium saccharin by 23-hour water-deprived rats (N = 4, two males and two females per group) after prior association of saccharin consumption and ip injections of saline, saline + 2 g/kg ethanol (10% w/v), 2 μg/kg of CCK-8 + saline, or ethanol + CCK-8, given immediately after saccharin access. Rats received one-bottle access for 30 minutes. Unpublished observations of M. Andrews and P. Kulkosky.

Conditioned Taste Aversion Procedure

Another experiment on the motivation of decreases in alcohol intake by CCK employed the conditioned taste aversion design (Kulkosky, 1985b). We gave water-deprived rats a 0.125% (w/v) saccharin solution for 30 minutes followed by either two saline injections, 2 μg/kg of CCK-8 + saline, saline + 2 g/kg of ethanol, or CCK-8 + ethanol (Andrews and Kulkosky, unpublished observations). After a recovery day, the rats were again presented with the single bottle of saccharin solution. Only the group that had combined injections of CCK and alcohol developed an aversion to saccharin (Fig. 10.6). A subsequent test showed that each treatment strongly suppressed food intake relative to saline when rats were food-deprived.

These results suggested that rats use CCK as a graded signal to stop drinking ethanol. When a satiating dose of CCK and a full bout's worth of ethanol were combined, an aversive signal of overconsumption (nimiety) was generated.

Earlier we Kulkosky et al. (1991) showed that repeatedly reducing alcohol intake by CCK injection resulted in increased alcohol consumption. Furthermore, Pérez and Sclafani (1991) showed that contingent low doses of CCK conditioned preferences for flavored saccharin solutions. From this viewpoint, the rat attends to the stimuli generated by CCK and alcohol to adjust intakes adaptively on the basis of experience. The result of this motivational process secures satiety and avoids nimiety.

Neural and Receptor Mechanisms of Peptidergic Satiation

Effect of Specific Antagonists

We have examined the effects of specific receptor antagonists of BN and CCK. The GRP-preferring antagonist D-Phe^6, Des-Met^{14} BN, (see Chapters 5 and 6) reduced the satiating effect of BN during the first 5 minutes of alcohol intake (Carr al., 1994). This suggests that the satiating effect of BN is mediated by GRP receptors (Fig. 10.7). When administered alone, the antagonist also increased intake during the first 5 minutes of the test (Fig. 10.7) (Vallejos et al., 1992). This suggests that endogenous GRP exerts inhibitory control of alcohol intake early in the drinking session.

The inhibitory effect of BN on drinking-associated feeding was also blocked by the antagonist. Furthermore, when the antagonist was administered alone, feeding increased at 20–25 minutes (Fig. 10.8).

Taken together, these results are consistent with a specific GRP receptor mechanism for the satiating effects of BN on intakes of alcohol and food. They also suggest that these satiating actions can be produced by endogenous GRP under these conditions (see Chapters 5 and 6).

Administration of MK-329 (devazepide), a specific antagonist of CCK_A receptors (Wang and Schoenfeld, 1988; see also Chapters 5 and 6), reduced the satiating effect of CCK on alcohol intake (Carr et al., 1993). The minimal effective dose for blocking the effect of CCK was 100 μg/kg, while 400 μg/kg was required to increase the intake of alcohol when administered alone (Fig. 10.9).

Similar results with MK-329 were observed on food intake during access to alcohol except that 100 μg/kg increased food intake when administered alone. Larger doses had no effect (Fig. 10.10).

These results indicate that exogenous CCK acts at CCK_A receptors to inhibit alcohol intake and feeding, and this is consistent with the presence of CCK_A receptors on subdiaphragmatic vagal afferent terminals and on smooth muscle in the pyloric region of the stomach (Chapter 6). The increases in alcohol intake and drinking-associated feeding produced by

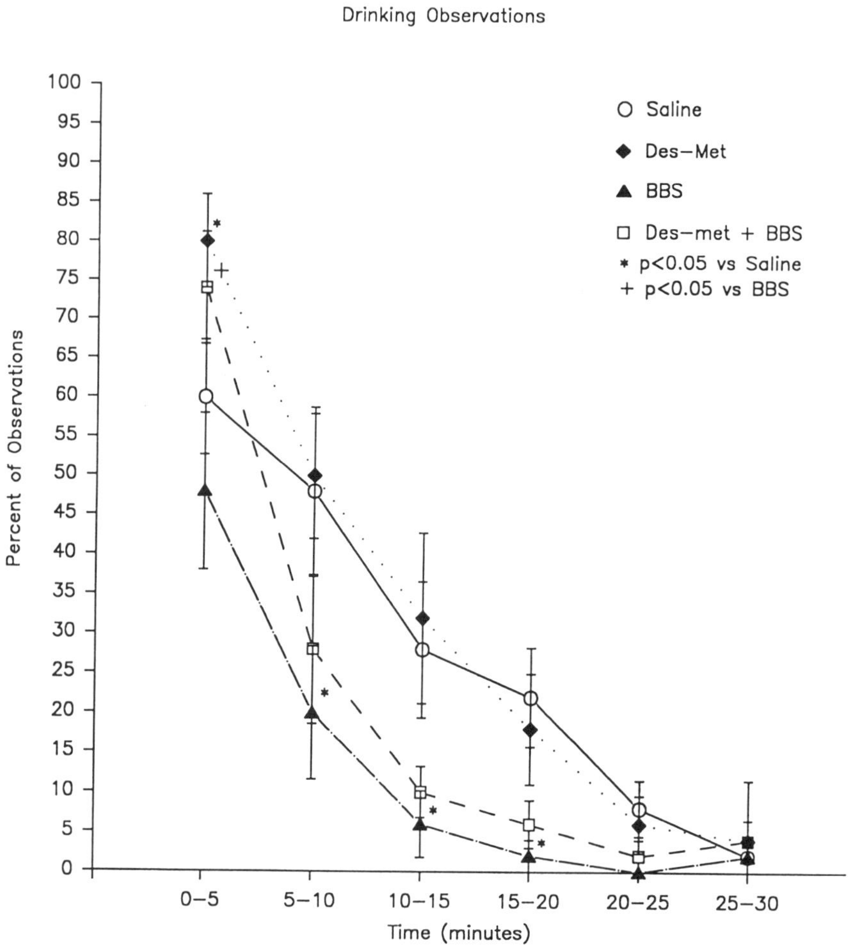

FIGURE 10.7. Mean (± SE) percentage of observations made in the category of drinking 5% w/v ethanol in 23-hour water-deprived rats (N = 10) after ip injection of saline, bombesin (4 μg/kg), bombesin antagonist (Des-Met analogue, 200 μg/kg), or bombesin + antagonist, 20 minutes prior to ethanol access. From Carr et al. (1994), with permission of the publisher.

MK-329 alone suggest that endogenous CCK also has a satiating effect on alcohol and food intake. The effect on alcohol intake is new; the effect on food intake is consistent with previous reports (Smith and Gibbs, 1992; see also Chapter 5).

These initial results with specific peptide receptor antagonists support the hypothesis of a structurally specific, peripheral peptide–receptor interaction as the initiating mechanism for peptidergic satiation of alcohol in-

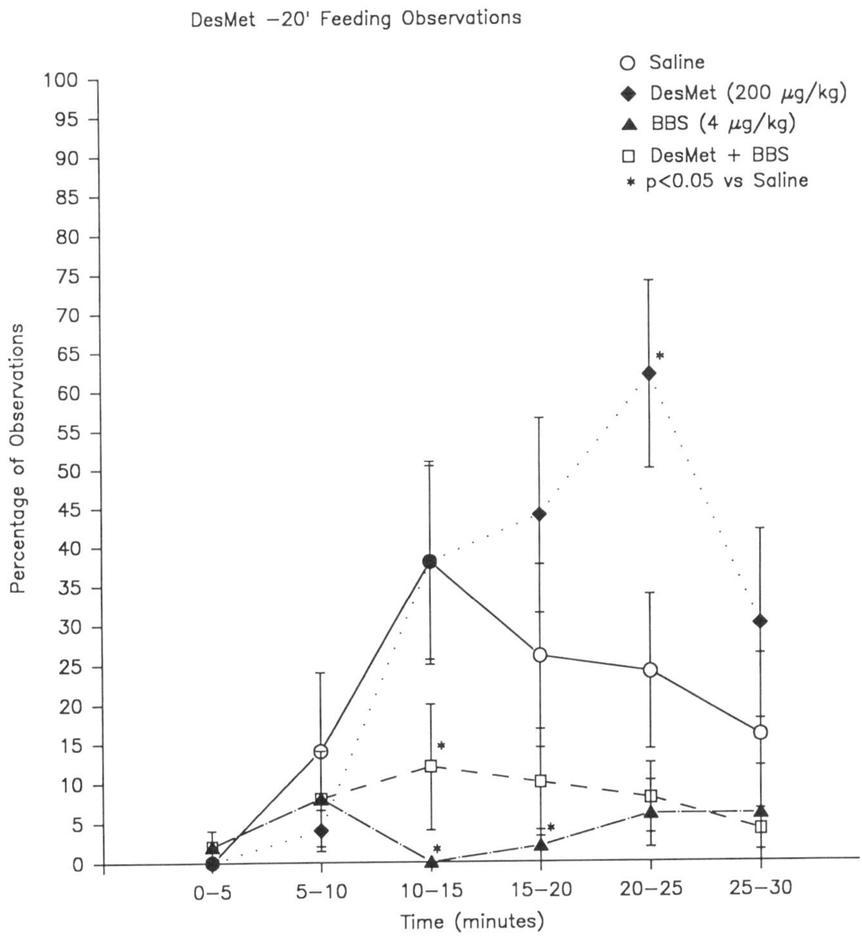

FIGURE 10.8. Mean (± SE) percentage of observations made in the category of feeding of rats described in Figure. 10.7. From Carr et al. (1994), with permission.

take. Furthermore, the results suggest that endogenous GRP and CCK have satiating effects on alcohol intake, but these observations must be replicated under a number of conditions before this action of the peptides can be considered proven.

Although both CCK and BN decrease intake of alcohol, investigation of dose additivity of CCK and BN on ethanol intake suggested that the effects of the peptides were not mediated by the same mechanisms (Kulkosky and Glazner, 1988). A similar result had been obtained with combinations of doses of CCK and BN on the inhibition of food intake (Stein and Woods, 1981). Furthermore, when both results were analyzed together, there was a

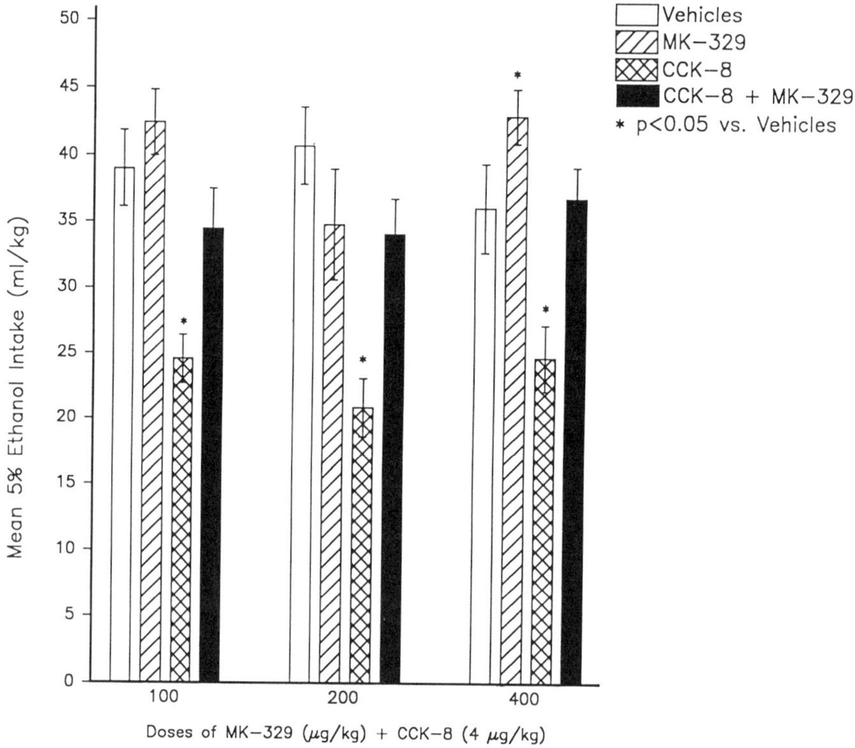

FIGURE 10.9. Mean (± SE) 5% ethanol intake of 23-hour water-deprived rats (N = 10) after ip injection of vehicles, MK-329 (100, 200, 400 μg/kg), CCK-8 (4 μg/kg), or MK-329 + CCK-8, 30 minutes (MK-329) and immediately (CCK-8) prior to one-bottle ethanol access for 30 minutes. From Carr et al. (1993), with permission of the publisher.

high correlation between the inhibitory effects of the combinations of CCK and BN on alcohol and food intake (Kulkosky and Glazner, 1988).

Vagus Nerve

Abdominal vagotomy, specifically the afferent fibers (Smith et al., 1985), blocked the satiating effect of CCK (Smith et al., 1981a; Lorenz and Goldman, 1982; Morley et al., 1982) or ceruletide (Smith et al., 1984) on food intake, but it did not change the satiating effect of BN (Smith et al., 1981b). To block the satiating effect of BN, abdominal vagotomy had to be combined with cord section at the level of T_2. This suggested that lesion of the vagal and spinal visceral afferents was required (Chapters 5 and 6).

To determine if the role of the vagus in mediating the satiating effect of

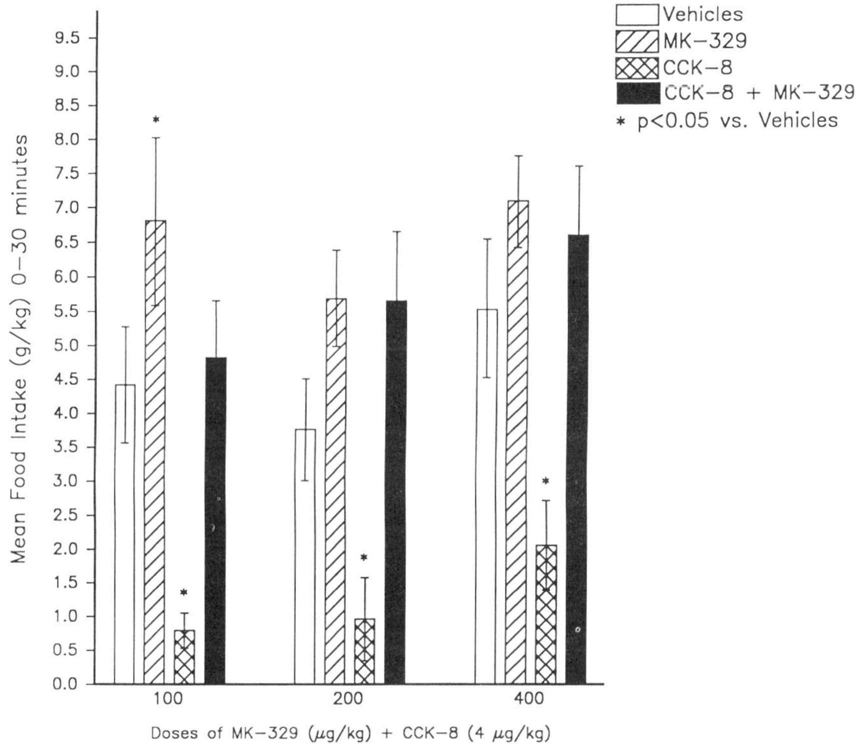

FIGURE 10.10. Mean (± SE) food intake (Purina chow) during ethanol access of rats described in Figure 10.9. From Carr et al. (1993), with permission of the publisher.

ceruletide and BN on alcohol intake was the same as for food intake, we administered ceruletide (1 μg/kg) and BN (4 μg/kg) to intact and vagotomized rats. The results were the same as for food intakes: Abdominal vagotomy abolished the satiating effect of ceruletide on alcohol intake, but did not change the satiating effect of BN (Kulkosky and Chiu, 1986; Kulkosky et al., 1987).

Furthermore, abdominal, gastric, or hepatic vagotomy also reduced alcohol intake (Kulkosky et al., 1987; Toth et al., 1990a) The decreased intake after gastric vagotomy was not explained by the decreased rate of absorption and increased rate of metabolism observed in rats with this lesion because these changes would tend to increase intake by decreasing blood concentration of ethanol (Toth et al., 1990a).

Toth et al. (1990a) noted that the decreased intake after selective or total abdominal vagotomy was similar to the effect of these lesions on food intake. This is consistent with our hypothesis (Kulkosky, 1985a).

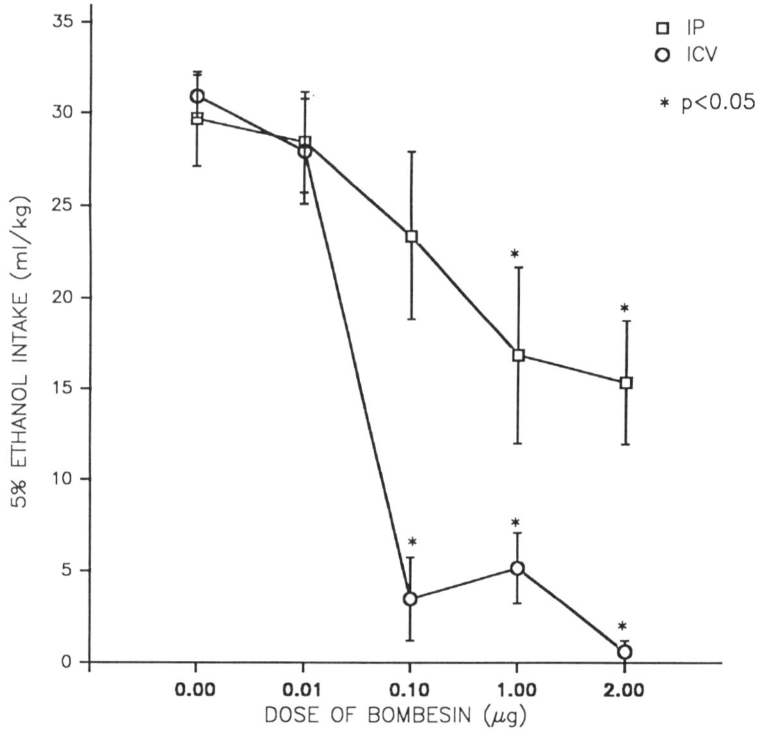

FIGURE 10.11. Mean (± SE) 5% w/v ethanol intake of 23-hour water-deprived rats after ip (N = 5) or icv (N = 5) injection of bombesin. Rats received one-bottle access to ethanol for 45 minutes. From Kulkosky et al. (1987), with permission of the publisher.

The reduction in alcohol intake also suggests that the abdominal vagus may be required to sustain alcohol intake as well as to mediate its suppression by CCK and ceruletide. This is consistent with our speculation that low levels of vagal activity elicit appetite, moderate levels signal satiety, and high levels cause nimiety.

Cerebroventricles

Administration of BN into the lateral cerebral ventricle inhibited alcohol intake more potently and completely than intraperitoneal injection (Fig. 10.11) (Kulkosky et al., 1987). The suppression of alcohol intake by intracerebroventricular (icv) BN, however, was accompanied by excessive grooming and scratching quite unlike the typical ethogram of ethanol satiation described after peripheral administration of BN (Fig. 10.1). We had previously described this grooming effect of central BN in feeding experiments (Gibbs et al., 1981; Kulkosky et al., 1982).

We concluded that BN may act directly on the brain to elicit satiation, but this action could not be mediated solely by penetration of circulating BN into the cerebrospinal fluid through the lateral ventricle. A more site-specific penetration of BN into structures adjacent to the fourth ventricle in the caudal hindbrain that produce a satiating effect for food (Flynn, 1989, 1992; Johnston and Merali, 1988; Kyrkouli et al., 1987; Ladenheim and Ritter, 1988; Willis et al., 1984) should be explored as central sites of action of BN-like peptides in satiation of alcohol intake.

It is possible that CCK released from neurons in the brain has a satiating effect on alcohol intake, but no relevant reports have appeared. The satiating effect of peripherally administered CCK appears to act peripherally in the abdomen at a site innervated by vagal afferent terminals (see above).

Other Neuropeptides and Satiation

In addition to CCK, BN, and their respective antagonists, we tested other neuropeptides for their effects on alcohol intake (Table 10.1). Except for ceruletide, an analogue of CCK, and litorin, an analogue of BN, none of the peptides decreased intake significantly. Neurotensin (Sandoval et al., 1992) and β-endorphin (Sandoval et al., 1991) had no effect, and neuropeptide Y (Kulkosky et al., 1989b) and vasoactive intestinal peptide (Kulkosky et al., 1989a) produced small increases of intake.

Naloxone is the only antagonist we have tested besides the antagonists of CCK and BN. In contrast to the increase in intake produced by the antagonists of CCK and BN, naloxone decreased alcohol intake. Because naloxone antagonizes opioid receptors, the decreased intake suggests that opioids, presumably in the brain, are necessary for the maintenance of alcohol intake once it has begun. Note that this inhibitory effect of naloxone is not a satiating effect; it is the blockade of a stimulating effect.

These results with naloxone suggest that opioid peptides may be endogenous stimulatory controls of alcohol intake that interact with CCK's satiating effect (Kulkosky et al., 1988). When naloxone and CCK were administered together, we observed an infra-additive dose interaction for the inhibition of alcohol intake (Fig. 10.12). This suggests a common process accessed by stimulation of CCK_A and blockade of opioid receptors as theorized earlier (Baile et al., 1986). In contrast to these results, we found a purely dose-additive interaction of naloxone and BN, suggesting separate mechanisms (Kulkosky et al., 1988).

TABLE 10.1.
Effects of Neuropeptides and Antagonists on Alcohol Intake

Peptide	Dose	Effect	Other Effects
Sulphated CCK octapeptide	2–4 μg/kg ip	Decrease	None on water intake
Ceruletide	1 μg/kg ip	Decrease	Blocked by vagotomy
Desulfated CCK octapeptide	1–16 μg/kg ip	None	None
BN tetradecapeptide	4–16 μg/kg ip; 0.1–2 μg icv	Decrease	icv: excessive grooming
BN nonapeptide	4 μg/kg ip	Decrease	Decreased blood ethanol levels
Litorin	16–32 μg/kg ip	Decrease	Equally decreased glucose intake
Neuropeptide Y	0.37 μg icv	Increase	Golden hamster; increased feeding
Vasoactive intestinal peptide	100 μg/kg ip	Increase	Decreased feeding; blocked by CCK
β-endorphin	0.1–500 μg/kg ip	None	Increased active behaviors
Neurotensin	0.1–100 μg/kg ip	None	Decreased feeding, increased water intake
Naloxone	0.1–10 mg/kg ip	Decrease	Infra-dose-additive with CCK-8
[Tyr^4, D-Phe^{12}] BN	20 μg/kg ip	Increase	Blocked effect of BN
[D-Phe^6, Des-Met^{14}] BN	200 μg/kg ip	Increase	Increased feeding; blocked effect of BN
L-364,718 (MK-329 or devazepide)	100–400 μg/kg ip	Increase	Increased feeding; blocked effect of CCK

Critique and Prospectus

Our experiments and those of others reviewed here demonstrate reliable inhibitory effects of CCK, BN, and naloxone on the intake of alcohol. We characterized the motivational and paradigmatic constraints on these inhibitory effects and interpreted them as satiating effects of CCK and BN and antagonism of opioid stimulation of intake with naloxone. The identification of excitatory and satiating effects of these peptides suggests an integrative neuropeptide control system for the regulation of alcohol intake, preference, and blood levels.

The repeated observations of the importance of experience with alcohol in the development of peptide control of intake support an interpretation that peptide controls are learned by natural conditioning. E.C. Tolman's functionalistic interpretation (1967) of the "docility" or trainability of feed-

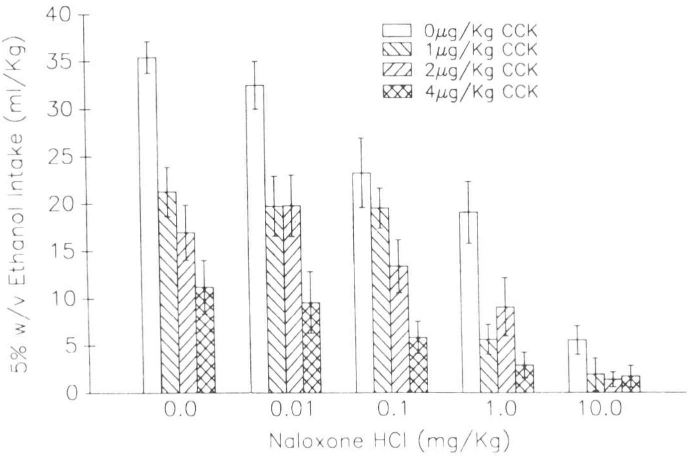

FIGURE 10.12. Mean (± SE) 5% w/v ethanol intake of 23-hour water-deprived rats (N = 11) after ip injections of combinations of naloxone and CCK-8. Rats received one-bottle ethanol access for 30 minutes. From Kulkosky et al. (1989c), with permission of the publisher.

ing accords with the notion that CCK is processed by the nervous system as a "sign" that stands for a later "significate," the consequences of ingestion. An adaptive, flexible process allows the organism to use peptide signals of appetite, satiation, and nimiety as informative tools to plot a path to optimal reinforcement and avoidance of overconsumption.

Criteria for proving that a peptide is a genuine signal of satiation are stringent and extensive and are not yet satisfied by available results with these neuropeptides and alcohol. In particular, more experiments are clearly required on neuropeptide secretion and action during alcohol intake and for the delineation of the central and peripheral pathways of the nervous system involved in alcohol satiation. Likewise, receptor mechanisms and the interacting roles of traditional neurotransmitters, such as serotonin, plainly await further research. Criteria related to dose- and repletion-dependent conditioning of satiety are also not fully met. Thus, CCK and BN-like peptides are "putative" satiating signals for alcohol intake.

Nevertheless, initial applications of the hypothesis of peptide control of alcohol intake have proven successful in the treatment of human alcoholism and alcohol abuse. Peniston and Kulkosky (1989) showed that an innovative alpha–theta brainwave biofeedback therapy for chronic alcoholism prevented a rise in radioimmunoreactive β-endorphin levels associated with relapse in traditionally treated alcoholics. This clinical finding was interpreted with reference to the potent inhibition of alcohol intake by naloxone in rats. For example, if endogenous opioid peptides are stimulatory controls of alcohol intake as suggested by the potent inhibition of alcohol intake in rats by naloxone, preventing their action should facilitate abstinence in the recovering alcoholic. We found a combination of relaxation therapies, prominently including brainwave training, stopped the increase in serum β-endorphin levels and resulted in prolonged prevention of relapse (Peniston and Kulkosky, 1989). These results have since been replicated in a variety of alcoholic patients and reported in the clinical literature (Peniston and Kulkosky, 1991). Recent findings of the successful use of naltrexone, a longer acting opioid antagonist, in the treatment of alcohol dependence and prevention of relapse (O'Malley et al., 1992; Volpicelli et al., 1992) are also consistent with our model of peptide control of alcohol intake. On the basis of these results, the U.S. Food and Drug Administration approved a wider clinical use of naltrexone (Revia) for the treatment of alcoholism in 1995. Because exploration of the role of neuropeptides in alcohol satiation has proven helpful in the clinical approach to human alcoholism (Litten and Allen, 1991), further application and research on the unsolved problems of neuropeptidergic control of alcohol intake and preference are likely to lead to further improvements in therapy of alcoholism.

ACKNOWLEDGEMENT

This research was supported by a Minority Biomedical Research support grant (2S06 GM-08197) from the National Institutes of Health. The author thanks the colleagues who inspired and contributed to this research and Brian Carr for his further assistance in preparation of the manuscript.

REFERENCES

Adolph, E. F. Intakes are limited; Satieties. *Appetite* 1:337–342, 1980.

Adolph, E. F. Regulations of intakes: Clearances. *Am. J. Physiol.* 240:R356–R363, 1981.

Amit, Z., and M. H. Stern. Alcohol ingestion without oropharyngeal sensations. *Psychon. Sci.* 15:162–163, 1969.

Antin, J., J. Gibbs, J. Holt, R. C. Young, and G. P. Smith. Cholecystokinin elicits the complete behavioral sequence of satiety in rats. *J. Comp. Physiol. Psychol.* 89:784–790, 1975.

Antin, J., J. Gibbs, and G. P. Smith. Intestinal satiety requires pregastric food stimulation. *Physiol. Behav.* 18:421–425, 1977.

Antin, J., J. Gibbs, and G. P. Smith. Cholecystokinin interacts with pregastric food stimulation to elicit satiety in the rat. *Physiol. Behav.* 20:67–70, 1978.

Baile, C.A., C. [illegible] McLaughlin, and M. A. Della-Fera. Role of cholecystokinin and opioid pe[illegible]des in control of food intake. *Physiol. Rev.* 66:172–234, 1986.

Bolles, R. C. Introduction: Associative processes in the formation of conditioned food aversions-an emerging functionalism? *Ann. NY Acad. Sci.* 443:1–7, 1985.

Booth, D. A. The physiology of appetite. *Br. Med. Bull.* 37:135–140, 1981.

Booth, D. A. Food-conditioned eating preferences and aversions with interoceptive elements: Conditioned appetites and satieties. *Ann. NY Acad. Sci.* 443:22–41, 1985.

Campbell, C. S., and J. D. Davis. Licking rate of rats is reduced by intraduodenal and intraportal glucose infusion. *Physiol. Behav.* 12:357–365, 1974.

Carr, B. A., J. D. Ballou, D. A. Marrinan, and P. J. Kulkosky. Desmethionine-bombesin receptor antagonist blocks bombesin-induced inhibition of alcohol intake. *Alcohol* 11:125–131, 1994.

Carr, B. A., J. D. Ballou, J. C. Snell, and P. J. Kulkosky. MK-329 blocks the inhibition of alcohol intake by CCK-8. *Peptides* 14:1193–1197, 1993.

Cornell, N. W., K. E. Crow, M. G. Leadbetter, and R. L. Veech. Rate-determining factors for ethanol oxidation in vivo and in isolated hepatocytes. In Li, T.-K., S. Schenker, and L. Lumeng (eds.): *NIAAA Research Monograph No. 2: Alcohol and Nutrition.* Washington, DC: U.S. Government Printing Office, 1979, pp. 315–335.

Fedorchak, P. M., and R. C. Bolles. Nutritive expectancies mediate cholecystokinin's suppression-of-intake effect. *Behav. Neurosci.* 102:451-455, 1988.

Flynn, F. W. Fourth ventricle bombesin injection suppresses ingestive behavior in rats. *Am. J. Physiol.* 256:R590–R596, 1989.

Flynn, F. W. Caudal brain stem systems mediate effects of bombesin-like peptides on intake in rats. *Am. J. Physiol.* 262:R39–R45, 1992.

Gewiss, M., Ch. Heidbreder, and Ph. De Witte. Differential effects of caerulein and bombesin on the ethanol intake in the rat. *Arch. Int. Physiol. Biochim. Biophys.* 99:151–153, 1991.

Gibbs, J., D. J. Fauser, E. A. Rowe, B. J. Rolls, E. T. Rolls, and S. P. Maddison. Bombesin suppresses feeding in rats. *Nature* 282:208–210, 1979.

Gibbs, J., P. J. Kulkosky, and G. P. Smith. Effects of peripheral and central bombesin on feeding behavior of rats. *Peptides* 2(Suppl. 2):179–183, 1981.

Gibbs, J., R. C. Young, and G. P. Smith. Cholecystokinin decreases food intake in rats. *J. Comp. Physiol. Psychol.* 84:488–495, 1973.

Glazner, G. W., R. L. Cannon, and P. J. Kulkosky. Effect of bombesin on behaviors associated with ethanol satiation and blood ethanol levels. *Alcohol* 5:325–330, 1988.

Grossman, M. I. Satiety signals. *Am. J. Clin. Nutr.* 8:562–568, 1960.

Johnston, S. A., and Z. Merali. Specific neuroanatomical and neurochemical correlates of grooming and satiety effects of bombesin. *Peptides* 9 (Suppl. 1):233–244, 1988.

Kulkosky, P. J. Brain-gut neuropeptides and the limitation of ethanol consumption. *Neurosci. Biobehav. Rev.* 9:179–190, 1985a.

Kulkosky, P. J. Conditioned food aversions and satiety signals. *Ann. N.Y. Acad. Sci.* 443:330–347, 1985b.

Kulkosky, P. J., and N. Chiu. Effect of vagotomy on peptide control of ethanol intake. *Appetite* 7:273, 1986.

Kulkosky, P. J., J. S. Doyle, V. I. Cook, G. W. Glazner, and M. A. Foderaro,

Vasoactive intestinal peptide: Behavioral effects in the rat and hamster. *Pharmacol. Biochem. Behav.* 34:387–393, 1989a.

Kulkosky, P. J., M. A. Foderaro, S. L. Sandoval, S. S. Cesar, and D. A. Marrinan. Cholecystokinin-induced satiation with ethanol: Effects of lighting cycle and limited access procedures. *Alcohol* 8:223–227, 1991.

Kulkosky, P. J., J. Gibbs, and G. P. Smith. Behavioral effects of bombesin administration in rats. *Physiol. Behav.* 28:505–512, 1982.

Kulkosky, P. J., and G. W. Glazner. Dose-additive inhibition of intake of ethanol by cholecystokinin and bombesin. *Alcoholism Clin. Exp. Res.* 12:277–281, 1988.

Kulkosky, P. J., G. W. Glazner, H. D. Moore, C. A. Low, and S. C. Woods. Neuropeptide Y: Behavioral effects in the golden hamster. *Peptides* 9:1389–1393, 1989b.

Kulkosky, P. J., H. D. Moore, R. L. Cannon, and N. Chiu. Naloxone, cholecystokinin, and bombesin interact to inhibit ethanol intake. *Soc. Neurosci. Abstr.* 14:295, 1988.

Kulkosky, P. J., M. R. Sanchez, N. Chiu, and G. W. Glazner. Characteristics of bombesin-induced inhibition of intake of ethanol. *Neuropharmacology* 26: 1211–1216, 1987.

Kulkosky, P. J., M. R. Sanchez, M. A. Foderaro, and N. Chiu. Cholecystokinin and satiation with alcohol. *Alcohol* 6:395–402, 1989c.

Kulkosky, P. J., M. R. Sanchez, and G. W. Glazner. Cholecystokinin octapeptide: Effect on the ethogram of ethanol consumption and blood ethanol levels in the rat. *Physiol. Psychol.* 14:23–30, 1986.

Kulkosky, P. J., M. R. Sanchez, and D. A. Marrinan. Bombesin reduces alcohol choice in nutritive expectancy and limited-access procedures. *Alcohol* 9:123–127, 1992.

Kyrkouli, S. E., B. G. Stanley, and S. F. Leibowitz. Bombesin-induced anorexia: Sites of action in the rat brain. *Peptides* 8:237–241, 1987.

Ladenheim, E. E., and R. C. Ritter. Low-dose fourth ventricular bombesin selectively suppresses food intake. *Am. J. Physiol.* 255:R988–R992, 1988.

Liddle, R. A., I. D. Goldfine, and J. A. Williams. Bioassay of plasma cholecystokinin in rats: Effects of food, trypsin inhibitor, and alcohol. *Gastroenterology* 87:542–549, 1984.

Linseman, M. A. Alcohol consumption in free-feeding rats: Procedural, genetic and pharmacokinetic factors. *Psychopharmacology* 92:254–261, 1987.

Litten, R. Z., and J. P. Allen. Pharmacotherapies for alcoholism: Promising agents and clinical issues. *Alcoholism Clin. Exp. Res.* 15:620–633, 1991.

Lorenz, D. N., and S. A. Goldman. Vagal mediation of the cholecystokinin satiety effect in rats. *Physiol. Behav.* 29:599–604, 1982.

Macdonall, J. S., and H. Marcucella. Increasing the rate of ethanol consumption in food- and water-satiated rats. *Pharmacol. Biochem. Behav.* 10:211–216, 1979.

Morley, J. E., A. S. Levine, J. Kneip, and M. Grace. The effect of vagotomy on the satiety effects of neuropeptides and naloxone. *Life Sci.* 30:1943–1947, 1982.

O'Malley, S. S., A. J. Jaffe, G. Chang, R. S. Schottenfeld, R. E. Meyer, and B. Rounsaville. Naltrexone and coping skills therapy for alcohol dependence: A controlled study. *Arch. Gen. Psychiatry* 49:881–887, 1992.

Peniston, E. G., and P. J. Kulkosky. α-θ brainwave training and β-endorphin levels in alcoholics. *Alcoholism Clin. Exp. Res.* 13:271–279, 1989.

Peniston, E. G., and P. J. Kulkosky. Alpha-theta EEG biofeedback training in alcoholism and post-traumatic stress disorder. *Newslett. ISSSEEM* 2:5–7, 1991.

Pérez, C., and A. Sclafani. Cholecystokinin conditions flavor preferences in rats. *Am. J. Physiol.* 260:R179–R185, 1991.

Richter, C. P. A study of the effect of moderate doses of alcohol on the growth and behavior of the rat. *J. Exp. Zool.* 44:397–418, 1926.

Rowland, N. E., and M. Barnett. Sham ingestion of alcohol in rats. *Alcohol* 9:75–77, 1991.

Sandoval, S. L., and P. J. Kulkosky. Effects of peripheral neurotensin on behavior of the rat. *Pharmacol. Biochem. Behav.* 41:385–390, 1992.

Sandoval, S. L., D. A. Marrinan, and P. J. Kulkosky. Beta endorphin injection and ingestive behaviors in the rat. *Proc. Fifth Natl. Conf. Undergrad. Res.* 2:859–864, 1991.

Smith, G. P. The place of gut peptides in the treatment of obesity. In Curtis-Prior, P. B. (ed.): *Biochemical Pharmacology of Obesity.* New York: Elsevier, 1983, pp. 407–419.

Smith, G. P., and J. Gibbs. The development and proof of the CCK hypothesis of satiety. In *Multiple Cholecystokinin Receptors in the CNS.* Dourish, C. T., S. J. Cooper, S. D. Iversen and L. L. Iversen. (eds.): New York: Oxford University Press, 1992, pp. 166–182.

Smith, G. P., C. Jerome, B. J. Cushin, R. Eterno, and K. J. Simansky. Abdominal vagotomy blocks the satiety effect of cholecystokinin in the rat. *Science* 213:1036–1037, 1981a.

Smith, G. P., C. Jerome, and J. Gibbs. Abdominal vagotomy does not block the satiety effect of bombesin in the rat. *Peptides* 2:409–411, 1981b.

Smith, G. P., C. Jerome, P. Kulkosky, and K. J. Simansky. Ceruletide acts in the abdomen, not in the brain, to produce satiety. *Peptides* 5:1149–1157, 1984.

Smith, G. P., C. Jerome, and R. Norgren. Afferent axons in abdominal vagus mediate satiety effect of cholecystokinin in rats. *Am. J. Physiol.* 249:R638–R641, 1985.

Stein, L. J., and S. C. Woods. Cholecystokinin and bombesin act independently to decrease food intake in the rat. *Peptides* 2:431–436, 1981.

Tolman, E. C. *Purposive Behavior in Animals and Men.* New York: Appleton-Century-Crofts, 1967.

Toth, P., M. A. Linseman, E. Perlanski, and L. A. Grupp. The role of the gastric and hepatic vagus in voluntary alcohol intake. *Pharmacol. Biochem. Behav.* 36:69–76, 1990a.

Toth, P., C. Shaw, E. Perlanski, and L. A. Grupp. Cholecystokinin octapeptide reduces ethanol intake in food- and water-sated rats. *Pharmacol. Biochem. Behav.* 35:493–495, 1990b.

Vallejos, J. J., J. D. Ballou, B. S. Butler, and P. J. Kulkosky. Bombesin antagonist increases alcohol intake. *Proc. SW Rocky Mt. Div. AAAS* 9:39, 1992.

Volpicelli, J. R., A. I. Alterman, M. Hayashida, and C. P. O'Brien. Naltrexone in the treatment of alcohol dependence. *Arch. Gen. Psychiatry* 49:876–880, 1992.

Wang, R. Y., and R. Schoenfeld (eds.): *Cholecystokinin Antagonists.* New York: Alan R. Liss, 1988.

Willis, G. L., J. Hansky, and G. C. Smith. Ventricular, paraventricular and circumventricular structures involved in peptide-induced satiety. *Regul. Pept.* 9:87–99, 1984.

Wolf, A. V. *Thirst: Physiology of the Urge to Drink and Problems of Water Lack.* Springfield, IL: Thomas, 1958.

Woods, S. C. The eating paradox: How we tolerate food. *Psychol. Rev.* 98:488–505, 1991.

INDEX